Anthrax:
History, Biology, Global Distribution, Clinical Aspects, Immunology, and Molecular Biology

Authored By

Robert E. Levin

University of Massachusetts
Amherst
USA

CONTENTS

FOREWORD

This work consists of the latest presentation of our knowledge of *Bacillus anthracis*, the causative agent of anthrax in a single volume. There are few better illustrations of the importance of "pure" research than the elucidation of the genetic factors involved in the virulence of *B. anthracis* in view of its recent emergence as a major potential instrument of biological warfare. An understanding of the relationship of *B. anthracis* to other members of the *B. cereus* group is essential for anyone working with *B. anthracis* and has been fully described in chapter two, and thoroughly elucidated with the application of molecular techniques presented in chapter six. The symptomatic aspects of the three forms of anthrax: cutaneous, gastrointestinal, and inhalation, manifest distinct clinical symptoms, and are thoroughly presented with appropriate illustrations in chapter three. Although several human vaccines have been developed and used over the years with reported success, the development of a reliably effective human vaccine devoid of adverse reactions and not requiring periodic boosting is still elusive, particularly a vaccine against inhalation anthrax. The highly detailed presentation of developmental vaccine efforts over the years in chapter five is comprehensive and should serve as a guide for future efforts in this area. The student and others will find this volume highly informative and replete with extensive references to the published literature.

Ronald G. Labbe
University of Massachusetts
Amherst
USA

PREFACE

Following the post 9/11/2001 distribution of anthrax spores through the U.S. mail, and the resulting deaths of five individuals, primarily due to initial misdiagnosis, there has been a renewed interest in anthrax, particularly as a biological warfare agent. This volume is an attempt to bring forth all essential aspects of anthrax. Included are its early history, non-natural outbreaks of anthrax, characteristics of the causative organism *Bacillus anthracis* and its relationship to other members of the *B. cereus* family. Also extensively included are clinical aspects, mechanisms of anthrax virulence and genetics of virulence factors. The extensive studies over the years regarding the development of veterinary and human vaccines, genetics of virulence, and molecular studies including conventional PCR and real-time PCR are dealt with in complete detail, with inclusion of a multitude of primers and probes. The author hopes that this text will serve as an advanced presentation and reference work for individuals pursuing introductory and advanced information regarding anthrax and as a guide for individuals contemplating studies on anthrax. The format, organization, tables, figures, and extensive references to the published literature should facilitate its use as an advanced text.

ACKNOWLEDGEMENT

The author wishes to acknowledge the assistance of Ruth Whitkowski in proof reading certain portions of the text.

CONFLICT OF INTEREST

The author confirms that the ebook contents have no conflict of interest.

Robert E. Levin
University of Massachusetts
Amherst
USA
E-mail: relevin@foodfci.umass.edu

2

Send Orders for Reprints to reprints@benthamscience.net
Anthrax: History, Biology, Global Distribution, Clinical Aspects, Immunology, and Molecular Biology, 2014, 3-23   3

CHAPTER 1

History, Global Distribution, and Non-Natural Incidents of Anthrax

Abstract: This chapter deals with the early history of anthrax, its industrial incidence and association with animal products such a wool and hair, and effective methods for destroying anthrax spores. The global distribution and prevalence of anthrax along with geographic areas considered endemic for anthrax are presented. Major non natural anthrax incidents that have occurred during the past century are discussed in detail.

Keywords: Cutaneous anthrax, inhalation anthrax, gastrointestinal anthrax, wool sorters disease, formaldehyde, Russia, Soviet Union, Caucasus, Spain, Canada, America (U.S.A.), Haiti, Africa, Central America, Europe, Australia, Central Asia, Turkey, Albania, Zimbabwe.

INTRODUCTION

Anthrax has been an important disease throughout human history. The earliest recorded description of anthrax is in the book of Exodus, in which the "sixth plague" is reported to have killed the Egyptians' cattle and affected people with black spot (cutaneous anthrax). Writings of the ancient Hindus, Greeks, and Romans, especially Virgil, contain many colorful descriptions of anthrax. Vergil wrote of the shortage of animals caused by what may have been anthrax and also noted that the disease spread to humans, not by human to human contact but by human contact with infected animals, and contact with hides and wool from infected animals which constitutes a well recognized epidemiological aspect of anthrax.

The industrialization of Europe in the 1800s was accompanied by outbreaks of human anthrax in commercial plants where imported animal hides, wool, and hair were processed. The designations wool-sorters disease and rag-pickers disease, reflect the relationship of anthrax and such occupations. By 1844, it was well established in the medical literature that shepherds, butchers, tanners, and woolworkers suffered from occupationally acquired "charbon" (anthrax) as it was referred to at that time.

Robert E. Levin

The widespread recognition of anthrax in Europe is reflected in the following designations found in the early literature of the 1800's: French: *charbon* Italian: *carbonchio* German: *milzbrand* (spleen fire), Dutch: *mildzvuur* (spleen fire), Spanish: *pustula maligna.*

Eurich (1913) was the first to apply formaldehyde for destruction of *B. anthracis* spores in wool and presented a highly detailed description of a typical wool processing plant and the incidence of anthrax associated with the industry at the time. He also discussed the problems of efficacy in using formaldehyde to destroy spores of *B. anthracis* associated with wool and reported that 2% formaldehyde was quite effective with unprotected spores (not trapped in blood clots or dirt). Elmhirst Duckering in 1914 described the first successful chemical treatment of wool. The Duckering process consisted of bathing wool successively in troughs filled with solutions of sodium carbonate and of sodium hydroxide then immersing in warm (100° F) formaldehyde. The sodium hydroxide presumably facilitated the dissolution of blood clots adhering to the wool so as to enhance the penetration of the formaldehyde.

Historically anthrax in humans has been transmitted from products not only derived from hides, and wool but also from imported shaving brushes derived from animal hair. Anthrax can be transmitted from animal to animal or from animals to humans. Human to human transmission has only recently been documented (Yakupogullari and Koroglu, 2007). Currently, very few cases of anthrax occur naturally in developed countries as a result of vaccination of high-risk people and animals, in addition to improvements in industrial hygiene. In contrast, anthrax spores remain endemic in certain rural areas of the world, such as Africa and Asia, where vaccination programs are difficult to implement.

In 1850, the French physician Rayer and the French parasitologist Casimir-Joseph Davaine reported observing nonmotile microscopic rods in the blood of sheep, but failed to associate the rods with anthrax. In 1863, stemming from bacterial fermentation studies published by Pasteur in 1861, Davaine applied the name "bacteridies" to emphasize that these rods were always associated with anthrax.

It was Ferdinand Cohn, after observing Kochs microscopic blood slides from patients stricken with anthrax who applied the Linnean name, *Bacillus anthracis*;

the genus designation being derived from the Latin word "bacillus" (a small staff or rod), to deliberately distinguish it from "bacterium" (a large stave). The species designation, *anthracis* is derived from the Greek word for coal (*anthrakas*), reflecting the clinical appearance of typical black cutaneous anthrax lesions or eschars. It was Cohn who discovered that *B. anthracis* formed spores.

Historically, anthrax was the first disease where a specific bacterium was proven to be the cause of an infectious disease. In 1877, Robert Koch was able to culture the organism on the aqueous humor of the ox's eye, described its life history, and reproduced the disease with a pure culture of the organism. Louis Pasteur developed the first anthrax vaccine in 1881.

In nature anthrax is a zoonosis, that is almost always associated with grazing animals especially sheep, cattle, goats, and horses that acquire the infection from soil contaminated with spores of *B. anthracis*, usually from a prior burial site of an infected animal. Horses, goats, and swine are less frequently infected. Rabbits, guinea pigs, and white mice are also susceptible. Mice are extremely susceptible to lethal infections derived from highly virulent strains. Rats are notably resistant to infection by *B. anthracis*. As a zoonosis anthrax can be transferred from animal to animal and from animals to humans, but is rarely transferred from human to human.

GLOBAL PREVALENCE OF ANTHRAX

Human anthrax is now very infrequent to rare in Canada and the U.S.A however, it is still a significant problem in West Africa, Spain, Greece, Turkey, Albania, Romania, areas of eastern Europe and Central Asia. A tabulation of the annual number of cases reported for various countries is given in Table **1**.

NATURAL GEOGRAPHIC DISTRIBUTION OF ANTHRAX

Anthrax in Russia and the Former Soviet Union

Anthrax in Russia has for a long time posed a serious problem for pubic health and veterinary services. Anthrax is well known to be endemic in various parts of Russia, particularly Siberia. In Russian "Sibirskaya yazva" translates to "Siberia ulcer", a common name for cutaneous anthrax in Russia. In czarist times, the

Russian incidence of anthrax reportedly was about 50,000 animal deaths per year, 11,500 human cutaneous infections, and 3,500 human deaths. In 1649, after a terrible anthrax epidemic Czar Aleksey decreed that no more animal carcasses could be dumped in Moscow streets and anthrax was looked upon as permanently endemic to Russia. By the 1930s Soviet morbidity and mortality rates were halved

Table 1. Number of annual cases of human anthrax reported among countries of Europe, North America and Japan.[a]

Countries	1997	1998	1999	2000	2001	2002	2003	2004	2005	2006
Belgium	0	0	0	0	0	0	0	0	0	0
Bulgaria	8	8	3	13	6	2	2	0	1	1
Czech Rep.	0	0	0	0	0	0	0	0	0	0
Denmark	0	0	0	0	0	0	0	0	0	0
Germany	0	0	0	0	0	0	0	0	0	0
Estonia	0	0	0	0	0	0	0	0	0	0
Ireland	0	0	0	0	0	0	0	0	0	0
Greece	4	1	4	0	2	1	2	4	3	1
Spain	70	38	28	34	25	19	17	26	10	12
France	3	0	0	0	0	0	1	0	0	0
Italy	1	0	3	5	1	3	0	0	0	0
Cyprus	0	0	0	0	0	0	0	0	0	0
Latvia	0	0	0	0	0	0	0	0	0	0
Lithuania	0	0	0	0	0	0	0	0	0	0
Luxembourg	0	0	0	0	0	0	0	0	0	0
Hungary	0	0	0	2	0	0	0	0	0	0
Malta	0	0	0	0	0	0	0	0	0	0
The Netherlands	0	0	0	0	0	0	0	0	0	0
Austria	0	0	0	0	0	0	0	0	0	0
Poland	0	2	1	0	0	1	0	1	2	0
Portugal	0	4	0	3	0	4	0	0	0	0
Romania		12	8	51	7		1		2	1
Slovenia	0	0	0	0	0	0	0	0	0	0
Slovakia	0	0	0	0	0	0	1	0	0	0
Finland	0	0	0	0	0	0	0	0	0	0
Sweden	0	0	0	0	0	0	0	0	0	0
United Kingdom	0	0	1	1	2	1	0	0	0	1

Table 1: contd...

Croatia	1				1	1	1	0	10	1
Macedonia	9	3	3	8	6	1	2	0	4	0
Turkey	690	404	460	396	532	368	325	268	320	272
Iceland	0	0	0	0	0	0	0	0	0	0
Norway	0	0	0	0	0	0	0	0	0	0
Switzerland	0	0	0	0	0	0	0	0	0	0
Albania	75	95	64	62	56	52	47	36	47	
Bosnia & Herzegovina	2	1	2	0	3	3	0	0	0	0
Montenegro										0
Serbia	9		0	0	1	0	1	0	0	0
Canada	0	0	0	0	0	0	0	0	0	1
United States	0	0	0	1	23	2	0	0	0	0
Japan	0	0	0	0	0	0	0	0	0	0

[a]Compiled from the European Centre for disease Prevention and control, Pubic Health Agency of Canada, U.S. CDC, and Japanese Infectious Disease Surveillance Center.

as a result of extensive animal vaccination programs. However, just prior to World war II, human anthrax rates in the more remote soviet republics such as Azerbaijan, Bessarabiak, and Kasakhtan still resembled those of the czarist era with animal and human cases in the thousands (Guillemin, 1999). By the 1950's animal anthrax cases in the Soviet Union were reduced to several thousand per year, human infections to twelve hundred, with human deaths in the low hundreds by the 1960s. By the 1960s, more than ten million animals and two million people were being vaccinated annually with the Soviet STI vaccine which reportedly has given rise to a notable incidence of side effects (lesion at vaccination site and temporary immobilization of the arm vaccinated) and is therefore not administered to young children, the elderly, and the infirm.

From 1995-1996 a number of outbreaks occurred in Russia involving occupational anthrax in addition to bovine cases resulting in the death of cattle and several human deaths. In the Russian Federation the registration of anthrax loci is obligatory for veterinary and sanitary-epidemiological services. In Russia, an outbreak involving one human death and eight people hospitalized from occupational exposure and enteric anthrax associated with dead cattle was reported south-east of Moscow in Tambov in 1995. In 1996 another outbreak, virtually identical, broke out in another village in the same general area, with one death, 23 hospitalized and 1500 vaccinated. In 1996 a very similar outbreak with human and bovine cases was reported near Stavropol in Southern Russia. In 1995 epidemiologically identical outbreaks occurred in Tblisi, Georgia, and in and around Buka and Azerbaijan. There have also been multiple human cases in Krasnodar, Stavropol, Kalnynya, and Sanaov in the late 1990's. All of these events reflected the poor quality of public health and veterinary services in their respective regions, an over-reliance on human sentinel cases, and on the economic pressure in the communities to buy and eat meat from infected dead and dying cattle (Hugh-Jones, 1999).

Anthrax is considered endemic in the Caucasus and the frequency of outbreaks increased dramatically from 1996 to 2006. Georgia is one of the countries most affected by these outbreaks, including human cases (Imnadze *et al.*, 2002). In the Southern caucasion region, which is a border area between Europe and the Middle East, there was an increase in anthrax following the dissolution of the FSU. Anthrax is considered epizootic in the Southern Caucases, is endemic in Georgia, and is sporadic in Armenia (Merabishvili *et al.*, 2006). From 1990 to 2003, 196 human cases of anthrax were recorded in Georgia.

Merabishvili *et al.*, (2006) collected 239 soil samples from eight areas of Georgia known to have a history of anthrax outbreaks. The soil samples were derived from infected animal burial sites, abattoirs, and animal stalls in addition to blood samples from infected two humans and one infected goat. A total of 18 isolates were identified as *B. anthracis* and were subjected to multilocus variable number tandem repeat (MLVTR) typing according to Keim *et al.*, (2000) using primer pairs that amplified 2 plasmid (pXO1 and pXO2) loci and 6 chromosomal loci (see chapter 7). Three distinct MLVTR genotypes were found for the 18 Georgia isolates, and all belonged to cluster A3.a of Keim *et al.*, (2000). The most informative variable number tandem repeat (VNTR) markers were those from pXO1 and pXO2 which are considered the most discriminatory for *B. anthracis* (Keim *et al.*, 2000). Two genotypes were previously shown to include Turkish isolates, indicating that there is a regional pattern in the South Caucasian Turkish region. The authors presented a diagram depicting the evolutionary origin of the Georgia and Turkey isolates.

Anthrax in Kazakhstan

Aikimbayev *et al.,* (2007) genotyped 92 clinical and environmental isolates of *B. anthracis* from anthrax outbreaks over a 55 year period in Kazakhstan which is recognized as being an anthrax endemic area of central Asia. A total of 12 identified subtypes were identified, the majority of which (74 isolates) belonged to the A1a genetic cluster, 6 isolates belonged to the A3b cluster, and 2 isolates belonged to the A4 cluster (Keim *et al.,* 1999; Keim *et al.,* 2000). Over half of the A1a isolates belonged to previously described genotypes (38/74), including G3 (N = 15), G6 (N = 2), and G13 (N = 21). Two Kazakh types (N = 6) appeared to represent new sub lineages in the "A" branch.

Anthrax in Canada

During the first-half of the 20th century, the majority of anthrax outbreaks in Canada occurred in the Southern portions of Ontario and Quebec and were often associated with pastures contaminated by effluent from textile industries dealing with imported animal materials. Federal legislation introduced in 1952 requiring disinfection of these materials greatly reduced the incidence of anthrax in eastern Canada. Between 1962 and 1963, nine anthrax outbreaks occurred in bison herds of the Northwest Territories and Northern Alberta with at least 1309 bison succumbing. Interestingly, during these bison outbreaks the majority of carcasses were sexually mature bulls. These northern outbreaks occurred after a wet spring, during drought conditions in late summer, and ended with cooler weather. Environmental stress factors coupled with breeding stress during the late summer were hypothesized as predisposing the bulls to infection (Dragon *et al.,* 1999). Alternatively, climatic conditions involving the flow of surface waters may have concentrated anthrax spores into low lying wallows preferentially utilized by the bulls. Isolates from northern bison outbreaks form their own distinct strains although closely related to isolates from domestic outbreaks. In addition, the pre-rut period is associated with increased levels of testosterone in bulls. Testosterone has been shown to be immunosuppressive in other ungulates which may play an additional role in predisposing the bulls to anthrax infection (Folstad *et al.,* 1989).

During the week of July 13, 2009, anthrax was detected in 3 cattle herds in southeastern Manitoba, in the RM [rural municipality] of Franklin. One unvaccinated herd has lost a total of 16 cattle. Another cattle herd that was vaccinated for anthrax has had one animal die. The third unvaccinated cattle herd also has had one death. Overland flooding has occurred in this area in the spring which may have allowed anthrax spores to be available to livestock grazing on pasture. Canada's second 2009 anthrax outbreak claimed the lives of at least 234 cattle and other livestock in the Canadian Prairies. About 8,500 animals are estimated to have been vaccinated and both outbreaks were considered under control.

The Minnesota Board of Animal Health in the US has reported that Minnesota which borders Manitoba to the south had 68 cattle and other livestock dead due to anthrax since mid-June of 2009. The federal food safety agency placed farms under 21-day quarantines and on Thursday stated that 48 farms remained under quarantine as positive premises.

Anthrax in the U.S.A.

In the U.S.A. 224 cases of cutaneous anthrax in humans have been documented between 1944 and 1994 with an additional single case in 2000 (Thompson, 2003). The Arms Textile Mills in Manchester, New Hampshire, in 1957 employed 632 workers. Arms Mills imported goat hair from Pakistan, Iraq, and Iran for the manufacture of warm, waterproof linings of men's coats. The goat hair was not disinfected and between January 1, 1941 and June 30, 1957, there were 136 cases of cutaneous anthrax at the mill, only one of which was fatal. There were no cases of inhalation anthrax until 1957. Suddenly, during a ten-week period from August to November, nine employees developed anthrax infections: five inhalation and four cutaneous. Four of the inhalation cases were fatal. All nine infected individuals had handled a particular shipment of black goat hair from Pakistan. Three of the patients with inhalation anthrax succumbed due in part to misdiagnosis and two recovered following penicillin therapy. The four patients with cutaneous anthrax recovered following penicillin treatment. (Plotkin *et al.*, 2002; Thompson, 2003).

Between 1978 and 2001, additional cases of inhalation anthrax in the U.S.A. have occurred, mostly individuals employed in woolen mills. In 1988 a 42-year old male textile maintenance worker became the first case of human anthrax in the United States since 1984. He initially developed cutaneous lesions on his forearm with ensuing pain, fever, chills, and edema which responded to *i.v.* antibiotics. Interestingly, the plant had never had a previous case of anthrax during its 25-years of operation. The plant did use cashmere hair from China, Afghanistan, and Iran. Eight of twelve cashmere samples yielded spores of *B. anthracis*.

In the U.S. there have been natural sporadic cases in South Dakota, Nebraska, New Mexico, and Oklahoma since 1991 with animal deaths in California, Kansas, Mississippi, and Arkansas. All appeared to be possibly related to old anthrax graves (Hugh-Jones, 1999). In the summer of 1997 an epidemic occurred in SW Texas involving 22 counties resulting from long-lasting hyper-endemic anthrax in the area, primarily affecting white tailed deer and spreading to adjoining livestock herds as a result of large numbers of blowflies. In Texas, an anthrax outbreak in 2005 was notable not for the numbers but for the location. Anthrax in Texas livestock is reported nearly every year in the southwest region of the state, but discovery of the disease in 2005 at two Sutton County ranches marked the first occurrence in that west central area in more than 20 years. The state has experienced heavy losses of deer and cattle from anthrax in prior years, including severe outbreaks in 1997 and 2001. Anthrax is under-reported, because many automatically dispose of carcasses and vaccinate livestock, although anthrax is a reportable disease. In the summer of 1988 there was an outbreak in North Dakota, involving six herds and 27 cattle deaths in three separate areas.

In 2005 an outbreak of anthrax afflicted a 660-head of cattle and a bison herd in South Dakota with 155 deaths occurring in the herd. Anthrax was confirmed in five additional herds. Nearly 200 animals had died by the end of July in the State. These mortality figures are more than double the South Dakota figures from a severe outbreak in 2002, when 53 animals in three counties died. That year also saw a rare case of cutaneous anthrax in a veterinarian who had examined an infected carcass. His infection responded to antibiotic treatment.

North Dakota's producers also experienced extensive livestock losses in 2005 from the disease. Approximately 200 grazing animals died in 10 southeastern counties. A previous severe outbreak in North Dakota in 2000 killed nearly 150 animals.

Anthrax in Haiti

Anthrax has been endemic in Haiti for two-hundred years. On December 28, 1973, a 22-year old female member of the U.S. navy took shore leave from the hospital ship on which she was stationed which was docked at Port-au-Prince, Haiti. Medics on the hospital ship treated 30 Haitions for anthrax while docked. The woman returned to the ship with several bongo drums with goatskin drumheads and shortly developed cutaneous anthrax in one eye. The Centers for Disease Control subsequently found that 387 cases of anthrax (all cutaneous) had occurred in the Haitian district she had visited during 1973 and another 74 during the first four months of 1974. Between 1963 and 1974, 194 cases of anthrax (cutaneous and inhalation) were identified among Haitians producing goat skin products (voodoo dolls, skins, rugs, purses, and drums). An organized continuous veterinary anthrax vaccination program has never been permanently established in Haiti.

Anthrax in the United Kingdom

During the nineteenth century, in the textile mills of England, 20% of patients with inhalation anthrax were said to recover spontaneously (Plotkin *et al.,* 2002). During the past decade the United Kingdom has had a total of six cases of human anthrax (Table **1**) with no more than 2 in any one year.

Anthrax in Mexico and Central America

Up to 1999, veterinary anthrax has been viewed as raging with slight effective controls and extensive under-recognition of the disease in Mexico and Central America (Hugh-Jones, 1999). Veterinary anthrax is considered hyper endemic in El Salvador and Guatemala and endemic in Mexico, Honduras, Nicaragua, and Costa Rica. El Salvador ~1999 continued to suffer a series of outbreaks in horses.

Veterinary Anthrax in South America

In South America, anthrax as of 1999 has been sporadic in Brazil, Ecuador, Paraguay, Uruguay, and Venezuela (Hugh-Jones, 1999).

Anthrax in Africa

Africa is severely afflicted with anthrax. As of 1991 West Africa was the largest region in the World with hyper-endemic and epidemic anthrax (Hugh-Jones,

1991). Genotype studies suggest that the geographic origin of *B. anthracis* may be sub-Saharan Africa. Among world-wide isolates of *B. anthracis*, southern Africa contains the widest range of chromosomal genotypic diversity (Smith *et al.,* 1999). Representative isolates of the primary node branches have been found in the Kruger National Park (KNP). One of these node branches has given rise to other distinct geographically correlated lineages with world-wide distribution, and the other presently appears unique to the Southern African subcontinent. Certain areas of the KNP are anthrax endemic areas.

Smith *et al.,* (1999) undertook a pilot study involving the characterization of genetic diversity within two anthrax epidemics in the KNP between 1970 and 1997. The KNP consists of about five million acres of subtropical savanna woodland located in the north-eastern corner of the republic of South Africa. The entire area is surrounded by game-proof fencing and adjoins commercial farms. A total of 78 isolates were involved in the study. Eight highly pleomorphic loci were utilized for multilocus variable amplification (MLVA) PCR. The results revealed a diverse range of genotypes within the park allied with three genotype reference groups. Kruger (44 isolates), Ames (16 isolates) and Southern Africa (18 isolates). The Mautel test for time-space clustering (Mautel, 1967) revealed that there is time-space clustering present with respect to outbreaks occurring prior to 1990 and later than 1990. The relatively large amount of genetic diversity observed among the isolates was interpreted to be due to independent infective events rather than gene flow.

Anthrax in Europe

In Europe, the major regions affected with veterinary anthrax as of 1999 were Turkey and Greece (Hugh-Jones, 1999). However, significant numbers of outbreaks were still regularly occurring in Spain, Albania, Italy, and Romania. Central Spain suffered from an extraordinary number of human cases each year from 152 in 1990 to 50 in 1996, indicating a hyperendemic or even epidemic livestock situation. From 1997 to 2006 Turkey had the highest number of reported cases of human anthrax in Europe varying from 272 to 690 annual cases (Table **1**). Spain (10 - 70 annual cases) and Albania (36 - 70 annual cases) were next.

Pilo, Perreten, and Frey (2008) analyzed and compared strains of *B. anthracis* from animal husbandry and industrial anthrax cases in Switzerland between 1952 and 1981 using the MLVTR system of Keim *et al.,* (2000). Strains isolated from autochthonous cases of anthrax in cattle were found to belong to genotype B2, together with strains from continental Europe, while human anthrax strains clustered with genotype A4. These strains were traced back to outbreaks of human anthrax that occurred between 1978 and 1981 in a factory processing cashmere wool from the Indian subcontinent. The worldwide occurrence of *B. anthracis* strains of cluster A4 was interpreted to be due to the extensive global trade of untreated cashmere wool during the last century.

Outbreaks of anthrax occurred in two regions of France in 1997 where previous outbreaks had occurred within the previous two decades (Patra *et al.,* 1998). Ninety-four animals died and there were three nonfatal human cases. The first cases appeared in the Pyrenees mountains, 1000 feet above sea level, toward the end of May, in very humid areas. The outbreak began after a period of heavy rain that followed a dry spring, and it lasted until July, with infected animals occurring in several locals. Thirty-five cows died and 82 herds, involving 1,800 cattle were vaccinated. The second outbreak occurred in the Alpine mountain region and began at the end of July and lasted until October after a long period of drought followed by torrential rain at the beginning of July. The livestock of 15 farms in the Alps were affected. Fifty-eight of 381 cattle died, in addition to a one-month old calf. The vaccination program in this area involved 35 farms and 750 animals. Three nonfatal human cases of anthrax occurred in the Pyrenees involving a veterinarian, a farmer, and a young boy; all infected *via* cutaneous wounds after contact with infected animals. All three were treated and recovered. No human cases were reported in the Alps, but antibiotics were administered as a preventive measure. All 15 isolates *of B. anthracis* from the outbreaks were found to belong to the variable number tandem repeat (VNTR) group $VNTR_3$. Surprisingly, a penicillin resistant strain from an infected Alpine cow was among the 15 isolates. During the period 1998 to 2006 only one case of human anthrax occurred in France.

Anthrax in Central Asia

Central Asia suffers from a comparative weakness of veterinary services which is thought to have resulted in a rapid increase in human and livestock cases during the 1990's.

Anthrax in Australia

The outbreak of anthrax in North Central Victoria Australia in 1997 represents a classic example of collective resources brought to focus to effectively control and abate a large scale outbreak (Turner *et al.,* (1999)). A total of 83 farms were rapidly infected in a geographic area that had been free of anthrax since 1914, although anthrax did occur in the area in the 1880s and 1890s. Immediately after diagnosis, infected herds were quarantined with a prohibition of movement of stock, recent movements onto and off affected farms were traced, infected carcasses were disposed of by burning, the site where the animal died was disinfected with 5% formaldehyde after removal of the carcass wrapped in plastic, people entering infected properties were required to wear protective clothing and footwear which was disinfected before leaving the property, vaccination of livestock on each affected farm and on surrounding farm instituted, enhanced surveillance of stock by private and government veterinarians was undertaken, the local disaster resource program was activated, and health authorities alerted. Vaccination was very effective in minimizing losses. A vaccination zone of about 30 km East and West and 20 km North and South was established to include all infected farms and the surrounding 457 farms. The whole vaccination zone encompassed less than 0.04% of Victoria's land mass. The epidemic peeked in the third week and persisted for seven weeks.

Anthrax in India

Anthrax is endemic in certain cattle breeding areas of South India. In 2005, India had 119 anthrax outbreaks in which 668 animals were afflicted and 569 died (Narayan *et al.*, (2009). Meningoencephalitis (ME) due to *B. anthracis* is particularly lethal. Narayan *et al.,* (2009) assessed the number and case histories of anthrax ME patients admitted to a South India referral hospital over a 20 year period. Two hospital admission clusters occurred: four during 1992 – 1994 and six in 1998 – 2000. All patients were males with agricultural related occupations. Two of the patients exhibited a cutaneous ulcer, and three of the ten patients had no evidence of primary focus of infection. The majority was in a coma at admission and had documented evidence of septicemia. The cerebral spinal fluid was hemorrhagic and death was the uniform outcome in spite of high dose intravenous administration of penicillin G. The maximum duration of hospital survival was 48 hrs and the average was 14.4 hrs. After 2000 no further cases of anthrax ME were detected at this referral hospital. However, isolated cases of cutaneous anthrax continued to attend the outpatient clinics at 4 – 5 cases per year.

Global Isolates

Sue *et al.,* (2007) documented the genetic diversity of an historical collection of 124 isolates of *B. anthracis* collected from the 1950's to the 1980's from various global locations and sources. The *pagA* gene which encodes the protective antigen (PA) was sequenced for all 124 diverse isolates and all previously identified *B. anthracis pagA* types except type 4 were found. Sixty-three of the 124 strains were identified as *pagA* type 6, while 44 were *pagA* type 5, 12 were *pagA* type 1, and individual isolates were identified for types 2 and 3, respectively. Two new *pagA* genotypes were discovered among three environmental isolates within the historical collection. Two isolates had the same new genotype, a third isolate produced a second new genotype. Multilocus variable number tandem peat analysis (MLVA) as described by Keim *et al.,* (2000) detected 22 previously described genotypes in the collection. In addition, 33 new MLVA genotypes were found. For 11 isolates, an MLVA genotype cold not be assigned because one or more alleles did not amplify. The sequencing of *pagA* discriminated isolates with the same MLVA genotype. MLVA revealed that 39 of the 124 isolates were previously undocumented genotypes.

NON NATURAL ANTHRAX INCIDENTS

Germany

Redmond *et al.,* (1998) have described the first attempt to use *B. anthracis* as a biological warfare agent. The curator of a police museum in Norway discovered in 1997 in his archive collection a glass bottle containing two lumps of sugar photographically documented. A small hole in one of the lumps contained a sealed capillary tube. A descriptive note read "A piece of sugar containing anthrax bacilli, found in the luggage of Baron Otto Karl von Rosen, when he was apprehended in Karasjok in January 1917, suspected of espionage and sabotage". Direct agar cultivation from the capillary tube failed to yield *B. anthracis*, due presumable to a low number of viable spores after 80 years. However, enrichment for eight days at 37 °C resulted in the isolation of *B. anthracis* which was confirmed by polymerase chain reaction (PCR) assays using primers specific for the chromosomal Ba813 marker, the lethal factor (*lef*) on the pXO1 plasmid, and the capsule B gene (*capB*) on the pXO2 plasmid. Norway was neutral at that time but favored the allies. Considerable transport of war supplies from the allies to Finland across Norway *via* horses and reindeer was known to be taking place. Presumably the sugar cubes were to be fed to the horses and reindeer, whose teeth would rupture the vials, for the purpose of disrupting this transport. Since anthrax is not transmitted from horse to horse it is doubtful that significant disruption would have occurred.

Former Soviet Union

In 1953, at the Soviet Union's Microbiology Research Institute at Kirov, a defective reactor accidentally spilled live liquid anthrax spores into the city's sewer. Three years later, rats from the sewer system were captured, and one yielded a more virulent strain than the original escaped spores and was designated strain 836 which was then developed as a weaponized strain in the military complex at Sverdlosk. Sverdlovsk, was a military industrial city of 1.2 million people in the Ural Mountains 880 miles (1400 km) east of Moscow. All but 1% of the city's income came from military production. It was renamed Yekateringurg in late 1979. In April and May of 1979 an unusual anthrax epidemic that lasted six weeks in Sverdlovsk, occurred, which was initially attributed to the consumption of contaminated meat but which is presently considered the largest human outbreak of inhalation anthrax (Meselson *et al.*, 1994). Most victims worked or lived in a narrow zone extending from a military microbiology research facility to the southern city limit. Livestock died of anthrax farther south. The zone was aligned with the northerly wind that occurred shortly before the outbreak. A total of 96 cases of human anthrax were identified, with 79 gastrointestinal, 17 cutaneous with 64 deaths among the former and one among the latter for an overall death rate of 66.7%. Hospitalized patients were treated with cephalosporin, penicillin, chloramphenicol, corticosteroids, and anti-anthrax globulin, in addition to the administration of artificial respiration. The average hospitalization for fatal cases was 1 to 2 days and ~3 weeks for survivors. The ages of patients ranged from 24 to 72 with a mean of 42 for men and 55 for women. The absence of patients under the age of 24 is unexplained, although earlier reports from Russian rural communities of inhalation anthrax also indicated a lack of young people afflicted.

Subsequent investigations indicated that all of the fatal cases were due to inhalation anthrax, presumably resulting in a death rate of 81% for those afflicted with inhalation anthrax. The majority of cases involved men, with case histories available for 77 of the 96 total patients. In 1991, the Russian government revealed that the origin of the epidemic was a military research facility in the city. A plume of *B. anthracis* spores had been accidentally released and was carried south by a north wind. The resulting 4 km wide plume and ensuing human infections extended 4 km south from the point of origin, with no human cases reported beyond 4 km from the point of origin. However, animals infected on farms were as distant as 50 km from the point of origin along the southern dissemination plane of the plume. About 2 km south of the point of origin was a ceramics factory that reported 1 - 2% incidence of anthrax among employees. The attack rate for anthrax among livestock 50 km distant from the point of origin was estimated to be 2% for sheep and resulted in the deaths of seven sheep and 1 cow. The calculated dosage at 50 km was 10-fold less than at the ceramics factory. This suggests that sheep, who have been reported to be more susceptible to inhalation anthrax than monkeys (Meselson *et al.*, 1994) may also be more susceptible than humans. In the rural village of Abramova, on the outskirts of Sverdlovsk, in an attempt to encapsulate the spores, contaminated dirt roads were paved and the village became the only asphalted village in the Urals.

Japan

During an August 1945 sweep of Manchuria, the Soviet army reportedly captured a stockpile of Japanese fragmentation bombs loaded with 900 lbs of anthrax (Graysmith, 2003).

In June of 1993 a Japanese cult sprayed a liquid suspension of anthrax spores from the roof of a building in Kameido, Japan, near Tokyo (Keim *et al.*, 2001). Malodor complaints resulted in health authorities collecting fluid from the outside of the building which was stored at 4°C. Forty-eight resulting *B. anthracis* colonies were purified and subjected to MLVA genotyping. The pXO2 plasmid was found absent in all isolates and the MLVA genotype was consistent with the Sterne vaccine strain. Given the availability of the Sterne 34F2 vaccine strain in Japan for veterinary purposes, it was concluded that the cult had obtained a Sterne veterinary strain. Since the Sterne strain is nonencapsulated, it had little chance of endangering human life.

U.S.A.

Detrick Field, located in Frederick, Maryland, was originally established in 1931 as a small municipal airport. In 1941, it was converted to a U.S. Army Air Force squadron facility. In 1943 it was renamed Camp Detrick when it became the U.S. Government's biological warfare (BW) facility for the study of foreign plant pathogens, the development and purification of biological warfare agents, vaccines, toxoids, disinfectants, and antiseptics. *B. anthracis* was a major microorganism of research focus. The facility was renamed Fort Detrick in 1956. In 1969 President Richard Nixon signed an executive order outlawing offensive biological research by the U.S. Since then, research at Fort Detrich has been purely defensive in nature, focusing on diagnostics, preventives, and treatments for BW agent infections. At its peak strength in 1945, Camp Detrick had a total of 1,770 personnel encompassing a total of 1,200 acres. Two accidental anthrax deaths are known to have occurred at the facility involving individuals at risk during its first decade as a BW facility.

From October 4 to November 2, 2001, 11 confirmed cases of inhalation anthrax and 11 confirmed cases of cutaneous anthrax were reported among persons who worked in the district of Columbia, Florida, New Jersey, and New York (Jernigan *et al.,* 2001). Epidemiological investigation indicated the intentional distribution of weaponized *B. anthrac*is spores of the Ames strain through mailed letters and packages, presumably by a Federally employed scientist, leading to the term "Amerithrex". Among the 12 inhalation cases, seven were postal employees in New Jersey and the district of Columbia. Eight of the 11 patients with inhalation anthrax were in the initial phase of illness when they first sought care. Among these eight, six received antibiotics active against *B. anthracis* on the same day and all six survived. Four patients in addition to one with meningitis were exhibiting fulminant symptoms on initial treatment with antibiotics active against *B. anthracis* and all five succumbed. Chapter 6 presents a detailed account of the forensic microbiology applied and of the presumed source of the mailed spores. Before this outbreak only 18 spontaneous cases of human inhalation anthrax had been reported in the South in the 20th century. The most recent spontaneous case was in 1976. Most cases have been related to animal products in textile mills processing goat hair, goat skins, and wool (Jernigan *et al.,* 2001). However, weaponized anthrax spores should be assumed to be capable of resulting in over 90% levels of lethal infections resulting from inhalation in the absence of early antibiotic treatment.

As a result of this mail distribution incident Wood and Martin (2009) reported on the development of a mobile trailer (3 m x 14.6 m x 4 m) housing a chlorine dioxide (ClO_2) generating system for the decontamination of buildings contaminated with spores of *B. anthracis*. Separate support trailers contained the N_2 gas supply and a portable 60 kW electric generator to operate the scrubber fans and pumps. The initial design was based on generation of 54.43 kg/h of ClO_2, while maintaining the ClO_2 concentration in each generator line below 8%, and a relative humidity of ~80% in a 100,000 ft^2. building at 21 $^{\circ}$C. The ClO_2 generating procedure utilized 4% of Cl_2 gas, set at a maximum flow of 0.11 m^3/min. passed through a porous bed of thermally stable solid sodium chlorite where N_2 gas was the diluent, set at a minimum flow of 2.83 m^3/min. The ClO_2 generating reaction was:

$$2NaClO_2 \text{ (solid)} + Cl_2 \text{ (gas)} \longrightarrow 2ClO_2 \text{ (gas)} + 2NaCl \text{ (solid)}$$

The United Kingdom

Gruinnard Island, is a small, oval-shaped Scottish island about 2 miles long and 0.62 miles wide located in Gruinnard Bay, about 0.68 miles from the mainland. It has a total area of 0.76 square miles. The Scottish island was used in 1942 and 1943 to assess spores of *B. anthracis* as biological weapon. Most trials were conducted by placing a bomblet, containing a heavy suspension of spores of *B. anthracis*, on the ground and detonating it electronically. The effectiveness of the bomblets was assessed by exposing sheep, located downwind of the resulting airborne cloud of spores. The number of spores inhaled by each sheep was measured by taking air samples next to each animal. Many sheep were found to have inhaled infectious numbers of spores. The dispersal of the spore cloud was found to have caused only light surface contamination over much of the island. However, only a small percentage of the fluid spore suspension was dispersed as an aerosol in the respirable range of one to five microns. The remainder was scattered as large globules over the ground at the detonation point and for a short distance downwind. Annual soil sampling from 1946 to 1969 indicated that although the number of spores was slowly declining, contamination was likely to persist well into the 21st century. Core sampling covering the entire island in 1979 indicated that only a small area of the island (2.6 ha) was still contaminated and nearly all of the spores were confined to the top 10 cm of soil (Manchee *et al.*, 1981; Manchee *et al.*, 1983). Preliminary studies by Manchee *et al.,* (1994) indicated that among four biocides, the most effective was a 5% solution of formaldehyde and that a seven-day exposure was found to effectively inactivate all bacterial spores present in the soil (Manchee *et al.*, 1983). Large scale application of 5% formaldehyde to the contaminated area was conducted with some 1 m^2 areas receiving as much as 20 liters/m^2. Removal of ground water and vegetation was also considered desirable. On the basis of these preliminary results, full-scale decontamination of a 4.1 ha area was undertaken. Vegetation was first sprayed with a herbicide and one week later the dead vegetation was burnt. The treatment area was then covered with parallel runs of irrigation tubing (0.9m apart) which delivered the 5% formaldehyde as a fine surface spray. A total of 50 Km of tubing was required for delivery of 50 liters of 5% formaldehyde over a 10 min. period. The site of a suspected aerial bomb drop (with deeper penetration of *B. anthracis* spores into the soil) was treated by injecting 38% formaldehyde to the depth of the bedrock. Two months after formaldehyde treatment, core samples indicated that among 58 initially contaminated points, only three were still contaminated after formaldehyde treatment. Following formaldehyde injection at the aerial bombsite, only two of 88 core samples were still positive for spores of *B. anthracis* with 10^6 spores/g of soil at a depth of 50 cm. Additional injection was applied to this zone and 6 months later, cores samples were negative. The island was reseeded with grass and a flock of 40 sheep located on the island. After grazing for five months, all of the sheep were healthy. This study suggests that it may be possible to decontaminate localized soil areas with formaldehyde where animals have fallen and succumbed to anthrax with subsequent decay of the carcass over an extended period of time.

Zimbabwe

In the modern world the largest recorded outbreak of anthrax occurred in Zimbabwe (formerly Rhodesia). Before 1978, there was an average of 13 human cases of anthrax annually. From 1979 through 1980, the revolutionary period, there were about 10,000 cases of anthrax. Of these, 182 died (Graysmith, 2003). This massive outbreak is thought to have been initiated by the Rhodesian military.

REFERENCES

Aikimbayev, A., Lukhnova, L., Zakaryan, S., Temiraliyeva, G., Y. Pazylov, T., Meka- Mechenko, T., Easterday, Van Ert, M., Keim, P., Hadfield, S., Francesconi, S., Blackburn, J., Hugh-Jones, M. (2007). Molecular diversity of *Bacillus anthracis* in Kazakhstan. Abstract no. P1851, 17th ECCMID & 25th ICC Conf. March 31 – April 4, 2007, Munich, Germany. Cherkasskiy, B. (1999). A national register of historic and contemporary anthrax foci. *J. Appl. Microbiol.* 87:192-195.

Dragon, D., Elkin, B., Nishi, J., Ellsworth, T. (1999). A review of anthrax in Canada and implications for research on the disease in northern bison. J. Appl. Microbiol. 87:208- 213.

Graysmith, R. (2003). Amerithrax –The Hunt for the Anthrax Killer. The Berkeley Publishing Group, N.Y., N.Y. 454 Pg. Guillemin, J. (1999). Anthrax: The Investigation of a Deadly Outbreak. University of California Press, Berkeley, CA 339 Pg. Folstad, I., Nilssen, A., Halvorsen, O., Anderson, J. (1989). Why do male reindeer (*Rangifer t. tarandus*) have higher abundance of second and third instar larvae of *Hypoderma tarandi* than females. Oikos. 55:87-92.

Hugh-Jones, M. (1999). 1996-97 global anthrax report. J. Appl. Microbiol. 87:189-191.

Imnadze, P., Bakinidze, M., Manrikyan, M., Kukhalashvili, E., Zangaladze, M., Kelelidze, M., Tsreteli, D., Nadiradze, M., Tsanava, S., Tserstvadze. N. (2002). Anthrax in southern caucasus. Antibiot. Monit. 18:17-18.

Jernigan, T., Stephens, D., Ashford, D., Omenaca, C., Topiel, M., Galbraith, M., Tapper, M., Fisk, T., Zaki, S., Popovic, T., Meyer, R., Quinn, C., Harper, S., Fridkin, S. Sejvar, J., Shepard, C., McConnell, M., Guarner, J., Shieh, W., Malecki, J., Gerberding, J., Hughes, M, Keim, P., Kalif, A., Schupp, J., Hill, K., Travis, S., Richmond, K., Adair, D., Hugh-Jones, M., Kuske, C., Jackson, P. (1997). Molecular evolution and diversity in *Bacillus anthracis* as detected by amplified fragment length polymorphism markers. J. Bacteriol. 179:818-825.

Keim, P., Klevytska, A., Price, L., Schupp, J., Zinser, G., Smith,K., Hugh-Jones, M., Okinaka, R, Hill, K, Jackson, P. (1999). Molecular diversity in *Bacillus anthracis*. J. Appl. Microbiol. 87:215-217.

Keim, P., Price, I., Klevytska, A, Smith, K, Schupp, J, Okinaka, R, Jackson, P, Hugh-Jones, (2000). Multiple-locus variable-number tandem repeat analysis reveals genetic relationships within *Bacillus anthracis*. J. Bacteriol. 182:2928-2936.

Keim, P., Smith, K., Keys, C., Takahashi, H., Kurata, T., Kauafman, A. (2001). Molecular investigation of the Aum Shinrikyo anthrax release in Kameido, Japan. J. Clin. Microbiol. 39:4566-4567.

Manchee, R., Broster, M., Anderson, I., Henstridge, R., Melling, J. (1983). Decontam- ination of *Bacillus anthracis* on Gruinard Island? Nature (London). 303:239-240.

Manchee, R., Broster, M., R., Melling, J., Henstridge, R., Stagg, A. (1981). *Bacillus anthracis* on Gruinard Island. Nature (London). 294:254-255.

Manchee, R., Broster, M., Stagg, A., Hibbs, S. (1994). Formaldehyde solution effectively inactivate spores of *Bacillus anthracis* on the Scottish island of Gruinnard. Appl. Environ. Microbiol. 60:4167-4171.

Mautel, N. (1967). The detection of disease clustering and a generalized regression approach. Cancer Res. 27:209-220.

Merabishvili, M., Natidze, M., Rigvava, S., Brusetti, L., Raddadi, N., Borin, S., Chanishvili, N., Tediashvili, M., Sharp, R., Barbesci, M., Visca, P., Daffonchio, D. (2006). Diversity of *Bacillus anthracis* strains in Georgia and of vaccine strains from the former Soviet Union. Appl. Environ. Microbiol. 2006. 72:5631-5636.

Meselson, M., Guillemin, J., Hugh-Jones, M., Langmjir, A., Popova, I., Shelokov, A. Yampolskaya, O. (1994). The Sverdlovsk anthrax outbreak of 1979. Science. 266:1202-1208.

Narayan, S., Sreelakshmi, M., Sujatha, S., Dutta, T. (2009). Anthrax meningoencephalitis — Declining trends in an uncommon but catastrophic CNS infection in rural Tamil Nadu, South India. *J. Neurol. Sci.* 281:244-248.

Patra, G., Vaissaire, J., Weber-Levy, M., Le dooujet, C., Mock, M. (1998). Molecular characterization of *Bacillus* strains involved in outbreaks of anthrax in France in 1997. J. Clin. Microbiol. 36:3412-3414.

Pilo, P., Perreten, V., Frey, J. (2008). Molecular epidemiology of *Bacillus anthracis*: determining the correct origin. Appl. Environ. Microbiol. 74:2928-2931.

Plotkin, S., Brachman, P., Utell, M., Bumford, F., Atchison, M. (2002). An epidemic of inhalation anthrax, the first in the twntieth century: I. clinical features. *Amer. J. Med.* 112:4-12.

Redmond, C, Pearce, M., Manchee, R., Berdal, B. (1998). Deadly relic of the great war. Nature. 393:747-748.

Smith, K., de Vos, V., Bryden, H., Hugh-Jones, M., Klevgtska, A., Price, L. (1999). Meso- scale ecology of anthrax in Southern Africa: a pilot study of diversity and clustering. *J.* Appl. Microbiol. 87:204-207.

Sue, D., Marston, C., Hoffmaster, A., Wilkins, P. (2007). Genetic diversity in a *Bacillus anthracis* historical collection (1954-1988*).* J. Clin. Microbiol. 45:1777-1782.

Thompson, M. (2003). The killer strain. Harper Collins, Publ. N.Y., 246 pages. Turner, A., Galvin, J., Rubira, R., Miller, T. (1999). Anthrax explodes in an Australian Summer. J. Appl. Microbiol. 87:196-199.

Wood, J., Martin, G. (2009). Development and field testing of a mobile chlorine dioxide generation system for the decontamination of buildings contaminated with *Bacillus anthracis. J. Hazard. Matls.* 164:1460-1467.

Yakupogullari, Y., Koroglu, M. (2007). Nosocomial spread of *Bacillus anthracis*. J. Hosp. Infect. 65:401-402.

CHAPTER 2

Characteristics of *Bacillus anthracis*

Abstract: *Bacillus anthracis,* the causative organism of anthrax is a member of the *B. cereus* group of bacilli. The stained organism exhibits a unique and characteristic "Boxcar" appearance microscopically. The three forms of anthrax: (1) cutaneous, (2) inhalation, and (3) gastrointestinal are presented with clinical details. The bacteriology of *B. anthracis* is presented in terms of its minimal diagnostic characteristics, cultivation, colony appearance, along with the use of diagnostic bacteriophage. All of the major outer structural components of *B. anthracis* (capsule, cell wall, S layer) are pictorially illustrated and discussed. The extracellular enzymes such as hemolysins and phospholipases, in addition to the intracellular superoxide dismutases that presumably influence virulence are presented in detail. Studies on the long-term persistence of *B. anthracis* spores in soil are described along with stability of the 2 major virulence determining plasmids.

Keywords: Cutaneous anthrax, inhalation anthrax, gastrointestinal anthrax, bacteriology, morphology, cell wall, Capsule, S-layer, boxcar shape, clinical aspects, subclinical infections, cultivation, selective media, hemolysins, phospholipases, superoxide dismutases, γ bacteriophage, Immunoassays, Rhizosphere, Soil.

INTRODUCTION

Bacillus anthracis has uniquely evolved genetically from presumably *B. cereus* by acquiring two plasmids, pXO1 encodes the protective antigen (PA), the lethal toxin (LT), and the edema toxin (ET). These interactive proteins along with the novel nonimmunogenic capsule that is formed only under conditions of elevated CO_2 as occurs in the infected mammalian body, and whose genes are encoded by the pXO2 plasmid, constitute the extremely lethal components of the anthrax bacillus. The organism is considered a member of the *B. cereus* group of closely related spore forming bacilli. In addition the production of spores allows the organism to be notably resistant to desiccation and other adverse environmental factors and to persist for many decades in contaminated soil. This chapter deals specifically with these unique characteristics of *B. anthracis*.

MORPHOLOGY

B. anthracis is one of the largest pathogenic bacteria and ranges from 4.5 to 10μ in length. The ends of the rods are often concave and somewhat swollen resulting in chains of cells often described as resembling a "jointed bamboo fishing rod" or a chain of "boxcars" (Fig. **1**). In the body, cells appear singly, in pairs, or in short chains, while in pure culture long chains are formed. The rods are non-motile except for certain strains isolated in China (Liang and Yu, 1999). Capsules are usually present in smears from infected tissue but are usually absent on conventional media. The capsule is not polysaccharide as with most bacteria but is composed of poly-γ-D-glutamic acid and is the first demonstrated natural occurrence of D (-) glutamic acid constituting a polypeptide composed of a single amino acid. Spores are formed as microscopic refractile bodies centrally located in vegetative cells and may also be seen free. Spores are formed most abundantly at 32 to 35 °C and only under aerobic conditions and are not observed in infected blood or tissue. The bacillus is Gram-positive, however the spores are stained with difficulty and require hot carbol fuchsin and are difficult to decolorize. Colonies are irregular, and have a curled or hair-like structure, and consist of tangled coils of long chains of bacilli (Fig. **2**).

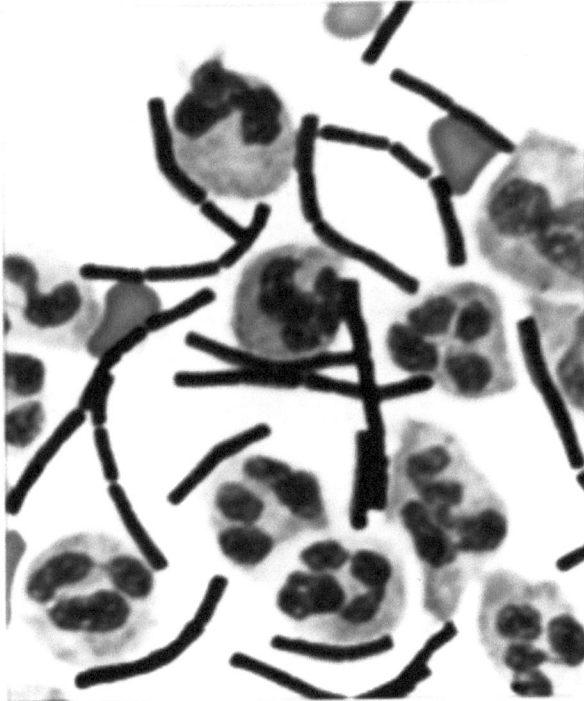

Figure 1: Gram stain of cerebrospinal fluid containing *B. anthracis*, displaying the Characteristic "boxcar-shaped" morphology. From Sejvar, Tenover, and Stephens (2005). With permission.

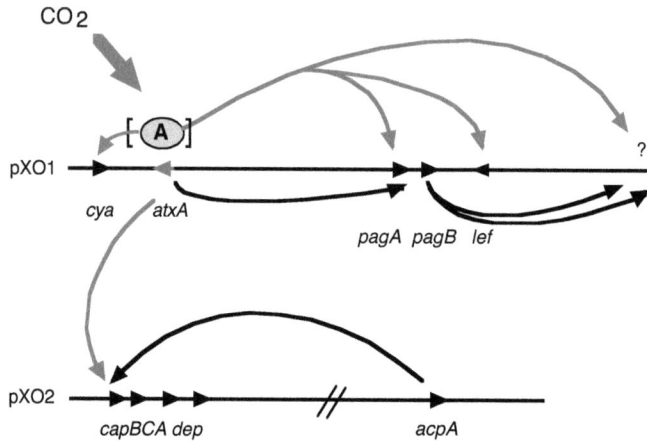

Figure 2: Gram stain of chains of *B. anthracis*. From CDC, Public Health Image Library at: http://www.cbwinfo.com/Images/1898.jpg. bThallous acetate is toxic and should be handled with care; avoid skin contact or inhalation of the powder.

B. Anthracis as a Member of the *B. Cereus* Group

B. anthracis is a member of the *B. cereus* group consisting of six species of the spore forming genus *Bacillus*, *B. cereus*, *B. anthracis*, *B. thuringiensis*, *B. mycoides*, *B. pseudomycoides*, *B. weihenstephanensis*, all of which are considered closely related. *B. anthracis* is therefore phenotypically difficult to distinguish from these other bacilli. 16S rDNA sequence analysis is unable to distinguish isolates of *B. anthacis, B. cereus,* and *B. thuringiensis* (Ash *et al.*, 1991). The major genetic distinctions between *B. anthracis* and the other members of the *B. cereus* group involve the unique poly–γ-D-glutamic acid capsule and the toxin genes. Major phenotypic characteristics of *B. anthracis* isolates that distinguish them from other members of the *Bacillus cereus* group are presented in Table **1**.

Table 1: Major phenotypic characteristics of *B. anthracis* isolates that distinguish them from other members of the *Bacillus cereus* group

Characteristic	Appearance or reaction of *B. anthracis* isolates
Colonies on blood agar	White or grey-white, 2-4 mm diam., tacky
Hemolysis (after 24 hrs. incubation at 37 °C)	Negative or very weakly hemolytic
Gamma phage	Plaques produced
Penicillin	Sensitive
Motility	Negative except for certain Chinese strains
PCR for capsule & toxin genes	Positive

THE INFECTIOUS PROCESS

Clinical Aspects of Human Anthrax

The pathogenicity of anthrax is mediated by the presence of two virulence plasmids pXO1 and pXO2. pXO1 encodes three proteins designated PA (protective antigen), LF (lethal factor), and EF (edema factor). Both LF and EF are able to bind to PA forming what is known is the "anthrax toxin". PA is required for the formation of a membrane penetrating pore allowing translocation of LF and EF to the cytoplasm of macrophages. LF when bound to PA is referred to as the "lethal toxin" or LF; EF bound to PA is referred to as the "edema toxin" or ET. PA, LF and EF are encoded by the *pagA, lef,* and *cya* genes on the pXO1plasmid, respectively. The genes encoding capsule synthesis are located on the pXO2 plasmid and are found in the *capBCAD* operon. Capsule synthesis occurs only in the presence of elevated levels of CO_2 as is found in mammalian tissues and when supplemented during cultivation in media. The presence of the capsule is considered essential for establishment of a systemic infection in most mammals.

Among the three forms of anthrax in humans, cutaneous, inhalation, and alimentary tract, the cutaneous form is the most frequent, accounting for over 95% of all recorded cases (Turnbull, 1999a). In cutaneous anthrax, approximately three to five days after entry of the organism through a lesion in the skin, a small pustule appears which soon becomes a vesicle containing a brownish fluid. Over the next two to three days, the center of the pustule ulcerates and becomes a dry dark brown or black scab surrounded by a ring of vesicles. This is the characteristic black anthrax eschar (Chapter 3, Fig. **1**) which is usually 2 cm or more across and is accompanied by significant edema surrounding the lesion. With most cases the eschar begins to resolve in about 10 days after the initial appearance of the pustule and resolves in two to six weeks.

There are two clinical forms of alimentary tract anthrax: (1) oropharyngeal in which the major symptoms consist of soar throat, dysphagea (inability to swallow), fever, regional lymphadenopathy in the neck, and toxemia, and (2) intestinal anthrax resulting in nausea, vomiting fever, abdominal pain, haematemesis (vomiting blood), and bloody diarrhea, leading to shock, collapse, and death. Historically, intestinal anthrax has rarely occurred in humans although it has frequently been the path of infection in cattle that have grazed in infected pastures.

The distribution of weaponized anthrax spores through the mail in late 2001 led to classic symptoms of inhalation anthrax. The average incubation period was four days, with patients exhibiting fever or chills, sweating, fatigue, minimal or non-productive cough, dyspnea (difficulty in breathing), nausea and vomiting. Chest-x-rays indicated infiltrates and pleural effusion (Jernigan *et al.,* 2001). Untreated initial mild symptoms end in the rapid onset shock, collapse, and death with meningitis being a potential complication.

Infections leading to septicemia result in enormous multiplication of bacterial cells occurring in the blood and internal organs. The spleen is a deep-red color and is greatly enlarged, hence the name "splenic fever". The U.S. Department of Defense estimates that the lethal dose for humans is approximately 8,000 - 10,000 spores (Inglesby, Henderson, and Bartlett, 1999).

The feeding of spore-free cells to highly susceptible animal species is usually innocuous due to the destruction of the vegetative cells by the acidity of the stomach. The feeding of spores leads to infection of the more susceptible species.

Subclinical Infections

Asymptomatic *B. anthracis* bacteremia has been observed in chimpanzees (Albrink and Goodlow, 1959). In addition, serological evidence among goat hair workers suggests that subclinical or mild forms of anthrax occur in humans (Norman *et al.,* 1960). Wattiau *et al.,* (2008) studied the microbial flora of a Belgian factory that processes wool and goat hair from all over the world. Viable spores of *B. anthracis* were found in goat hair fibers, air dust, and unprocessed waste water produced from goat hair scouring. No clinical cases of anthrax were recorded among the employees except for a possible cutaneous lesion, reported by a worker in 2002. None of the workers were vaccinated against *B. anthracis*. In a subsequent study, Wattiau *et al.,* (2009) obtained blood samples from 66 of these 67 employees in this Belgium factory and an ELISA assay was used to detect anti-PA IgG. Among the 66 blood samples obtained the first year (2006), 3 employees had protective IgG levels and the titer for a fourth was considered borderline protection. A year later, in 2007, 54 employees yielded similar ELISA results except for three additional borderline individuals. Western dot blot analysis yielded 4 individuals positive for IgG to PA and to LF the first year of sampling and 6 individuals positive the second year of sampling. The authors concluded that seroconversion most likely resulted from asymptomatic *B. anthracis* infections.

Until recently if an isolate was unable to produce anthrax in a laboratory animal it was discarded without being given a species designation, or it was called *B. cereus*, or given a name such as *B. anthracis* similis. In view of the present emphasis on rapid identification of *B. anthracis* with respect to bioterrorism it has become of greater importance to achieve absolute identification as rapidly as possible, particularly with respect to national security. Isolates with characteristic colony morphology, are markedly tacky, white or grayish white, non-hemolytic on sheep blood, with curling of colony edges, non-motile (with the exception of certain Chinese strains), sensitive to penicillin, and the diagnostic γ phage, able to produce the capsule on agar containing 0.7% $NaHCO_3$ following incubation in a $5 - 20\%$ CO_2 atmosphere can be considered virulent *B. anthracis* (Turnbull, 1999b). In addition, the PCR can be used to confirm the presence of the capsule and the toxin genes. The only time a problem arises in identification of potential isolates involves the isolation of rare strains that are Cap^-/tox^+. Such strains would certainly not be isolated from cases of anthrax nor would they be encountered in "weaponized" *B. anthracis*.

Cultivation of *B. anthracis*

The organism can be grown on simple synthetic media. Thiamine, Mg^{++}, Fe^{++}, and Ca^{++}, are required in addition to a source of energy and nitrogen. Uracil, adenine, guanine, and Mn^{++} markedly stimulate growth (Brewer *et al.*, 1946). Growth occurs at temperatures as high as 41 to 43 °C with an optimum at 37 °C. The organism is facultatively anaerobic. Dextrose and trehalose are fermented rapidly but without gas production. Lactose, galactose, mannitol, dulcitol, rhamnose, and xylose are not fermented.

When *B. anthracis* is grown on Nutrient Agar or in broth tubes in air, the cells appear as endless filaments with no capsule. With agar cultures, 0.7% sodium bicarbonate and an atmosphere of 5 - 20% CO_2 results in capsule formation. Liquid cultures require at least 50% horse serum to yield capsules. Overnight incubation on blood agar yields white or grey-white colonies, 2 to 4 mm in diameter with little or no hemolysis. Colonies have a tacky texture and can be made to stand up from the colony with a loop (Fig. **3**), which results from the long chains. Aerobic growth is required for sporulation but not for germination. Protease, amylase, catalase, lecithinase, and collagenase activities have been found in culture filtrates. Confirmatory tests following these presumptive observations consist of lysis by the diagnostic "gamma" phage, sensitivity to penicillin, and capsule production when grown in blood or serum on bicarbonate agar with 5 to 20% CO_2. Only the lack of motility, hemolysis, and capsule production under requisite conditions are considered helpful in distinguishing *B. anthracis* from other members of the *B. cereus* group. In addition, the sequences of primer pairs have been identified for PCR amplification of gene sequences derived from the toxin and capsule genes (see chapter 6).

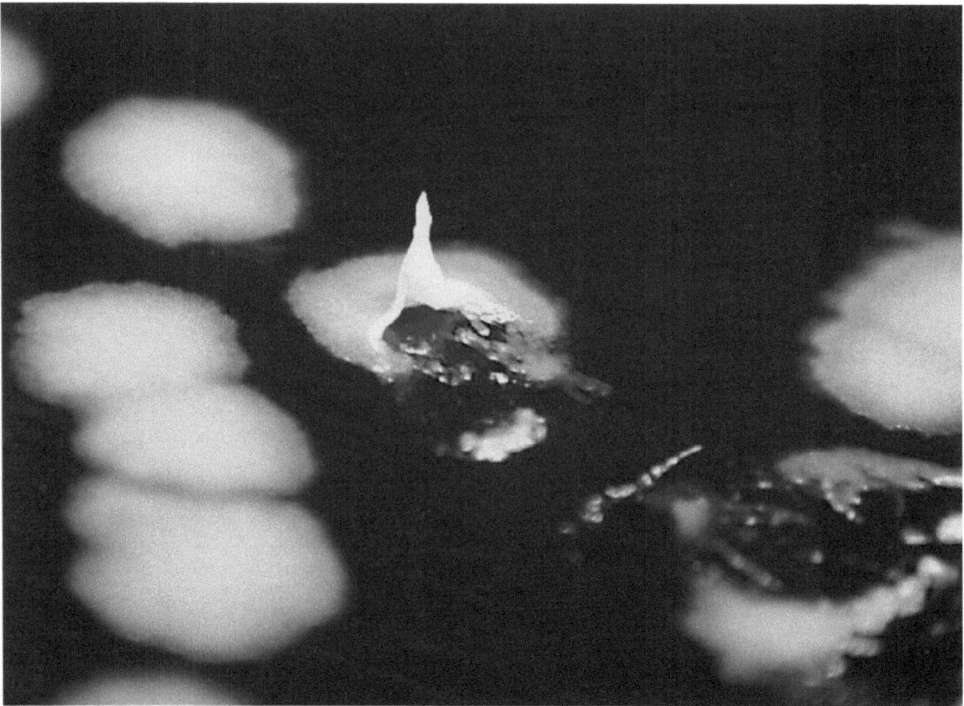

Figure 3: Characteristic vertical spike of *B. anthracis* colony after being teased due to long chains of bacilli. From CDC, Public Health Image Library at: http://www.cbwinfo.com/Images/1898.jpg.

Selective Isolation Media for *B. anthracis*

No selective enrichment broth has been developed as yet for *B. anthracis*. However, among selective agar media, the most successful historically is polymyxin-lysozyme-EDTA thallous acetate (PLET) agar (Table **2**) developed by Knisely (1966). It is considered the best selective agar for isolation of *B. anthracis* from processed or decomposed animal specimens, or food or environmental samples contaminated with other organisms including *Bacillus* species (WHO, 2008). However, its performance is notably dependant on how it is prepared, and the source of the Heart Infusion Agar (HIA) used and adequate cooling of the molten agar before addition of polymyxin and lysozyme. The final unadjusted pH was 7.35. Good recovery of vegetative cells was obtained in 24 hrs. at 37 °C with 20 of 21 strains of *B. anthracis*. One strain (ATCC 944) required 48 hrs. to yield a recovery of 29%. Recovery of vegetative cells from the other 20 strains *of B. anthracis* varied from 42 to 135% based on colony counts on HIA without selective components. A spore suspension of the Vollum 1B strain was recovered without difficulty. Ten of eleven strains of *B. cereus* and 5 strains of *B. mycoides* were inhibited. No inhibition occurred with the *B. cereus* strain ATCC 7064. On PLET, the colonies of this strain were very minute after 24 hrs at 37 °C, and after 48 hrs, the colonies were small and approximately the same size as 24 h-hr. *B. anthracis* colonies. The majority of other *Bacillus* species were also inhibited on the PLET medium with the exception of one lysozyme resistant *B. megaterium* strain and several *B. subtilis* strains. A wide variety of Gram-negative bacteria were inhibited. *Staphylococcus aureus* and *Streptococcus faecalis* were not inhibited.

Table 2: Polymyxin-Lysozyme-EDTA-Thallous Acetate (PLET) agar.[a]

Ingredient	Grams or units per liter
Heart Infusion agar	40.0
Thallous acetate[b]	0.04
$Na_2.EDTA$	0.30
Polymyxin B	30,000 U
Lysozyme	300,000 U

[a]Filter sterilized polymyxin B (0.5 ml) as a stock solution of 60,000 U/ml and filter sterilized lysozyme (0.5 ml) as a stock solution of 600,000 U/ml should be added to the autoclaved agar at the lowest possible temperature (45 - 47 °C) and rapidly poured into plates before thermal denaturation occurs.

Although PLET agar has been traditionally used for the isolation of *B. anthracis*, it contains highly toxic thallium acetate at a high concentration, which excludes its use in many countries. Tomasco *et al.,* (2006) compared the growth characteristics of 92 *B. anthracis* isolates and 132 other *Bacillus* strains representing 30 species on Cereus Identification Agar™ (CEI) and Anthrax Blood Agar™ (ABA). Thirty-seven of the non-*B. anthracis* isolates yielded on CEI Agar similar growth to *B. anthracis*. Thirty-six of the non-*B. anthracis* isolates yielded on ABA Agar similar growth to *B. anthracis.* The combination of CEI and ABA yielded only 10 of the non-*B. anthracis* isolates, with growth undistinguishable from *B. anthracis.*

In contrast to *B. anthracis*, the majority of *Bacillus* spp. enzymatically released a chromophore on CEI, thus acquiring a turquoise coloration in the center of the colonies. On ABA, *B. anthracis* colonies were gray, white or cream colored, with a diameter of 1-10 mm and were non-hemolytic. A "thread-like" consistency and resulting "Spikes" are considered typical of *B. anthracis* colonies.

B. anthracis data bases have been developed for the commercial API and Biolog systems facilitating species identification of *B. anthracis*. The API SCHB system for the identification of *Bacillus* species is depends on acidification of 49 carbohydrates and should be used along with the API 20 test strip. The Biolog system is based on "carbon source metabolic fingerprints" in 96 well-well trays. Results should be confirmed with the tests listed in Table **1** and/or the PCR (see chapter 6 for primers).

Use of Diagnostic Bacteriophage with Presumptive Isolates of *B. anthracis*

Among the standard diagnostic tests for identification of *B. anthracis* strains recommended by the U.S. Centers for Disease Control and Prevention (CDC) is testing for γ phage sensitivity. Although gamma phage exhibits a fairly narrow host range, being primarily limited to strains of *B. anthracis,* several *B. cereus* strains have been shown to be sensitive to infection by this phage (Abshire, Brown, and Ezzell, 2005; Schuch, Nelson, and Fischetti, 2002). The wild type AP50 phage, was originally found to have a narrow host range, with only one-third of 34 *B. anthracis* strains and none of the 52 strains belonging to six different *Bacillus* spp. susceptible to infection by AP50 (Nagy and Ivanovics, 1977). The AP50 wild type phage designated AO50t produces turbid plaques characteristic of lysogenic phages, and spontaneous clear plaque variants designated AP50c. The clear plaques are larger after incubation at 25 °C than after incubation at 37 °C and are larger in phage assay agar (Thorne, 1968) than on Luria-Bertani or blood agar plates. Sozhamannan *et al.,* (2008) found that phage AP50c infected 111 of 115 *B. anthracis* strains (~97%) and none of 100 *B. cereus* sensu lato strains. In contrast, 10 out of 115 *B. anthracis* strains were resistant to the γ phage, and among these, 9 strains were infected by the AP50c phage while 3 of the 4 AP50c-resistant strains were infected by the γ phage. The authors speculated that phages AP50c and γ probably have different target bacterial receptors and concluded that a combination of these two phages might be a better alternative for phage-based diagnostics. It should be noted that gamma phages do not lyse encapsulated *B. cereus* cultures (McCloy, 1951).

Diagnostic Immunoassays for *B. anthracis*

Immunoassays along with real-time PCR (Rti-PCR) can be used for verification of viability and confirmation of virulence. This requires recovery of spore DNA after a positive immuno-assay. The strong denaturant and lysis reagents usually used for dissociation of spores from antibodies can interfere with the PCR.

Hang *et al.,* (2008) described the development of methodology involving (1) a rapid immunoassay procedure using capillary tubes and an Integrated Waveguide Biosensor (IWB) followed by (2) germination and outgrowth of spores in Brain Heart Infusion (BHI) to determine viability, and (3) lysis of resulting vegetative cells and the PCR for confirmation. A "liquid phase" capture immunoassay was also described where spore suspensions in 1 ml of 0.05% Tween-20 and 20% BSA in PBS (PBSTB) were incubated together with both biotinylated and Cy5-labeled anti-*B. anthracis* antibodies at room temperature for 1 hr. Unbound antibodies were then removed from the spores with bound antibodies using an Ultrafree-MC 0.1 micron porosity centrifugal filter. Spores with bound antibodies retained on the filter were then resuspended in PBSTB to remove any non-specific bound antibody. Spores with bound antibodies were then resuspended in PBSTB and added to capillary tubes coated with neutravidin and blocked with 2% BSA and then incubated for 2 hrs. at room temperature. The capillary tubes were then washed with PBST to remove any non-specifically bound spores and fluorescence intensity determined with an IWB unit. The capillary tubes were then drained and filled with Brain heart Infusion (BHI) which the authors found would result in bound spores germinating) in 5 min. (loss of heat resistance, 70 °C for 30 min.). Incubation was then at 37 °C for 50 min. to allow outgrowth of spores to vegetative cells. The capillary tubes were then drained and vegetative cells lysed for PCR. The limit of detection was 10^3 spores per capillary tube.

STRUCTURAL COMPONENTS OF *B. anthracis*

Cell Wall

The cytoplasmic membrane of *B. anthracis* is surrounded by a thick peptidoglycan cell wall (Fig. **4**). Only one associated polymer, a polysaccharide composed of N-acetylmannosamine, galactose, and N-acetylglucosamine, is covalently linked to the peptidoglycan.

Guanidine extracts of crude *B. anthracis* cell wall material from vegetative cells were used by (Ezzel *et al.*, 1990) to develop two IgM monoclonal antibodies (MAbs) to cell surface antigens. The MAbs appeared to be directed to an epitope associated with the galactose-N-acetyl-D-glucoseamine polysaccharide. Both demonstrated specificity in binding to purified *B. anthracis* cell walls, intact nonencapsulated vegetative cells and *O*-stearoyl-polysaccharide conjugates. The interaction of the MAbs with purified cell wall polysaccharide from *B. anthracis* was inhibited by 0.5M lactose and galactose, but not by glycine, glycerol, glutamate, or N-acetylglucosamine. Electron microscopy revealed that both MAbs interacted with the the cortex of spores and the cell walls of vegetative cells. Neither MAb reacted with encapsulated vegetative cells or with intact spores. The MAbs conjugated to fluorescein isothyocyanate stained all 96 strains of *B. anthracis* and none of the strains of 19 other *Bacillus* spp. However, 2 of 31 *B. cereus* strains were stained. These two strains were considered transition forms between *B. cereus* and *B. anthracis* and could be distinguished from *B. anthracis* by being b-hemolytic and failing to produce PA and capsules under requisite conditions.

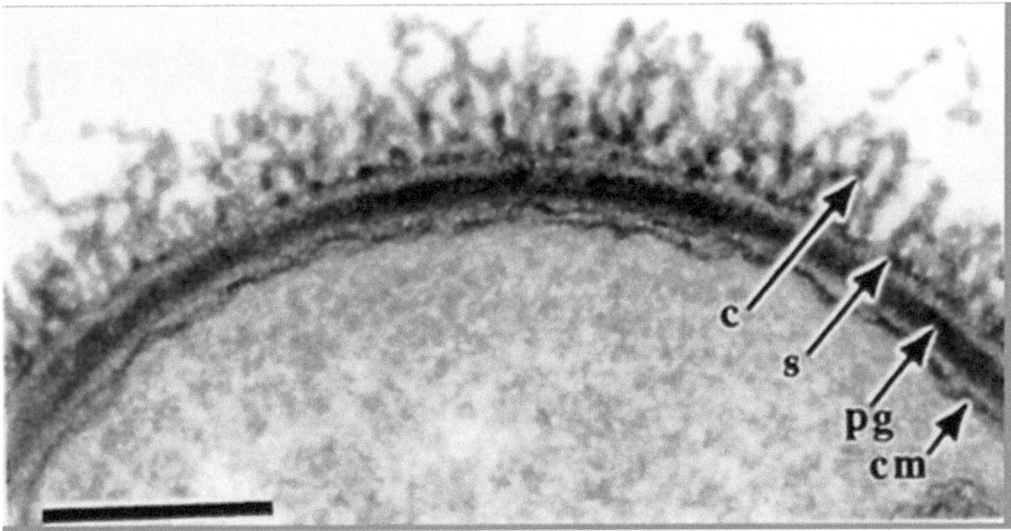

Figure 4: Electron microscopic visualization of envelope components on an encapsulated *B. anthracis* cell by negative staining of a thin section. *s*, S layer; *pg*, peptidoglycan cell wall; *cm*, cytoplasmic membrane; *c*, capsule. Bar 250 nm. From Fouet and Mesnage (2002) with permission.

S-Layer

Immediately outside the cell wall is an S-layer (surface layer, Fig. **4**), composed of two proteins designated Sap (surface array protein) encoded on the chromosome by the *sap* gene and EA1 (extractable antigen 1) encoded on the chromosome by the *eag* gene (Etienne-Toumelin *et al.,* 1995; Ezzell Jr., and Abshire, 1988). Fouet *et al.,* (1999) found that sera from infected mice recognized Sap and EA1. Mutants revealed that EA1 and Sap can each independently form an organized layer and are thus each S-layer components. There is no other S-layer component in that mutants doubly deleted for both *eag* and *sap* did not form structured cell surface layers. In addition, capsule synthesis occurred in double mutants (*eag⁻ sap*) indicating that capsule formation is independent of the S-layer.

The *sap* gene encoding the s-layer protein has been cloned and sequenced (Etiennne-Toumelin *et al.*, 1995). The SAP protein, is composed of 814 amino acid residues, including a classical prokaryotic 299-amino-acid signal peptide. The mature form has a calculated molecular mass of 83.7 kDa, is weakly acidic, contains only 0.9% methionine and no cysteine residues, and possesses many charged residues.

Capsule

A poly-γ-D-glutamate (PGA) capsule surrounds the S-layer (Fig. **4**) and is antiphagocytic. Its synthesis is dependant on environmental factors mimicking the mammalian host, such as bicarbonate and growth at 37 °C. An early observation indicated that an aqueous solution of methylene blue stains the capsule violet and is still used in the field to identify *B. anthracis* from infectious material (Fouet and Mesnage, 1999). The capsule is considered one of the two major virulence factors for *B. anthracis*. The genes encoding capsule synthesis are located on the 60-MDa pXO2 plasmid (Green *et al.*, 1985). The cistrons are arranged contiguously in the order *capB, capC, capA* and encode for proteins of 44, 16, and 46 kDa, respectively. The synthesis of the capsule and toxins is in part, under bicarbonate regulation by the protein product of the *trans*-acting *atxA* gene on the 100-MDa pXO1 toxin encoding plasmid and the *acpA* gene on the pXO2 plasmid. An additional protein is encoded by the *dep* gene located down stream from the *cap* region and appears to function as a depolymerase, hydrolyzing poly-D-glutamic acid into lower molecular weight polyglutamates that are presumed to inhibit host defense mechanisms (Ezell Jr. and Welkos, 1999). Strains that do not synthesize the capsule exhibit extremely attenuated pathogenicity. The presence of the capsule allows the organism to evade the host's immune system by inhibiting phagocytosis resulting in septicemia. In addition, the capsule is very weakly immunogenic. The presence of the capsule prevents accessibility of cellular antibodies to the cell surface or interior antigens.

Purified PGA has been found to rapidly clear from the blood of mice and to accumulate in the liver and spleen (Sutherland *et al.*, 2008). This rapid clearance from the blood contrasts with the much slower *in vivo* clearance of capsular polysaccharides which remain in the blood of mice for several days (Grinsell *et al.*, 2001; Sutherland *et al.*, 2008). Sutherland and Kozel (2009) found that PGA binds to and is internalized by J774.2 macrophage-like cells and accumulates in CD71 transferrin-receptor positive endosomes. The receptor mediated endocytosis inhibitors phenylarsine oxide and amantadine inhibited the uptake and binding of PGA in the cells. In addition PGA was found to be degraded in J774.2 cells starting four hours after uptake, with continuing degradation for at least 24 hrs. The authors concluded that the degradation of PGA may be responsible for its rapid clearance compared to the slow clearance of capsular polysaccharides.

Proteins for Identification of *B. Anthracis*

Krishnamurthy *et al.,* (2007) subjected five distinguishable strains of *B. anthracis* to mass-spectroscopic analysis of proteins in cell lysates following enzymatic cleavage to peptides. Resulting peptides were resolved on an HPLC column linked to a mass spectrometer. A total of 170 common proteins were found present among the five strains of *B. anthracis* out of a total of 266 proteins. The percent of matching proteins of four of the strains to the Vollum strain was used to identify each strain.

EXTRACELLULAR AND INTRACELLULAR ENZYMES OF *B. Anthracis*

Hemolysins of *B. Anthracis*

Read *et al.,* (2003) analyzed the complete sequence of the chromosome of the Ames strain of *B. anthracis* (~5.23 megabases) and found several chromosomally encoded proteins that presumably contribute to pathogenicity, including hemolysins, phospholipases, and iron acquisition functions. Almost all of the putative chromosomal virulence proteins were found to have homologues in *B. cereus.*

B. anthracis has earlier been considered non-hemolytic and the hemolytic genes transcriptionally silent (Klichko *et al.,* 2003). It was previously proposed that in the *B. anthracis* genome operons controlled by a pleotropic controlling regulator (PlcR) are silent, as a result of evolutionary-counter selection of a truncated and therefore inactive PlcR protein (Agaisse *et al.,* 1999). Earlier speculation suggested that one or more chromosomal factors of virulence may be produced by *B. anthracis* in addition to the LF and EF encoded on plasmid pXO1.

Klichko *et al.,* (2003) found that the hemolytic genes of *B. anthracis* encoding enzymes involved in red blood cell (RBC) lysis and designated anthralysins (Anls), could be induced by strict anaerobiasis. The Anl genes were found to be expressed *in vitro* within macrophages at the early stages of infection by vegetative bacilli after spore germination. Cooperative enhancement of pore-forming and phospholipase C activities of the Anls was found in hemolytic tests with human but not sheep RBCs. The authors began by using known sequences of hemolytic proteins in *B. cereus* and other bacteria as queries, and identifying homologous sequences in the genome *of B. anthracis*. Six open reading frames (ORFs) corresponding to putative membrane damaging proteins were found. All six were highly homologous to known phospholipases C and hemolysins from *B. cereus* as orthologs. Noncontiguous locations of the Anl genes are shown in Fig. **5**. Three of the genes encode for the enzymes phospotidylcholine-preferring phospholipase C (PC-PLC), sphingo-myelinase (SMase), and phosphatidylinositol-specific phospholipase C (PI-PLC) and have been designated *anlA, anlB,* and *anlC*, respectively. The other three Anls were identified as pore-forming hemolysins. AnlO is a thiol-activated, cholesterol-binding hemolytic factor. The nucleotide sequence of the putative *anl II* gene is 99% homologous to that of the *B. cereus hly II* gene, coding for hemolysin II. However it has two mutational differences compared to the *B. cereus* hemolysin II. The amino acid sequence of Anl III is identical to its ortholog in *B. cereus* and shares a high degree of similarity with the sequences of pore-forming hemolysins of other bacilli.

Under conditions of aerobic growth, *B. anthracis* displayed low, but detectable hemolysis on agar plates with human RBCs, visible solely below the colonies (Klichko *et al.*, 2003). In contrast, notably strong hemolytic activity occurred when colonies developed under strictly anaerobic conditions with sizeable zones of hemolysis of human RBCs surrounding each colony. No hemolysis of sheep RBCs occurred, even under anaerobic conditions. To identify the hemolytic properties of individual Anls the recombinant proteins were expressed in *E. coli* and purified. Neither AnlA nor AnlB alone demonstrated hemolysis of human RBCs. Together AnlA and AnlB exhibited strong cooperative lysis of human RBCs. Strong synergistic hemolysis of human but not sheep RBCs was also observed when AnlO was combined with AnlA. In addition, human RBCs were slowly lysed by Anl II, but no synergistic activity was observed when Anl II was combined with other Anls. AnlC did not exhibit hemolytic activity or synergism with other Anls. Primer pairs for five of the six Anls with the exclusion of primers for Anl II in addition to primers utilized for *eag, sap,* and *pag* and are given in Table **1** of chapter 6.

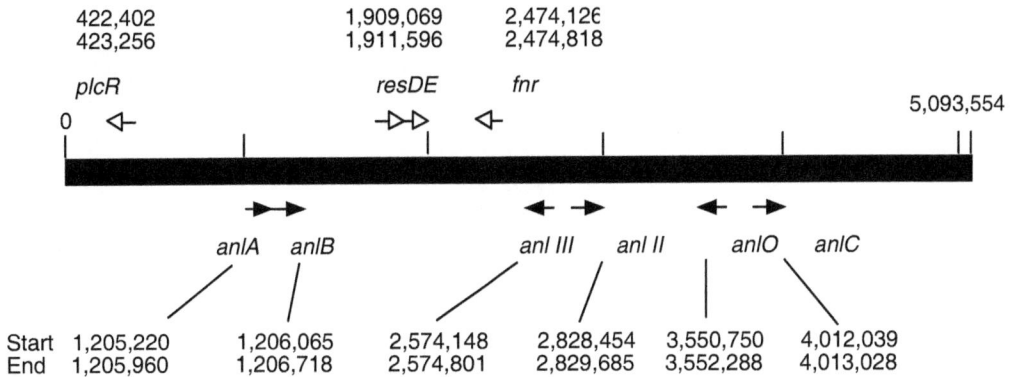

Figure 5: Location of the putative hemolysin Anl genes and their regulators in the genome of *B. anthracis* strain A2012. Solid arrows correspond to the Anl genes. Open arrows indicate possible transcription factors directly or indirectly involved in Anl gene regulation. *resdE* genes are homologous to the corresponding genes of *B. subtilis* involved in aerobic and anaerobic respiration. *fnr* encodes for a transcriptional regulator of the fumarate-nitrate reductase family. From Klichko *et al.,* (2003) with permission.

Although overt hemolysis is not a typical feature of human anthrax, signs of hemolytic anemia have been reported at the terminal stages of disease in guinea pigs (Ward *et al.,* 1965) and with some patients (Jernigan *et al.,* 2001). These observations suggest the *in vivo* production of hemolysins by *B. anthracis*. The use of blood agar plates prepared with human *versus* sheep blood incubated both aerobically and anaerobically may possibly serve as a highly definitive method for phenotypically confirming the identity of *B. anthracis* isolates.

Phospholipases

Phospolipase C's hydrolyze the polar head groups of phospholipids and exhibit a broad range of specificities dependant on recognition of the specific identity of the polar head group and the hydrophobic moiety. PC-PLC encoded by *placB* is able to act on phosphatidylcholine and a variety of other phospholipids. In contrast, SMase encoded by *smcA* and PI-PLC encoded by *plcA* are restricted to activity on sphingomyelin and phosphatidylinositol, respectively (Titball, 1998). Phospholipases contribute to bacterial virulence by involvement in deregulation of cellular signaling, phosphate acquisition, degradation of mucus layers, and tissue destruction (Schmiel and Miller, 1999) in addition to disrupting phagosomal membranes (Goldfine, *et al.,* 1998). PC-PLC, PI-PLC, and SMase of *B. cereus* a expressed as part of a regulon under the control of the transcriptional activator PlcR (Gohar *et al.,* 2002).

B. anthracis encodes a PlcR homolog that has a C-terminal truncation resulting in weak expression of the regulon leading to the absence of hemolysis on sheep blood agar. However, low levels of hemolysis have been reported. Hefferman *et al.,* (2006) characterized the respective contribution of three phospolipase C's (Plop's) to the pathogenesis of *B. anthracis* in DBA/25 mice. Deletion of all three PLC genes was required for attenuation of virulence in mice after intratracheal inoculation with spores. The authors attributed this attenuation to the possible inability of the PLC-null strain to grow in association with the macrophage and to cooperation among the PLC's to enhance virulence.

Hefferman *et al.,* (2006) also reported beta-hemolysis on sheep blood agar after 36 hrs. when plates were supplemented with 1 mM $CaCl_2$ and $MgCl_2$. Hemolysis was further enhanced when blood agar plates were incubated at 40 $^{\circ}$C. The hemolysis was attributable to SMase since the absence of hemolysis coincided with the deletion of *smcA*, whereas the disruption of *plcB* and *plcA* had no effect on hemolysis. PC-PLC and PI-PLC activities were detected respectively, on egg-yolk agar or BHI agar supplemented with a chromogenic derivative of myoinositol phosphate (X-PI). Deletion of the *plcB* gene resulted in the loss of the opaque zones beneath colonies on egg-yolk agar, reflecting the absence of PC-PLC. Strains containing a deleted *plcA* gene were white on X-PI plates due to the inability to release the blue chromogen from the chromogenic myoinositol derivative due to loss of PF-PLC activity. Function was restored in each null strain by complementation of the disrupted gene. The PlcR regulon was found to be nonfunctional in *B. anthracis*. Expression of the three PLC's was found to be independent of the PlcR regulon and pXO1.

Superoxide Dismutases

Cybulski *et al.,* (2009) found that four superoxide dismutases (SODs) SOD15, SODA1, SODC, and SODA2 are encoded by the *B. anthracis* genome which are capable of converting the superoxide radical (O_2^-) to molecular oxygen and hydrogen peroxide. Two (SOD15 and SODA1) are known to be located in the outer layers or exosporium of the *B. anthracis* spore (Liu *et al.*, 2004; Steichen *et al.*, 2003). Double mutants (*sod15⁻* and *sodA1⁻*) were found to have the same LD_{50} spore dose as the wild type strain for A/J mice in an intranasal infection model. A quadruple mutant strain (*sod15⁻, sodA1⁻, sodc⁻, and sodA2⁻*) resulted in more than a 40-fold increase in the LD_{50} spore dose for A/J mice infected intranasally. The authors concluded that the SOD molecules within the spore afford *B. anthracis* protection against oxidative stress and enhance the pathogenicity of *B. anthracis* in the lungs. The presence of the four alleles within the genome was considered to provide functional redundancy of this enzyme for enhanced virulence.

LONG TERM PERSISTENCE OF SPORES IN THE ENVIRONMENT

Longevity of Spores and Interaction of *B. anthracis* and The Rhizosphere

The multidecade persistence of *B. anthracis* spores in the soil and the environment constitutes a major epidemiological source in endemic anthrax areas of the world. Wilson and Russell (1964) reported that anthrax spores survived in dry soil for 60 years. Jacotot and Virate (1954) found that anthrax spores prepared by Pasteur in 1888 were viable 68 years later. In 1992 viable *B. anthracis* spores were found in the plaster and lagging of London's Kings Cross railway station roof space attributed to the use of contaminated horse hair to bind the plaster when the building was constructed a century earlier (Marston *et al.*, 2005). The longest report for survival of *B. anthracis* spores involved bones from an infected and buried animal carcass recovered during an archaeological excavation in Kruger National Park in South Africa which were estimated by carbon dating to be 200 + 50 years old (De Vos, 1990).

Anthrax spores are usually destroyed by boiling for 10 minutes and by dry heat at 140 °C for three hours. A 20% solution of formaldehyde will destroy spores in about ten minutes while a 5% solution requires about one hour of contact. The latter has been found effective for destruction of spores on the surface of soil derived from infected carcasses of livestock.

Saile and Koehler (2006) investigated the ability of *B. anthracis* spores to germinate in the rhizosphere and to establish horizontal gene transfer in the soil. Seeding a grass-plant soil system with spores of *B. anthracis* resulted in germination occurring on and around roots. From two to six days post inoculation, about one-half of the *B. anthracis* CFU recovered from soil containing grass seedlings were found to have arisen from heat sensitive cells. In contrast, CFU from soil seeded without plants consisted primarily of heat-resistant spores. These results clearly indicated that the rhizosphere greatly stimulates spore germination.

Co-inoculation of the plant-soil system with spores of a fertile *B. anthracis* strain carrying a tetracycline resistance plasmid and a *B. anthracis* selectable recipient strain resulted in the transfer of the plasmid as early as 3 days post inoculation Saile and Koehler (2006). These results suggest that horizontal gene transfer in the rhizosphere of grass plants may play a role in the evolution of the species comprising the *B. cereus* group.

Kahn *et al.,* (2009) investigated the survival and inactivation kinetics of the Sterne strain of *B. anthracis* in whole egg (WE), egg white, (EW), egg yolk plus 10% sucrose (YSU) and egg yolk plus 10% NaCl (YSA). In general, spore viability decreased at low and high temperatures and increased at moderate temperatures. At 0 and 5 °C, a 60 - 100% reduction in spore viability was observed within 2 - 3 weeks in WE and YSU, 0 - 30% in YSA, and 50-100% in EW. At -20 °C, no drop in viable spore counts occurred in YSU and EW but a 20% drop occurred in YSA and 50% in WE within 2 -3 weeks. At high temperatures (45 - 60 °C), WE, EW, and YSA produced a 20- 50% drop in viable spores within 1 - 4 hrs, whereas YSU resulted in 100% inactivation of spores within 0.75 hrs., From 15 to 40 °C spore were completely inactivated within 1 – 6 hrs. in EW. Inoculation of vegetative cells into YSU held at 45 °C resulted in an initial decrease in CFU followed by a period of recovery and exponential growth. Resulting colonies on TSA plates after growth at 45 °C yielded a variety of colony sizes from pin-head to small, to regular size colonies. These smaller colonies were considered to represent various heat-resistant polymorphic types of *B. anthracis.* At the moderate temperatures of 15 to 40 °C, EW resulted in inactivation of vegetative cells. As the temperature increased from 20 °C to 35 °C the rate of *B. anthracis* vegetative cell inactivation increased from -0.897 to -4.423 log CFU/hr. This was attributed to the presence of lysozyme in egg white. With YSU that did not contain lysozyme, complete inactivation of spores occurred within 3 - 40 min., 50 - 55 °C. This was attributable to sucrose induced spore germination and elevated temperature induced permeability of the vegetative cells which were then more susceptible to increased osmotic stress.

PLASMID STABILITY

The plasmid stability among 619 isolates of *B. anthracis* collected and stored from 1954 to 1989 was determined by Marston *et al.*, (2005). There was a much higher percentage of strains cured of both plasmids collected in the 19560's and 1960's (43%) compared to strains collected and stored in later decades (26% for 1970's isolates). When data was analyzed by decade, as storage time increased, isolates containing both plasmids decreased linearly. All isolates had been stored at room temperature on Tryptic Soy Agar slants overlaid with mineral oil in glass screw-cap tube. It is possible that many of the slants were overlayed with mineral oil before the formation of large number of spores. In addition, spores are formed most abundantly at 32 - 35 $^{\circ}$C so that incubation at ~20 $^{\circ}$C might have resulted in relatively low numbers of spores on the slants. The incubation temperature of the slants was not presented. The mechanism involving the loss of plasmids from spores is not known, however the authors speculated that it is possible that genetic damage over time may affect plasmid replication upon spore germination and vegetative growth.

REFERENCES

Abshire, T., Brown, J., Ezzell, J. (2005.) Production and validation of the use of gamma phage for identification of *Bacillus anthracis*. J. Clin. Microbiol. 43:47809-4788.

Agaisse, H., Gominet, M., Okstad, O., Kolsto, A., Lereclus, D. (1999). PlcR is a Pleiotropic regulator of extracelllar virulence factor gene expression in *Bacillus thuringiensis*. Mol. Microbiol. 32: 104310-104353.

Albrink, W., Goodlow, R. (1959). Experimental inhalation anthrax in the chimpanzee. Am. J. Path. 1959. 35:1055.

Ash, C., Farrow, J., Dorsch, M., Stackebrandt, Collins, M. (1991). Comparative analysis of *Bacillus anthracis, Bacillus cereus*, and related, species on the basis of reverse transcriptase sequencing of 16S rRNA. Int. J. Syst. Bacteriol. 41:343-349.

Brewer, C., McCullough, W., Mills, R., Roessler, W., Herbst, E. (1946). Application of nutritional studies for development of practical culture media for *Bacillus anthracis*. Arch. Biochem. 19:77-80.

Cybulski, Jr., R., Sanz, P., Alem, F., Stibitz, S., Bull, R., O'Brien, A. (2009). Four superoxide dismutases contribute to *Bacillus anthracis* virulence and provide spores with redundant protection from oxidative stress. Infect. Immun. 77:274-285.

De Vos, V. (1990). The ecology of anthrax in the Kruger National Park, South Africa Salisbury Med. Bull. 68 (Spec. Suppl.):19-23.

Etienne-Toumelin, I., Sirard, J., Duflot, E., Mock, M., Fouet, A. (1995). Characterization of the *Bacillus anthracis* S-layer: cloning and sequencing of the structural gene. *J. Bacteriol.* 177:614-620.

Eurich, F. (1913). Anthrax in the woollen industry, with special reference to Bradford. Proc. Royal acad. Med. 6:219-233.

Ezzell, Jr., J., Abshire, W. (1988). Immunological analysis of cell-associated antigens of *Bacillus anthracis*. Infect. Immun. 56:349-356.

Ezzel, Jr. J., Absire, T., Little, S., Lidgerding, C., Brown, C. (1990). Identification of *Bacillus anthracis* by using monoclonal antibody to cell wall galactose-N-acetyl- glucosamine polysaccharide. J. Clin. Microbiol. 28:223-231.

Ezell, Jr. J., Welkos, S. (1999). The capsule of *Bacillus anthracis*, a review. J. Appl. Microbiol. 87:250.

Fouet, A., Mesnage, S. (2002). *Bacillus anthracis* cell envelope components. Curr. Top. Microbiol. Immunol. 271:86-113.

Fouet, A., Mesnage, S., Tosi-Couture, E., Gounon, P., Mock, M. (1999). *Bacillus anthracis* surface: capsule and S-layer. J. Appl. Microbiol. 87:251-255.

Fraser, C. (2003). The genome sequence of *Bacillus anthracis* Ames and comparison to closely related bacteria. Nature. 423:81-86.

Gohar, M., Oksard, O., Gilois, N., Sanchis, V., Kolso, A., Lereclus, D. (2002). Two- dimensional electrophoresis analysis of the extracellular proteome of *Bacillus cereus* reveals the importance of the PlcR regulon. Proteomics. 2:784-791.

Goldfine, H., Bannam, T., Johson, N., Zückert, W. (1998). Bacterial phospholipases and intracellular growth: the two distinct phospholipases C of *Listeria monocytogenes*. J. Appl. Microbiol. 33:510-523.

Green, B., Battisti, L., Koehler, T., Throne, C., Ivins, B. (1985). Demonstration of a capsule plasmid in *Bacillus anthracis*. Infect. Immun. 49:291-297.

Grinsell, M., Weinhold, L., Cutler, J., Han, Y., Kozel, T. (2001). *In vivo* clearance of luconooxylomannan, he major capsular polysaccharide of *Cryptococcus neoformans*: a critical role for tissue macrophages. J. Infect. Dis. 184:479-487.

Hang, J., Sundaram, A., Zhu, P., Shelton, D., Karns, J., Martn, P., Li, S, Amstutz, P, Tang, (2008). Development of a rapid and sensitive immunoassay for detection and subsequent recovery of *Bacillus anthracis* spores in environmental samples. *J.* Microbiol. Meth. 73:242-246.

Hefferman, B., Thomason, B., Herring-Palmer, A., Shaughnessy, L., McDonald, R, Fisher, N., Huffnagle, G., Hanna, P. (2006). *Bacillus anthracis* phospholipases C facilitate macrophage-associated growth and contribute to virulence in a murine model of inhalation anthrax. Infect. Immun. 74:3756-3764.

Inglesby, T, Henderson, D., Bartlett, J. (1999). Anthrax as a biological weapon: medical and pubic health management. JAMA. 281:1735-1745.

Jacotot, H., Virate, B (1954). La longevite des spore de *B. anthracis* (premier vacc de Pasteur). Ann. Inst. Pasteur. 87:215-217.

Jernigan, T., Stephens, D., Ashford, D., Omenaca, C., Topiel, M., Galbraith, M., Tapper, M., Fisk, T., Zaki, S., Popovic, T., Meyer, R., Quinn, C., Harper, S., Fridkin, S. Sejvar, J., Shepard, C., McConnell, M., Guarner, J., Shieh, W., Malecki, J., Gerberding, J., Hughes, J., Perkins,

B. (2001). Bioterrorism-related inhalation anthrax: the first 10 cases reported in the United States. Emerg. Infect. Dis. 7:933-944.

Khan, S., Sung, K., Nawaz,M., Cerniglia, C., Tamplin, M., Phillips, R., Kelley, L. (2009). The survivability of *Bacillus anthracis* (Stern strain) in processed liquid eggs. Food Microbiol. 26:123-127.

Klichko, V., Miller, J., Wu, A., Popov, S., Alibek, K. (2003). Anaerobic induction of *Bacillus anthracis* hemolytic activity. Biochem Biophys. Res. Comm. 303:855-862.

Knisely, R. (1966). Selective medium for *Bacillus anthracis*. J. Bacteriol. 92:784-786.

Krishnamurthy, T., Deshpande, S., Hewel, J., Liu, Hongbin, L., Wick, C., Yates, J. (2007). Specific identification of *Bacillus anthracis* strains. *Int. J. Mass Spec.* 259:140-146.

Liang, X., Yu, D. (1999). Identification of *Bacillus anthracis* strains in China. J. Appl. Microbiol. 87:200-203.

Liu, H., Bergman, N, Thomason, B., Shallom, S., Hazen, A., Crossno, J., Rasko, D., Ravel, J., Read, T., Peterson, S., Yates III, J., Hanna, P. (2004). Formation and composition of the *Bacillus anthracis* endospore. J. Bacteriol. 186:164-178.

Marston, C., Hoffmaster, A., Wilson, K., Bragg, S., Plikaytis, B., Brachman, P., Johnson, S., Kaufman, A., Popovic, T. (2005). Effects of long-term storage on plasmid stability in *Bacillus anthracis*. Appl. Environ. Microbiol. 71:7778-7780.

McCloy, E. (1951). Studies on a lysogenic *Bacillus* strain. 1. A bacteriophage s pecific for *Bacillus anthracis*. J. Hygiene. 49:114-125.

Nagy, E., Ivanovics, G. (1977). Association of probable defective phage particles with lysis by bacteriophage AP50 in *Bacillus anthracis*. J. Gen. Microbiol. 102:215-219.

Norman, P., Ray, J., Brachman, P., Potkin, S., Pagano, J., (1960). Testing for anthrax antibodies in workers in a goat hair processing mill. *Am. J. Hyg.* 72:32.

Read, T., Peterson, S., Tourasse, N., Baillie,L., Paulsen, I., Nelson, K., Tettelin, H., Fouts, D., Eisen, J., Gill, S. Holtzapple, E., Okstad, O., Helgason, E., Rilstone, J., Wu, M., Kolonay, J., Beanan, J., Dodson, R., Brinkac, L., Gwinn, M., DeBoy, R., Madpu, R, Daugherty, S., Durkin, A., Haft, D., Nelson, W., Peterson, J., Pop, M., Khouri, H., Radune, D, Benton,J., Mahamoud, Y., Jiang, L., Hance, I., Weidman, J., Berry, K, Plaut, R, Wolf, A., Watkns, K., Nierman, W., Hazen, A., Cline, A., Redmond, C, Thwaite, J, White, O., Salzberg, S., Thomason, B., Friedlander, A., Koehler, T., Hanna, P., Kolsto, A.,Schmiel, D., Miller, V. (1999). Bacterial phospholipases and pathogenesis. Microbes Infect. 1:1103-1112.

Saile, E. and Koehler, T. (2006). *Bacillus anthracis* multiplication, persistence and genetic exchange in the rhizosphere of grassplants. *Appl. Environ. Microbiol.* 72:3168-3174.

Schuch, R., Nelson, D., Fischetti, V. (2002). A bacteriolytic agent that detects and kills *Bacillus anthracis*. *Nature.* 418:884-889.

Sejvar, J., Tenover, F., Stephens, D. (2005). Management of anthrax meningitis. Lancet Infect. Dis. 5:287-295.

Sozhamannan, S., KcKinstry, M., Lentz, S., Jalasvuori, M., McAfee, F., Smith, A., Dabbs, J., Ackermann, H., Banford, J., Mateczun, A., Read, T. (2008). Molecular Characterization of a variant of *Bacillus anthracis*-specific Phage AP50 with improved bacteriolytic activity. *App. Environ. Microbiol.* 74:6792-6796.

Steichen, C., Chen, P., Kearney, J., Turnbough, Jr., C. (2003). Identification of the immunodominant protein and other proteins of the *Bacillus anthracis* exosporium. *J. Bacteriol.* 185:1903-1910.

Sutherland, M., Thorkildson, P., Parks, S., Kozel, T. (2008). In *vivo* fate and distribution of poly-g-glutamic acid, the capsular antigen from *Bacillus anthracis. Infect. Immun.* 76:899-906.

Sutherland, M., Kozel, T. (2009). Macrophage uptake, intracellular localization, and degradation of poly-g-D-glutamic acid, the capsular antigen of *Bacillus anthracis. Infect. Immun.* 77:532-538.

Thorne, C. (1968). Transducing bacteriophage for *Bacillus cereus. J. virol.* 2:657-662.

Titball, R. (1998). Bacterial phospholipases. *J. Appl. Microbiol.* 84:127S-137s. Tomaso, H., Bartling, C., Al Dahouk, S., Hagen, R., Scholz, C., Beyer, W., Neubauer, H. (2006). Growth characteristics of *Bacillus anthracis* compared to other *Bacillus* spp. On the selective nutrient media Anthrax Blood Agar® and cereus Ident Agar®. Syst. Appl. Microbiol. 29:24-28.

Turnbull, P. (1999a) Anthrax history, disease, and ecology. Curr. Top. Microbiol. Immunol. 271:1-19.

Turnbull, P. (1999b.). Definitive identification of *Bacillus anthracis* — a review. *J. Appl. Microbiol.* 87:237-240.

Ward, M., McGann, V., Hogge Jr., A., Huff, M., Kanode Jr., R., Roberts, E. (1965). Studies on anthrax infections in immunized guinea pigs. J. Infect. Dis. 115:53-67.

Wattiau, P., Klee, S., Fretin, D., Van Hessche, M., Menart, M., Franz, T., Chasseur, C., Butaye, P., Imberechts, H. (2008). Occurrence and genetic diversity of *Bacillus anthracis* strains isolated in an active wool-cleaning factory. Appl. Environ. Microbiol. 74:4005-4011.

Wattiau, P., Govaerts, M., Frangoulidis, D., Fretin, D., Kissling, E., Van Hessche, M., China, B., Pt workers exposed to anthrax, Belgium. Emerg. Infect. Dis. 15:1637- 1640.

Wilson, J., Russell;, K. (1964). Isolation of *Bacillus anthracis* from soil stored 60 years. J. Bacteriol. 87:237-238.

Send Orders for Reprints to reprints@benthamscience.net

CHAPTER 3

Clinical Aspects and The Infectious Process of Anthrax

Abstract: Systemic infections usually result in the patient reaching "the point of no return". Among the three forms of anthrax: cutaneous, gastrointestinal, and inhalation, the cutaneous form constitutes 95% of all reported human cases. The clinical aspects of all three forms are presented in notable detail with illustrations. Meningeal anthrax infections resulting in the "red Cardinal's cap" are nearly always fatal. The major known virulence factors in anthrax include the cell-surface-associated antiphagocytic poly-D-glutamic acid capsule, the protective antigen (PA), and the edema (ET) and lethal (LT) toxins. Penicillin G, doxycycline, and the fluoroquinolone ciprofloxacin are considered the drugs of choice for the treatment of anthrax.

Keywords: Cutaneous anthrax, gastrointestinal anthrax, inhalation anthrax, meningitis, meningealencephalitis, Cardinal's cap, penicillin G, doxycycline, ciprofloxacin, quinolones, eschar, intracellular spore germination, capsule, nosocomial infection, treatment, protective antigen edema toxin, lethal toxin, virulence factors, D-glutamic acid.

INTRODUCTION

One of the unique aspects of anthrax is known as the "point of no return", whereby a systemic infection results in a critical number of cells in the blood leading to the inability of normally effective antibiotics to prevent death of the patient. Among the three forms of anthrax, cutaneous anthrax occurs most frequently and is usually occupationally related. Several nonnatural outbreaks of anthrax have occurred during the past few decades which have resulted in detailed clinical observations being recorded, particularly with respect to inhalation anthrax. Meningeal anthrax results from the spread of *B. anthracis* to the central nervous system (CNS) and usually results in the red "Cardinals cap" of the meninges. The treatment of meningea lencephalitis involves the administration of several antibiotics cable of penetrating the CNS. The ability of *B. anthracis* to germinate and grow inside macrophages is a critical factor mediating infection. An interesting aspect of inhalation anthrax involves the unusual absence of infection of the lungs and the absence of pneumonic infection; the organism developing in local mediastinal lymph nodes that drain the lungs.

Robert E. Levin

CLINICAL ASPECTS OF ANTHRAX

Septicemia and the "Point of no Return"

The major known virulence factors in anthrax include the cell-surface-associated antiphagocytic poly-D-glutamic acid capsule, the protective antigen (PA), and the edema (ET) and lethal (LT) toxins. *In vivo* spore germination and outgrowth of vegetative cells are critical factors leading to septicemia (Guidi-Rontani and Mock, 2002). Once an infection has progressed to the septicemic state it has usually reached "the point of no return" in that extensive intravenous administration of antibiotics resulting in elimination of *B. anthracis* from the blood still results in death.

The Three Forms of Anthrax

There are three clinical forms of anthrax in humans: cutaneous (>95% of cases), gastrointestinal, and inhalation. The infectious form of the organism, the spore, enters the body and germinates within macrophages. The organism then synthesizes the antiphagocytic capsule under elevated CO_2 conditions of the hosts tissues and the lethal and edema toxins.

Cutaneous Anthrax

After an incubation period of 2 to 5 days, the cutaneous form begins as a small pruritic (itchy) papule or pustule. Within 1 or 2 days, the site develops one or more vesicles (Brachman and Friedlander, 1999) filled with a clear or serous fluid containing numerous anthrax bacilli. Within a week, the vesicle erodes leaving a necrotic ulcer. The individual may exhibit malaise, headache, and toxicity. Edema surrounding the ulcer can be massive due to the production of the edema toxin (ET) and is followed by the development of a characteristic black eschar or slough (Fig. **1**) which heals in two to three weeks (Friedlander, 1999). The localized infection however can become systemic. Death from cutaneous anthrax occurs in about 20% of untreated cases (Jami, 2002). Felek, Akbulut, and Kalkan (1999) photographically documented a non-fatal course of cutaneous anthrax infection (Fig. **2**). The infection involved a massive eschar of the neck of a 16 year old male involved in sheep slaughtering and resulted in extensive edema of the face and torso (Fig. **2**) in addition to visual hallucinations that responded to haloperidol.

Scrapings from the wound and blood culture and were positive for *B. anthracis*. Satisfactory treatment consisted of *i.v.* administration of penicillin G, and ciprofloxacin, and the additional administration of the corticosteroid methylprednisolone, along with H_2 receptor antagonists and antacids. The area of the eschar required a skin graft.

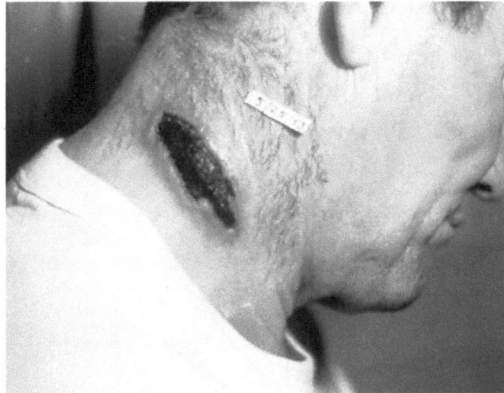

Figure 1: Cutaneous anthrax lesion (eschar). Source: Public Health Image Library, CDC at: http://www.bt.cdc.gov/agent/anthrax/anthrax- images/cutaneous.asp.

Cutaneous anthrax can also involve the eyelids from which infection can spread into the eye and cause panophthalmitis (inflammation of the entire eye).

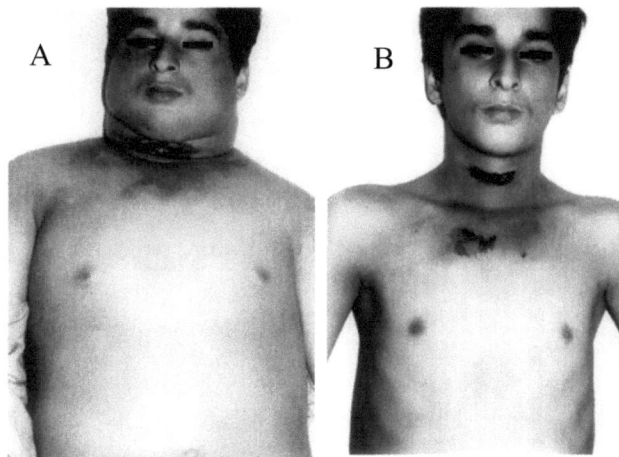

Figure 2: Cutaneous anthrax. A. Appearance of 16 year old patient with large black anthrax eschar in neck area and chest lesion exhibiting extreme edema of facial and torso areas, before effects of antibiotic treatment. B. Same patient after antibiotic treatment on day 16. From Felek, Akbulut, and Kalkan (1999) with permission.

Gastrointestinal Anthrax

The two forms of gastrointestinal anthrax are oropharyngeal and orogastric (abdominal). Oropharyngeal anthrax usually results in a severe sore throat or an ulcer in the oropharyngeal cavity associated with swelling of the neck, fever, toxicity, and dysphagea (inability to swallow). Gastrointestinal anthrax has never been reported in the United States. Most cases occur in developing countries, where disease results from consumption of meat from sick or dead animals. Abdominal anthrax begins with anorexia, nausea, vomiting, and abdominal pain. Diarrhea and ascites may occur with hemorrhaging (Turnbull, 2002). Untreated gastrointestinal anthrax can lead to lethal systemic infection and neuromengingitis (Berthier *et al*., 1996) in addition to lethal anthrax pneumonia (Meric *et al*., 2009). The mortality rate for untreated gastrointestinal anthrax approaches 100%. Berthier *et al.,* (1996) described a case of gastrointestinal anthrax that resulted in fulminant meningitis leading to the death of an individual in a French province where anthrax had not occurred in animals or humans for over 40 years. The patient was an 11-year old girl. Blood cultures were positive for *B. anthracis*. The girl's family was Muslim and followed the food precepts of Ramadan. A sheep was slaughtered in a clandestine slaughterhouse by the girl and her father without benefit of a veterinary inspection of the animal. The lungs were totally black, but the liver was only partly black. The offal was discarded, and the girl kept half the liver, which was only passed through the flame of a gas stove. She then used the undercooked liver for preparation of a kebab which she alone consumed. Afterward, the entire family ate the sheep which was cooked by boiling.

Inhalation Anthrax

Inhalation anthrax in humans, particularly anthrax derived from weaponized anthrax spores is considered 100% fatal without early treatment. However, during the nineteenth century in the textile mills of England, 20% of patients with inhalation anthrax were said to recover spontaneously (Plotkin *et al*., 2002). Once inhaled, the spores reach the respiratory broncheoli and alveoli and are phagocytised by alveolar macrophages that initiate innate immune responses in the lungs by releasing immunomodulatory products such as cytokines, chemokines prostoglandins, and leukotrienes. The secreted products act coordinately to activate and recruit and other leukocytes from the blood to the site of infection. The spore-bearing activated macrophages transport the engulfed spores along the lymphatic channels to the regional lymph nodes and to local mediastinal lymph nodes where they present antigens and secrete cytokines to facilitate activation of the adaptive immune system. From the lymph nodes the organism is transported to the blood stream (Guidi-Rantoni and Mock, 2002). The spores are able to germinate inside the macrophages. The macrophage is the target for the lethal toxin (Friedlander, 1986). The cytotoxicity of LF is due to its zinc-mediated protease activity.

Pathological investigations of human inhalation anthrax have indicated that the vegetative cells do not multiply in the lung itself but lead to infection of the mediastinal lymph nodes. The alveolar lining acts simply as a point of entry with multiplication and resulting bacteremia occurring only after infection of the lymph nodes that drain the lungs (Barnes, 1947). The lymph nodes act as foci for the multiplication and distribution of the vegetative cells resulting in septicemia and death.

An accidental outbreak of anthrax in 1979 at Sverdlovsk (now Ekaterinburg) Russia, from a military research facility, resulted in 96 cases and 64 deaths. Abramova *et al.,* (1993) presented the results of a highly detailed pathological study involving 42 autopsies of individuals succumbing from inhalation anthrax in this outbreak. Among the 42 autopsies, infection of the mesentery lymph nodes was observed in only nine which was interpreted to mean that observed signs of gastrointestinal infection were secondary to inhalation anthrax. In contrast, all 42 cases showed signs of massive infection in the chest area, with hemorrhagic thoracic nodes and with hemorrhaging in the mediastinum (area between the lungs). In most cases, the inside surface of the lungs appeared free of serious lesions, but in eleven cases there was a hemorrhagic lesion inside the lung. Twenty-one patients had cerebral hemorrhage, including sixteen with definite Cardinal's cap. These observations are particularly important in that the extent of pathological observations resulting from inhalation anthrax is unmatched by no other modern outbreak. It is now well recognized that in halation anthrax, inhaled spores are phagocytised and transported out of the lungs to mediastinal lymph nodes. Inhalation anthrax therefore does not adhere to the usual concept of a pulmonary infection or pneumonia, since the infecting organisms do not initiate growth in the lungs.

Plotkin *et al.,* (2002) have described in detail the classic symptoms of inhalation anthrax which results in two distinct stages. The initial stage is marked by the insidious onset of a mild fever (~101 °F), malaise, fatigue, myalgia (muscle pain), a non-productive cough, and frequently a sensation of precardial oppression. There are few objective findings which may lead the unsuspecting physician to a misdiagnosis. This initial stage typically lasts for several days. Misdiagnosis at this stage and the failure to administer penicillin or ciprofloxacin at this point is critical to the outcome. The second stage of the disease occurs suddenly with acute dyspnea (labored breathing) and subsequent cyanosis. The patient appears moribund, with accelerated pulse and respiration. The body temperature, although usually elevated to 102 °F or more, may be subnormal because of shock. Stridor (harsh respiratory sounds) occurs commonly, presumably due to partial obstruction of the trachea by enlarged mediastinal nodes. Profuse diaphoresis (sweating) is a frequent sign, and subcutaneous edema of the chest and neck may be present. Examination of the chest reveals moist crepitant rales and signs of pleural effusion are observed. Chest X-rays invariably indicate a notably widened mediastinum (Fig. **3**) caused by inhalation anthrax (Lindsey and Michie, 2002). The spleen is usually enlarged on autopsy. If meningeal involvement occurs, disorientation, coma, and meningismus occur. Premortem diagnosis of inhalation anthrax is considered difficult and the first stage is often mistaken for influenza or bronchitis, and the second stage resembles cardiac failure or cerebrovascular accident. Death usually follows in 24-36 hr from respiratory failure, sepsis, and shock. On the basis of animal studies, penicillin or ciprofloxacin can be expected to cure inhalation anthrax if administered before the onset of toxemia (Plotkin *et al.*, 2002).

Meningeal Anthrax

Meningeal infection is a rare complication of anthrax infections and results from hematogenous or lymphatic spread of bacilli to the central nervous system. Its symptoms have been described in detail by Jamie (2002). Symptoms include nuchal rigidity (rigidity of the neck), fever, headaches, seizures, agitation, and delirium. Pathologic findings are hemorrhagic meningitis, bloody cerebrospinal fluid containing inflammatory infiltrates and large numbers of bacilli, and dark red meninges at autopsy referred to as the "cardinals cap". Such meningeal infections are almost always fatal within 1-6 days, even with antibiotic therapy.

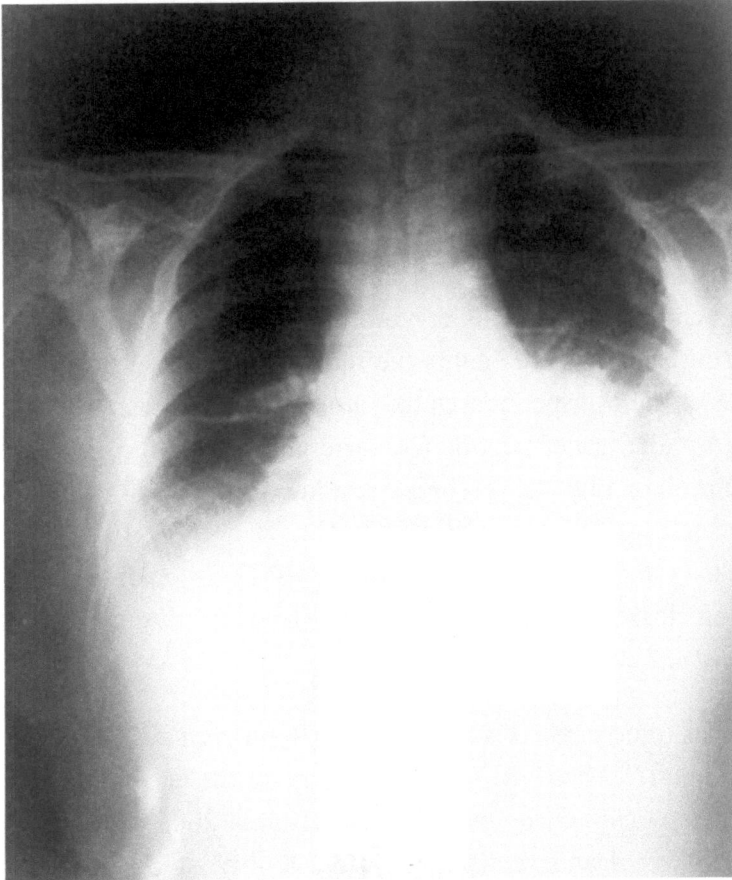

Figure 3: Chest radiograph showing widened mediastinum due to inhalation anthrax. From: Public Health Image Library, CDC. http://www.bt.cdc.gov/agent/anthrax/anthrax-images/cutaneous.asp.

THE INFECTIOUS PROCESS

Initiation of Infection and Intracellular Spore Germination

After entry of infectious anthrax spores into the body, host-specific signals induce spore germination, outgrowth of vegetative bacilli and the expression of lethal toxin and other virulence factors. Injection of sterile LT into test animals mimics the shock and sudden death observed during active bacterial infections. Once large levels of LT are produced within the body, antibiotic therapy is usually unsuccessful. Upon entering the cell cytoplasm of macrophages, the LF acts as a zinc-metalloprotease disrupting normal homeostatic functions. Removal of macrophages from mice renders the animals insensitive to LT challenge (Hanna, 1999). Low levels of LT induce macrophage production *in vitro* of the shock inducing cytokines, tumor necrosis factor (TNFα), and interleukin-1β (Il-1β) while higher levels of LT cause over-production of reactive oxygen intermediates, bursting of macrophages, and release of mediators of shock (Hanna, 1999).

One group of surface proteins common to Gram-positive bacteria are proteins that are attached to the peptidoglycan of the cell wall by a family of proteins known as "sortases". Such surface proteins contain an N-terminal signal peptide and a C-terminal sorting signal that contains the motif LPXTG that is recognized by the sortase. Sortases are transpeptidases that are anchored to the membrane *via* an N-terminal hydrophilic leader peptide (Navarre and Scheewind, 1999). They cleave the substrate protein after the threonine residue of the LPXTG motif and link the protein to the cell wall (Perry *et al.*, 2002; Ton-That *et al.*, 1999). Loss of sortase activity results in improper localization of the LPXTG-containing surface proteins with resulting loss of function and can lead to attenuation of virulence (Mazmanian *et al.*, 2000).

B. anthracis has three putative sortase genes (Read *et al.*, 2003). Two of these are named *srtA* and *srtB* based on their homology to the *srtA* and *srtB* genes of *S. aureus* and *List*eria *monocytogenes*. Zink and Burns (2005) examined the effect of mutations of the *srtA* and *srtB* genes on the ability of *B. anthracis* to grow in cultures of murine macrophage like 5774A.1 cells. No significant difference in the growth rate of either the *srtA⁻* or *srtB⁻* mutant strains occurred compared to the parent strain in BHI broth. In 57774A.1 cells the parent strain of *B. anthracis* Sterne 7702 increased in number for 8 hr. In contrast, the *srtA⁻* and *srtB⁻* mutants did not replicate intracellularly.

Complementation of the *srtA⁻ and srtB⁻* mutants with plasmids containing the intact *srtA* and *srtB* regions respectively restored intracellular growth. These observations indicate that the SrtA and SrtB proteins are essential for the intracellular growth of *B. anthracis* in 5774A.1 cells. The SrtB of *S. aureus* is known to be involved in tethering a protein involved in iron transport and is important for persistence of infections in mice. The SrtB of *B. anthracis* is therefore thought to be involved in the tethering of a protein possibly required for iron acquisition. Cendrowski, MacArthur, and Hanna (2004) demonstrated that disruption of a gene encoding a protein necessary for siderophore biosynthesis resulted in attenuation of the growth of *B. anthracis* within macrophages and notably reduced mouse virulence. These results suggest that sortases may be potentially suitable targets for therapeutic intervention in anthrax.

Ireland and Hanna (2002a) found that specific combinations of amino acids or purine nucleotides are required for efficient germination of endospores of Δ*B. anthracis*, a plasmid less strain. Alanine alone at 1 mM concentration had no effect on germination. Alanine (1mM) plus 100 mM histidine, proline, or alanine (1mM) with 1 mM tryptophan or 1 mM tyrosine yielded 31.0 to 56.1% germination. Interestingly, alanine alone at 100 mM yielded 98% germination. The co-amino acids alone at 100 mM had no effect on germination. Inosine alone at 1 mM had no effect on germination. However, inosine at 1 mM plus 100 mM of the individual amino acids: alanine, histidine, methionine, proline, serine, or valine yielded 46.3 to 84.5% germination. Inosine at 1 mM plus 1mM of tryptophan or tyrosine yielded 79.1% and 74.1% germination respectively. The *gerS*   operon (a tricistronic germination receptor operon) which when disrupted, completely eliminated the ability of *B. anthracis* spores to respond to amino-acid and inosine-dependant germination responses. The *gerS* disruption appeared to specifically affect use of aromatic amino acids as cogerminants with alanine and inosine. The authors concluded that efficient germination of *B. anthracis* endospores requires multipartite signals and that *gerS*-encoded proteins act as an aromatic-responsive germination receptor.

Spores of the Sterne 34F2 strain of *B. anthracis* which possess only the pXO1 and not the pXO2 plasmid failed to significantly germinate in rat, horse, rabbit, or swine serum (Ireland and Hanna, 2002b). In contrast, BHI, goat serum, and fetal bovine serum yielded 99.2, 89.4, and 36.6% germination, indicating that serum is not the sole germination factor in susceptible animals, nor the sole reason for variation in animal species resistance.

A dramatic increase in germination was observed when spores of *B. anthracis* were exposed to macrophage cultures, increasing from nearly 0 to 80% in the presence of macrophages in 10 min. Neither phagocytosis nor direct contact with the macrophages was found to be an absolute requirement for germination. Clearly, macrophages release one or more molecular species required for spore germination.

Among *B. anthracis, B. subtilis,* and *B. megatherium*, only *B. anthracis* spores exhibited a rapid germination response to macrophages. The *gerS* chromosomal locus was shown by Ireland and Hanna (2002a) to act as a *B. anthracis* germination receptor, recognizing aromatic ring-like germinant signals. A *gerS* chromosomal mutant was completely unable to germinate in macrophage cultured medium compared to the parent strain but was fully able to germinate when plated onto BHI agar. These results indicate that *gerS* is important for macrophage associated germination and suggest a role for aromatic amino acids and purines as potential host-supplied germinants.

Guidi-Rontani *et al.,* (1999a) determined that a germination locus designated *gerX* was organized as a tricistronic operon (*gerXB, gerXA,* and *gerXC*) between the *pag* and *atxA* genes on pXO1. Expression of the *gerX* operon using a *gerXB-lacZ* trancriptional fusion indicated that expression began 2.5-3 hr. after initiation of sporulation and was detected exclusively in the forespore compartment. A *gerX*-null mutant was less virulent for mice than the parental strain and did not germinate efficiently *in vivo* or *in vitro* within phagocytic cells. These results indicate that the virulence of *B. anthracis* spores is influenced by the *gerX*-encoded protein. Weiner and Hanna (2003) found that the *gerHABC* operon of *B. anthracis*, encoding a *ger-A*-like family member of germinant sensors, was required for spore germination in the presence of macrophages and in macrophage-conditioned media. Six tricistronic *gerA*-like chromosomal loci (*gerAA, gerAB, gerAC*) plus the *gerX* locus carried on pXO1 exist in *B. anthracis*. These loci encode germinant recognition proteins. The *gerS* operon in *B. anthracis* mediates germination in response to aromatic amino acids and is required for germination in the presence of cultured macrophages (Ireland and Hanna, 2002a,b). The gerHABC locus is required for germination with the following combinations of cogerminants: Ino-val, Ino-Met, Ino-His, Ino-Phe, and Ino-Tyr which together are referred to as the Ino-His germination response pathway (Weiner, Read, and Hanna, 2003). Weiner and Hanna (2003) found that *gerH* is required for *B. anthracis* spore germination in macrophage conditioned culture media. Spore germination appears to be a multifactorial event and like *gerSABC* (Ireland and Hanna, 2002a), the germination operon *gerHABC* is also required for macrophage related germination.

The Ino-His germination response of *B. anthracis* was found to require *gerH*, in which neither adenosine nor guanosine can substitute for Ino (Weiner, Read, and Hanna, 2003). In contrast, Ino can be replaced by adenosine and guanosine in the Ino-Ala germination response pathway. Therefore, Ino-His and Ino-Ala represent distinct germination pathways. In addition, his participates in two distinct germination pathways (Ireland and Hanna, 2002a; Weiner, Read, and Hanna, 2003) Ino-His and Ala-His. *gerH*-null spores were found to maintain an intact Ala-His germination response but lacked the Ino-His germination response (Weiner and Hanna, 2003). The authors concluded that Ino is likely to be one of the cogerminants secreted by macrophages because of the inability of other purines (adenosine and guanosine) to substitute for Ino in the Ino-His germination pathway.

The alveolar macrophage has been reported to be the primary site of *B. anthracis spore* germination in a murine inhalation infection model (Guidi-Rontani *et al.*, 1999b). Germinated spores were detected 30 min., after infection and appeared to be located within the perinuclear compartment of the macrophages. Germinated spores inside macrophages were found co-localized with F-actin-rich phagocytic cups. These results were interpreted to mean that the uptake of *B. anthracis* spores is dependant on F-actin assembly which is essential for the phagocytic process. The LT and ET have been found essential for the intracellular survival of vegetative cells of the Sterne 7702 strain (pXO1$^+$ pXO2$^-$) of *B. anthracis* in murine peritoneal macrophages and murine-like RAW264.7 macrophages (Guidi-Rontani *et al.*, 2001). Guidi-Rontani (2002) has diagrammatically presented the molecular and cellular events involved in the early phase of inhalation anthrax (Fig. **4**).

Glomski *et al.,* (2008) found that aerosolizing spores of a capsule deficient strain of *B. anthracis* administered to A/J mice at moderate doses (4.6 spores in lungs) initiated infection in the nasopharynx. Distribution beyond the nasopharynx was delayed for at least 24 hrs. and then targeted the kidneys. Interestingly, high intranasal doses ($10^{7.3}$ spores) led to spore germination in the alveoli. These results indicated that eliminating the capsule from subsequently developing vegetative cells while maintaining toxin production altered dissemination, but still allowed initiation of infection in the lungs.

Bergman *et al.*, (2005) reported on transcriptional changes within macrophages infected with *B. anthracis* (Sterne 34F2 strain). And the identification of several hundred host genes *via* microarray analysis that were differentially expressed during intracellular infection. These loci included genes that are known to be regulated differentially in response to *B. anthracis* but not to other bacterial species. The data obtained provided a transcriptional basis for a variety of physiological changes observed during infection, including the induction of apoptosis caused by the bacterial cells. Ornithine decarboxylase production by macrophages was found to play an important and previously unrecognized role in suppressing apoptosis in *B. anthracis*-infected cells. The transcriptional response to anthrax LT in activated macrophages revealed that many of the host inflammatory response pathways are dampened.

Bergman *et al.*, (2007) reported on the first genome-wide analysis of *B. anthracis* gene expression during infection of murine host cell macrophages using custom microarrays of *B. anthracis* to characterize the expression patterns occurring within intracellular vegetative bacteria throughout infection of the host macrophages starting with spores. Vegetative cells of *B. anthracis* (Sterne 34F2 strain) were found to adapt very quickly to the intracellular environment. Analysis identified metabolic pathways that appear to be important to the organism during intracellular growth in addition to individual genes that show significant induction *in vivo*. The microarray data obtained provided the first comprehensive view of how *B. anthracis* survives within the host macrophage. Nearly 1,100 genes (~19% of the Sterne genome) exhibited a statistically significant change in expression. These genes were separated into up- and down-regulated genes. In general, genes associated with prophage function and sporulation/germination were down-regulated 1 to 2 hrs. postinfection, while a large number of genes associated with energy metabolism (*e.g.*, the pentose phosphate pathway, the tricarboxylic acid cycle, and glycogen metabolism) were up-regulated.

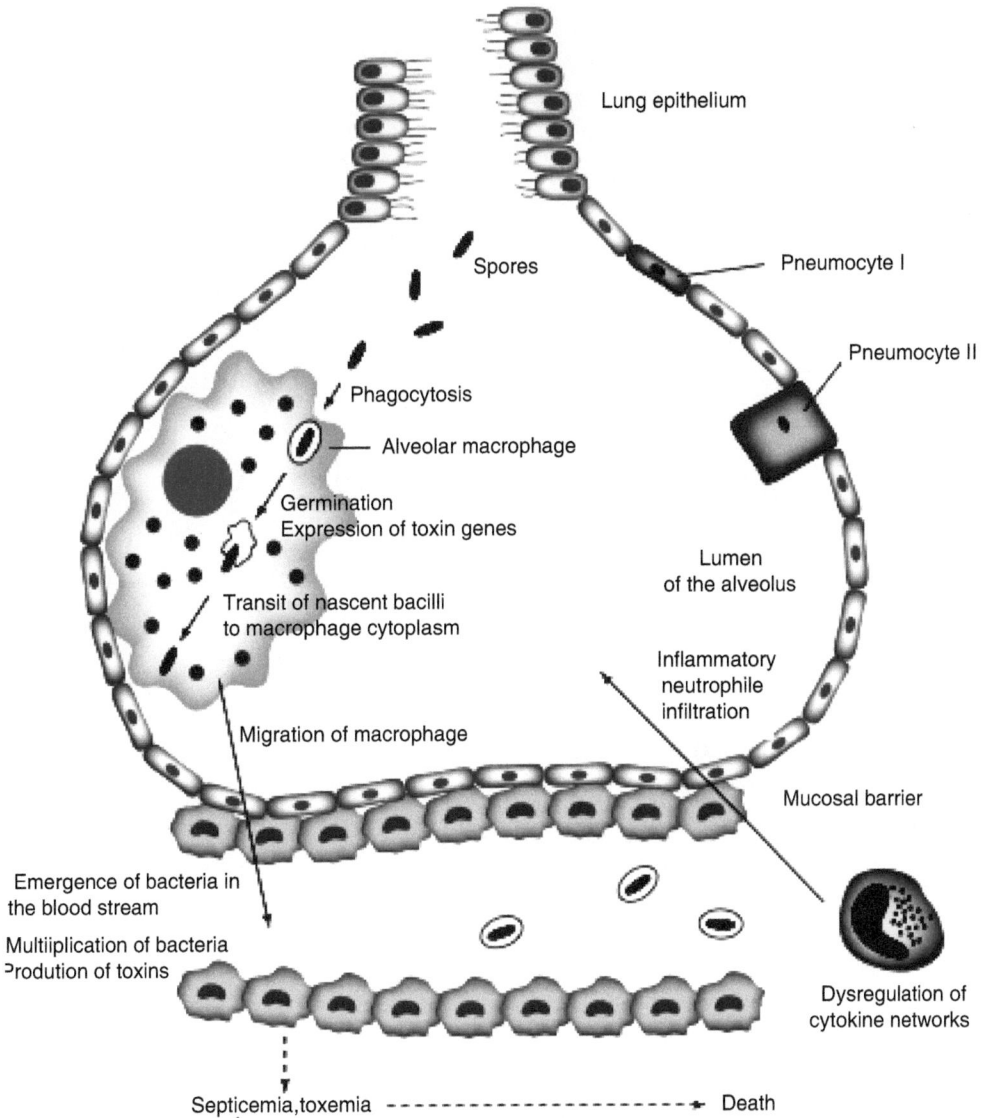

Figure 4: The molecular and cellular mechanisms involved in the early phase of *Bacillus anthracis* infection. *B. anthracis* spores are shown reaching the lumen of the alveoli. Once they reach the phagolysosome after phagocytosis by alveolar macrophages, the spores germinate and produce toxins (LT and ET). Disruption of the phagolysosome membrane (promoted by virulence factors with membrane-damaging activities) would allow the transit of nascent bacilli into the macrophage cytoplasm. Bacteria that have escaped from the macrophage phagosome are carried from the primary site of infection by the migrating macrophage. The survival of nascent bacteria is associated with a loss of macrophage cell viability. Once the bacteria emerge in the blood stream they can multiply. As the infection progresses bacteria spread through the blood and lymph causing bacteremia and the toxemic characteristics of a fatal anthrax infection. From Guidi--Rontani (2002) with permission.

Contribution of the Capsule to Infection

The sensitivity of A/J mice to infection with spores from toxigenic nonencapulated strains of *B. anthracis* has been correlated with the absence of the complement component C5 function (Welkos and Friedlander, 1988). Glomski *et al.,* (2007a) undertook studies to determine whether elimination of the capsule (Tox$^+$ Cap$^-$ strains) altered bacterial growth and characteristic infectious dissemination compared to encapsulated (Tox$^+$ Cap$^+$) strains in A/J mice. A bioluminescent Tox$^+$ *B. anthracis* strain was constructed to allow bioluminescent imaging for following *in vivo* dissemination of an infectious strain. In contrast to infection with a Tox$^+$ Cap$^+$ strains, the growth of a Tox$^+$ Cap$^-$ strain was (1) confined to the site of inoculation, (2) a minimal number of Tox$^+$ Cap$^-$ cells were found in the spleen throughout infection, (3) the Tox$^+$ Cap$^-$ strain infected the kidneys, and frequently, the gastrointestinal tract. Furthermore, PA immunization was found to allow sensitive strains of mice to confine Tox$^+$ Cap$^-$ infecting bacteria to the initial site of cutaneous infection without completely eliminating spore germination. When DBA/2J and BALB/c and A/J mice were injected cutaneously into the ears with spores of a bioluminescent Tox$^+$ Cap$^-$ strain, measurable bioluminescence occurred at the site of injection. With A/J mice bioluminescence of the Tox$^+$ Cap$^-$ strain disseminated to deeper tissues and ultimately killed the mice. At approximately 48 hrs., light became detectable in the superficial parotid lymph node, draining the ear. Luminescence then spread to the lungs and unexpectedly to the kidneys, and in 13 of 36 (36%) mice it also spread to the intestines and/or the pyloric region of the stomach. No luminescence and few CFU were observed in the spleen, in contrast to infection with luminescent Tox$^+$ Cap$^+$ bacteria (Glomski *et al.*, 2007b). Dissection late in the infection revealed that both the jejunum and ileum were transparent and distended with a watery liquid. By 48 hrs., edema was observed adjacent to the ear, and by 42 hrs., it encompassed most of the head and shoulder. The mouse strain DBA/2 resulted in a slower progression of infection with similar dissemination. In contrast, when naturally resistant BALB/c mice were infected, maximal luminescence at the site of injection was observed at 48 hrs., after which luminescence decreased and was eliminated by 144 hrs. postinfection. With a/J mice immunized with PA, luminescence was detected at the site of ear injection with spores within 8 hrs., and then diminished until no longer detectable, with luminescence not detected beyond the ear.

Nosocomial Infection

Although anthrax is not readily spread from person to person, the hospital spread of anthrax can readily occur if proper precautions are not taken in dealing with patients. Yakupogullari and Koroglu (2007) reported on a 20-day-old infant hospitalized with peri-umbilical edema with a few small papular lesions. High dose administration of infused penicillin G cured the infection. Inquiry revealed that two nurses and a child care worker subsequently developed cutaneous anthrax of the hands. Investigation indicated heavy contamination of the crib used to maintain the infant. Direct contact with the patient and inadequate use of gloves and ineffective disinfection of medical instruments in contact with the infant were considered to be the major reasons for this nosocomial spread of anthrax.

Treatment

Animal Studies

Early observations of infected and untreated guinea pigs indicated that they were in secondary shock resulting from pronounced loss of circulatory blood volume from extensive internal hemorrhaging leading to cessation of heart action. Guinea pigs dying two days after streptomycin treatment exhibited evidence of renal failure, a known sequel of secondary shock (Smith *et al.*, 1954, 1955).

With guinea pigs, following intradermal inoculation with *B. anthracis*, during the last 12 hrs. before death, the number of bacterial chains in the blood rises to 1×10^9/ml (Keppie, Smith and Harris-Smith., 1955). High doses of an effective antibiotic administered during the initial stage of bacteremic infection will inhibit bacterial growth and the blood is completely cleared of bacilli within 6 hrs., after treatment and the animals survive. However, if the blood count of *B. anthracis* chains exceeded 3×10^6/ml, even though high doses of an effective antibiotic were administered and *B. anthracis* was eliminated from the blood, the animals did not survive. This early observation suggested that the cause of death was due to a toxin. Such early studies revealed the "principle of no return" whereby once the infection reached a certain stage, the animal was irreversibly doomed, even after antibiotic destruction of the organism (Lincoln and Fish, 1970; Stephen, 1986). This effect is presently recognized as being due to the effects of the LT produced during systemic anthrax infections (Hanna, 1999).

The efficacy of postexposure antibiotic treatment was assessed in guinea pigs infected intranasally with either the Vollum strain of *B. anthracis* or ATCC 6605 spores at a dose of $75LD_{50}$ and $87LD_{50}$ respectively by Altboum *et al.,* (2002). Starting 24 hrs. post aerosol challenged animals were treated three times daily for 14 days with cefazolin, tetracycline, ciprofloxacin, erythromycin and trimethoprim-sulfamethoxazole (TMP-SMX). Cefazolin and TMP-AMX failed to protect the animals while erythromycin, tetracycline, and ciprofloxacin, and prevented death. Upon cessation of treatment, all erythromycin treated animals died. Among the tetracycline-treated animals, two of eight infected with the Vollum strain and one of nine infected with the ATCC 6605 strain survived; and among the ciprofloxacin group injected with either 10 or 20 mg/Kg body weight, five of nine and five of five animals respectively survived. To test the added value of extending the treatment time, Vollum infected animals were treated for 30 days with ciprofloxacin or tetracycline, resulting in protection of eight of nine animals respectively. Once treatment was discontinued, only four of eight and five of nine animals, respectively survived. In an attempt to impart protective immunity lasting beyond the termination of antibiotic administration, Vollum-infected animals were immunized with a PA-based vaccine simultaneously with treatment with either tetracycline or ciprofloxacin. The combined trimethoprim-sulfamethoxazole treatment protected eight of eight and nine of nine animals. Following cessation of antibiotic administration, seven of eight and eight of eight animals resisted rechallenge. These results indicated that a combined treatment of antibiotics together with a PA-based vaccine was able to provide long-term protection against anthrax. In addition, these results indicated that 14 days of injection of either tetracycline or erythromycin were not sufficient to result in a significant reduction in the number of viable spores in the infected lungs. Also, treatment with ciprofloxacin (15 mg/Kg) for 10 days, followed by 7.5 mg/Kg) for an additional 20 days resulted in only four survivors among 10 infected animals. All animals surviving. The CDC have recommended ciprofloxacin administration for at least 60 days following exposure to spores of *B. anthracis*. Ciprofloxacin is active only on actively growing cells and is bactericidal while tetracycline is a bacteriostatic agent, which explains the greater efficacy of ciprofloxacin.

Treatment of Human Anthrax

In developing countries, penicillin G is still the drug of choice because of low cost, abundant availability, and efficacy. In mild uncomplicated cases of cutaneous anthrax, the usual treatment is the intramuscular (*i.m.*) administration of 500 to 600 mg (80-100 million units) of procaine penicillin every 12 – 24 hrs for 3 - 7 days (WHO, 2008). Cutanous lesions usually become sterile in 24 to 48 hrs.

Penicillin G, doxycycline, and the fluoroquinolone ciprofloxacin are considered the drugs of choice for the treatment of anthrax (Pile *et al.*, 1998); Knudsen, 1986; Bartlett, 1998). Esel, Doganay, and Sumerkan (2003) found that the minimum inhibitory concentration for 90% of 40 isolates of *B. anthracis* (MIC_{90}) for penicillin G (PG), doxycycline (DC), ciprofloxacine (CP), gatifloxacin (GF), and levofloxacin (LF) were 0.0, 16, 0,03, 0.06, 0.06, and 0.12 mg/L, respectively. Susceptibilities indicated that the quinolones (DC, CD, GF, and LF) should also be included as alternatives to PG.

The Sterne strain of *B. anthracis* was used by Brook *et al.*, (2001) to test for the development of resistance to doxycycline and the three quinolones ciprofloxacin, alarofloxacin, and gatifloxacin after 21 sequential subcultures in sub-inhibitory concentrations of these antibiotics. In addition, 15 sequential subcultures in sub-inhibitory concentrations of the three macrolides clarithromycin, erythromycin, and azithromycin, were undertaken. After 21 subcultures the minimal inhibitory concentrations (MICs) increased from 0.1 to 1.6 mg/ml for ciprofloxacin, from 1.6 to 12.5 mg/L for alatrofloxacin, from 0.025 to 1.6 mg/ml for gatifloxacin and from 0.025 to 0.1 mg/ml for doxycycline. After 15 passages of sequential subculturing of macrolides, the MICs increased from 12.5 to12.5 or 50.0 mg/L for arizromycin, from 0.2 to 1.6 or 0.4 mg/L for clarithromycin and from 6.25 to 6.25 or 50 mg/L for erythromycin. After sequential passages with a single quinolone or doxycycline, each isolate was cross-tested for resistance to other drugs. All isolates selected for resistance to one quinolone were also resistant to the other two quinolones, but not to doxycycline. The doxycycline-resistant isolate was not resistant to any quinolone. The authors suggested that long-term treatment of anthrax with any one of the quinolones might result in a quinolone-class-based resistance which is particularly pertinent to the recommended course of ciprofoxacin of up to 60 days for inhalation anthrax.

Patients who have uncomplicated cutaneous anthrax can be treated as outpatients with oral antibiotics. The recommended treatment is with penicillin V (200 – 500 mg orally) four time a day for seven days (Jamie, 2002). For cutaneous anthrax with systemic symptoms, and all other forms of the disease, penicillin G (four million units *i.v.* every 4 – 6 hrs.) with a total daily infused dose of 20 - 24 million units until resolution of symptoms, should be administered intravenously in an intensive care setting (Jamie, 2002; WHO, 2008). At this point, *i.m.* procaine penicillin can be administered. Studies with rhesus monkeys have shown that penicillin together with streptomycin was more effective than either antibiotic alone (Lincoln *et al.,* 1964).

In severe cases, penicillin or another appropriate antibiotic such as ciprofloxacin may be combined with another agent, preferably one that penetrates well into the central nervous system (CNS). Penicillin G may be combined with clindamycin or clarithromycin in treating inhalation anthrax, or with an aminoglycoside such as streptomycin in gastrointestinal anthrax (WHO, 2008. When exposure to anthrax spores is suspected to have occurred, Jamie, 2002 has recommended treatment for 60 days with ciprofloxacin (500 mg orally) twice a day. An acceptable alternative is doxycycline (100 mg orally) twice a day. Vaccination is considered effective in preventing infections arising from dormant spores after antibiotic therapy. Bonn (2002) has suggested a 60 day period of antibiotic treatment may not be a long enough period, particularly if spores implanted in organ tissues do not all germinate within that time period and has indicated that vaccine if available should also be administered. Jamie (2002) has recommended that antibiotic treatment should be continued for at least two weeks after the third vaccination (eight weeks after the first vaccination).

Antwerpen *et al.,* (2007) developed a multiplex PCR targeting ten genes involved in resistance to vancomycin, ciprofloxacin, rifampin, doxycycline, rifampin, and vancomycin in *B. anthracis* and *B. cereus.* Enzymatic labeling of PCR products followed by hybridization to oligonucleotide probes on a DNA microarray enabled simultaneous detection of the 10 resistance genes. The assay allowed detection of clinically significant antibiotic resistance determinants within 4 hrs. to allow rapid diagnosis of antibiotic resistance.

Therapy with ciprofloxacin or doxycycline is recommended for adults and children for 60 days following exposure to spores of *B. anthracis* to prevent inhalation anthrax (Brook, 2002). Therapy with two or more antimicrobial agents that are predicted to be effective is recommended. Other agents with *in vitro* activity are also suggested for use in conjunction with ciprofloxacin or doxycycycline include rifampicin, vancomycin, imipenem, penicillin, ampicillin, clindamycin, and clarithromycin.

Treatment of Human Meningoencephalitis

In the U.S.A. the most up-to-date recommendation for anthrax meningoencephalitis is a fluoroquinolone such as ciprofloxacin administered *i.v.* with two additional agents that penetrate well into the CNS, such as a β-lactam (penicillin G, ampicillin, or meropenem), rifampicin, or vancomycin is recommended (Sejvar, Tenover, and Stephens, 2005).

REFERENCES

Abramova, F., Grinberg, L., Yampolskaya, O., Walker, D. (1993). Pathology of inhalation anthrax in 42 cases from Sverdlovk outbreak of 1979. Proc. Natl. Acad. Sci. USA. 90:2291-2294.

Altboum, Z., Gozes, Y., Barnea, A., Pass, A., White, M., Kobiler, D. (2002). Post ex- posure prophylaxis against anthrax: evaluation of various treatment regimens in intranasally infected guinea pigs. Infect. Immun. 70:6231-6241.

Antwerpen, M., Schellhase, M., Ehentreich-Forster, E., Bier, F., Witte, W., Nubel, U. (2007). DNA microarray for detection of antibiotic resistance determinants in *Bacillus anthracis* and closely related *Bacillus cereus*. Molec. Cell. Probes. 21:152-160.

Barnes, J. (1947). The development of anthrax following the administration of spores by inhalation. Br. J. Exp. Pathol. 28:385-394.

Bartlett, G. (1998). Pocket Book of Infectious Disease Therapy. Williams and Wilkins Baltimore, MD, P. 20.

Bergman, N., Anderson, E., Swenson, E., Janes, B., Fisher, N., Niemeyer, M., Miyoshi, A., Hanna, P. (2007). Transcriptional profiling of *Bacillus anthracis* during infection of host macrophages. Infect. Immun. 75:3434-3444.

Bergman, N., Passalacqua, K., Gaspard, R., Shetron-Rama, L., Quackenbush, J., Hanna, P. (2005). Murine macrophage transcriptional responses to *Bacillus anthracis* infection and intoxication. Infect. Immun. 73:1069-1080.

Berthier, M., Fauchère, J., Perrinin, J. Grignon, B., Oriot, D. (1996). Fulminant meningitis due to *Bacillus anrthacis* in 11-year-old girl during Ramadan. The Lancet. 347:828.

Bonn, D. (2002). Anthrax update. The Lancet. 2:64.

Brachman, P., Friedlander, A. (1999). Anthrax: In: S. A. Plotkin, W.A. Orenstein, eds. *Vaccines*, 3rd ed. W. B. Saunders Co., Philadelphia, PA. Pp. 629-637.

Brook, I., Elliot, T., Pryor, H., Sautter, T., Gnade, B., Thakar, J., Knudson, G. (2001). *In vitro* resistance of *Bacillus anthracis* Sterne to doxycycline macrolides and quinolones. Int. J. Antimicrobial Agents. 18:559-562.

Brook, I. (2002). The prophylaxis and treatment of anthrax. Antimicrob. Agents.20:320- 325.

Cendrowski, S., MacArthur, W., Hanna, P. (2004). *Bacillus anthracis* requires sid- erophore biosynthesis for growth in macrophages and mouse virulence. Mol. Microbiol. 51:407- 417.

Esel, D., Doganay, M., Sumerkan, B. (2003). Antimicrobial susceptibilities of 40 isolates of *Bacillus anthracis* isolated in Turkey. Int. J. Antimicrob. Agents. 22:70-72.

Felek, S., Akbulut, A., Kalkan, A. (1999). A case of anthrax sepsis: non-fatal course. J. Infect. 38:201-202.

Friedlander, A, (1986). Macrophages are sensitive to anthrax lethal toxin through an acid-dependant process. *J. Biol. Chem.* 261:7123-7126.

Friedlander, A. (1999). Clinical aspects, diagnosis and treatment of anthrax. J. Appl. Microbiol. 87:303.

Glomski, I., Dumeetz, F., Jouvion, G., Huerr, M., Mock, M., Goossens, P. (2008). Inhaled non-encapsulated *Bacillus anthracis* in A/J mice: nasopharynx and alveolar space as dual portals of entry, delayed dissemination, and specific organ targeting. Microbes & Infect. 10:1398-1404.

Glomski, I., Corr, J., Mock, M., Goossens, P. (2007a). Nonencapsulated toxigenic *Bacillus anthracis* presents a specific growth and dissemination pattern in naive and protective antigen-immune mice. Infect. Immun. 75:4754-4761.

Glomski, I., Piris-Gimenez, A., Huerre, M., Mock, M., Goossens, P. (2007b). Pimary involvement of pharynx and Peyer's pach in inhalation and intestinal anthrax. PLos Pathog. 3:e76.

Guidi-Rontani, C. (2002). The alveolar macrophage: the Trojan horse of *Bacillus anthracis*. Trends in Microbiol. 10:405-409.

Guidi-Rontani, C., Levy, M., Ohayon, H., Mock, M. (2001). Fate of germinated *Bacillus anthracis* spores in primary murine macrophages. Molec. Microbiol. 42:931- 928.

Guidi-Rontani, C., Mock, M. (2002). Macrophage interactions. Current topics in Microbiol. & Immunol. 271:113-141.

Guidi-Rontani, C., Preira, Y., Ruffie, S., Sirard, J., Weber-Levy, M., Mock, M. (1999a). Identification and characterization of a germination operon on the virulence plasmid pXO1 of *Bacillus anthracis*. Molec. Microbiol. 33:407-414.

Guidi-Rontani, C., Weber-Levy, M., Labruyère, E., Mock, M. (1999b). Germination of *Bacillus anthracis* spores within alveolar macrophges. Molec. Microbiol. 31:9-17.

Hanna, P. (1999). Lethal toxin actions and their consequences. J. Appl. Microbiol. 87: 285- 287.

Ireland, J., and Hanna, P. (2002a). Amino acid-and purine ribonucleoside induced germ- Ination of *Bacillus anthracis* Sterne endospores: *gerS* mediates responses to aromatic ring structures. J. Bacteriol. 184:1296-1303.

Ireland, J., Hanna, P. (2002b). Macrophage-enhanced germination of *Bacillus anthracis* endospores requires *gerS*. Infect. Immun. 70:5870-6872.

Jamie, W. Anthrax: diagnosis, treatment, prevention (2002). Prim. Care Update. 9:117- 121.

Keppie, J., Smith, H., Harris-Smith, P. (1955). The chemical basis of the virulence of *Bacillus anthracis*. II. The role of terminal bacteraemia in death of guinea pigs from anthrax. Brit. J. Exp. Pathol. 36:315-322.

Knudsen, G. (1986). Treatment of anthrax in man: history and current concepts. *Mil. Med.* 151:71-77.

Koehler, T., Hanna, P., Kolsto, A., and Fraser, C. (2003). The genome sequence of *Bacillus anthracis* Ames and comparison to closely related bacteria. *Nature* 423:81–86.

Lincoln, R., Klein, F., Walker, S., Haines, B., Jones, W., Mahlandt, B., Friedman, R. (1964). Successful treatment of rhesus monkeys for septicemic anthrax. Antimicrob. Agents & Chemother. 4:759-763.

Lincoln, R., Fish, D. (1970). Anthrax toxin. In T. C. Monte, S. Kadis, and S. J. Ajl (eds.) Microbial Toxins III ed. Pp. 361-414. Academic Press, New York. Lindsay, J. (2002). Three steps to targeting anthrax toxin. Trends in Molec. Med. 8:6.

Mazmanian, S., Liu, G., Jensen, E., Lenoy, E., Schneewind, O. (2000). *Staphylococcus aureus* sortase mutants defective in the display of surface proteins and in the path- ogenesis of animal infections. *Proc. Natl. Acad. Sci. USA.* 97:5510-5515.

Meric, M., Willke, A., Muezzinoglu, B., Karadnizli, A., Hosten, T. (2009). A case of pneumonia caused by *Bacillus anthracis* secondary to gastrointestinal anthrax. Int. J. Infect. Dis. 13:e456-458.

Navarre, W., Scheewind, O. (1999). Surface proteins of gram-positive bacteria and mechanisms of their targeting to the cell envelope. Microbiol. Mol. Biol. Rev. 63:174- 229.

Perry, A, Ton-That, H., Mazmanian, K, Schneewind, O. (2002). Anchoring of surface proteins to the cell wall of *Staphylococcus aureus*. III. Lipid II is an *in vivo* peptidoglycan substrate for sortase-catalyzed surface protein anchoring J. Biol. Chem. 277:16241- 16248.

Pile, J., Maline, J., Eitzen, E., Friedlander, A. (1998). Anthrax as a potential biological warfare agent. *Arch. Intern. Med.* 158:429-434.

Plotkin, S., Brachman, P., Utell, M., Bumford, F., Atchison, M. (2002). An epidemic of inhalation anthrax, the first in the twentieth century: I. clinical features. Amer. J. Med. 112:4-12.

Read, T., Peterson, S., Tourasse, N., Baillie, L., Paulsen, I., Nelson, K., Tettelin, H., Fouts, D., Eisen, J., Gill, S., Holtzapple, E., Okstad, O., Helgason, E., Rilstone, J., Wu, M., Kolonay, J., Beanan, R., Dodson, L., Brinkac, M., Gwinn, R., DeBoy, R. Madpu, S., Daugherty, A., Durkin, D., Haft, W., Nelson, J., Peterson, M., Pop, H., Khouri, D. Radune, J., Benton, Y., Mahamoud, L., Jiang, I., Hance, J., Weidman, K. J. Berry, R., Plaut, R., Wolf, A., Watkins, K., Nierman, W., Hazen, A., Cline, R., Redmond, C., Thwaite, J., White, O., Salzberg, S., Thomason, B., Friedlander, A., Koehler, T., Hanna, P., Kolsto, A., and Fraser, C. (2003). The genome sequence of *Bacillus anthracis* Ames and comparison to closely related bacteria. Nature. 423:81–86.

Sejvar, J., Tenover, F., Stephens, D. (2005). Management of anthrax meningitis. Lancet Infect. Dis. 5:287-295.

Smith, H., Keppie, J., Ross, J., Stanley, J. (1954). Observations on the cause of death in experimental anthrax. *The Lance*t. 263:474-476.

Smith, H., Keppie, J., Ross, J., Stanley, J., Harris-Smith, P. (1955). The chemical basis for virulence of *Bacillus anthracis*. IV: secondary shock as the major factor in death of guinea pigs from anthrax. Brit. J. Pathol. 36:323-335.

Stephen, J. (1986). Anthrax toxin. In F. Dorner and J. Drew (eds), *Pharmacology of* Bacterial Toxins. Pp. 381-395. Pergamon Press, Oxford, England. Ton-That, H., Liu, G., Mazmanian, S, Faull, K., Schneewind, O. (1999). Purification and characterization of sortase, the transpeptidase that cleaves surface proteins of *Staphylococcus* at the LPCTG motif. Proc. Natl., Acad. Sci. USA. 96:12424-12429.

Turnbull, P. (2002). Introduction: anthrax history, disease and ecology. Current Topics in Microbiol. Immunol. 271:1-19.

Weiner, M., Hanna, P. (2003). Macrophage-mediated germination of *Bacillus anthracis* endospores requires the *gerH* operon. Infect. Immun. 71:3954-3959.

Weiner, M., Read, T., Hanna, P. (2003). Identification and characterization of the *gerH* operon of *Bacillus anthracis* endospores: a differential role for purine nucleosides in germination. J. Bacteriol. 185:1462-1464.

Welkos, S., Friedlander, A., (1988). Pathogenesis and genetic control of resistance of the Sterne strain of *Bacillus anthracis*. Microb. Pathog. 4:53-69.

Yakupogullari, Y., Koroglu, M. (2007). Nosocomial spread of *Bacillus anthracis*. J. Hosp. Infect. 65:401-402.

Zink, S., Burns, D. (2005). Importance of *srtA* and *srtB* for growth of *Bacillus anthracis* in macrophages. Infect. Immun. 73:5222-5228.

Send Orders for Reprints to reprints@benthamscience.net

CHAPTER 4

Functionality of *Bacillus anthracis* Toxins

Abstract: Both the lethal toxin (LT) and edema toxin (ET) of *Bacillus anthracis* are binary toxins and each for activity, must be bound to the protective antigen (PA) to enter and exert toxicity on eukaryotic cells. Eukaryotic cells harbor specific PA membrane receptors. Proteolytic activation of PA results in the formation of ring shaped heptamers which form ion-conductive pores in cell membranes with receptors. Each prepore binds a total of three EF and/or LF peptides competitively, and after endocytosis and transport of the complex to an acidic, vesicular compartment, it undergoes membrane insertion and mediates translocation of EF/LF to the macrophage cytosol where it exerts its cytotoxic effect. The transmembrane pore houses an aromatic iris or disk that creates a structural bottleneck in the pore, requiring that the LF and EF peptides unfold in order to be translocated into the cellular cytosol. This restricted site has been termed the "ϕ-clamp" or phenylalanine clamp. The ϕ-clamp or hydrophobic ratchet is thought to grasp hydrophobic sequences as they unfold from the molten peptide globule to direct the translocating chain through the channel. Treatment of macrophages with LT has been found to influence the expression of 103 genes. LT is proteolytic towards MAPKKs and thereby disrupts intracellular signaling.

Keywords: Lethal toxin, LT, edema toxin, ET, transmembrane pore, protective antigen, PA, anthrax toxin receptor, ATR, heptamers, ratchet, MAP kinase kinases, MAPKKs, Macrophages, oxidative burst, apoptosis, autophagy, proteasome activity, cyclic AMP, cAMP.

INTRODUCTION

The lethality of anthrax is based on the antiphagocytic activity of the poly-γ-D-glutamic acid capsule and the ability of the lethal toxin (LT) and edema toxin (ET) to form binary toxins with the protective antigen (PA) which forms a transmembrane pore allowing these toxins to enter into the cytosol of phagocytes. The LT is a highly specific zinc metallo protease that cleaves members of the mitogen-activated kinase kinase (MAPKK) family, leading to inhibition of various signaling pathways and exerts other functions so as to inhibit various aspects of the innate immune system. The ET functions to increase the intracellular cyclic AMP (cAMP) resulting in typical edema in cutaneous anthrax patients and also increases interleukin-6 (IL-6) production that inhibits phagocytosis, thereby increasing host susceptibility to infection. A considerable

Robert E. Levin

body of experimental information has been generated during the past half-century resulting in a significant level of mechanistic understanding as to why anthrax, particularly inhalation anthrax is so highly lethal.

BINARY NATURE OF THE ANTHRAX TOXINS

Interaction of PA, LF and EF

The protective antigen PA, encoded by the *pag* gene on plasmid pXO1, is named for its ability to elicit a protective immune response and is required for the activity of both the edema factor (EF) and the lethal factor (LF). The combination of PA and EF produces a skin edema but not death, whereas PA with LF is lethal without producing skin edema (Stanley and Smith, 1961). Earlier literature referred to the production of three toxins, whereas contemporary literature now usually refers to two separate toxins (Lacy, D., Collier, R. 2002): edema toxin (ET, PA + EF) and lethal toxin (LT, PA +LF). ET and LT have been found to fall into the class of "binary" (bipartite) toxins or A-B toxins where the enzymatic activity (EF or LF, "A" domain) and binding activity (PA, "B" domain) are derived from separate proteins similar to the mechanisms involving the lethality of ricin and other binary toxins. With ET and LT, the A (EF and LF) and B (PA) proteins are released from *B. anthracis* as monomers and combine at the surface of eukaryote cell receptors to form the binary toxin, where PA is the binding protein.

THE PROTECTIVE ANTIGEN (PA) AND ITS INTERACTION WITH THE LETHAL FACTOR (LF) AND EDEMA FACTOR (EF)

Cell Membrane PA Receptors

Bradley *et al.,* (2001) cloned and sequenced the PA cellular membrane receptor termed ATR (anthrax toxin receptor), and found it to be a type I membrane protein with an extracellular von Willebrand factor A (VWA) domain that binds directly to PA. The DNA sequence encoded a 368-amino-acid protein. The first 364 amino acids of ATR were found to be identical to a protein encoded by the human *TEM8* gene. *TEM8* is upregulated in the colorectal cancer endothelium. VWA domains are present in the extracellular regions of a variety of cell surface proteins and are important for protein-protein interactions and constitute ligand-

ligand sites for integrins. At least three different ATR/TeM8 protein isoforms have been described that are produced from alternatively spliced mRNA transcripts, ATR/TEM8 sv1 and sv2 which function as ATRs, whereas the putative secreted sv3 protein does not (Bradley *et al.,* 2001; Liu and Leppler, 2003; Scobie *et al.,* 2003).

Scobie *et al.,* (2003) identified a second PA receptor encoded by the capillary morphogenesis gene 2 (CMG2), which has 60% amino acid identity to ATR/TEM8 within the VWA/I domain. The initial activity of PA therefore involves binding to either of the two anthrax toxin receptors, tumor endothelial marker 8 (ATR/TEM8) or capillary morphogenesis protein 2 (CMP2) on the host cell membrane.

Domains of PA, its Proteolytic Activation, and Requirement for Acidic pH for Pore Formation

Domains of PA

PA is an 83-kDa protein (735 amino acids) which has been found to consist of four distinct and functional independent domains. After PA binds to the cell membrane receptors it is cleaved *in vivo* to yield a 63-kDa protein. This 63-kDa protein is then able to bind LF or EF resulting in biological activity. Each domain is required for a specific step in the cytotoxic process. Domain 1 (residues 1 - 249) contains a furin cleavage site for activating proteases, and is the site of interaction of PA with LF and EF. Cleavage of the native PA_{83} protein, *in vitro* by trypsin or *in vivo* by cellular proteases, cleaves off the sub-domain 1a, corresponding to the N-terminal 20 kDa fragment designated PA_{20} (Leppla, 1995). The sub-domain 1b is part of the remaining 63kDa fragment (PA_{63}) which has the property to oligomerize to ring-shaped heptamers. The proteolytic activation of PA_{83} yielding PA_{63} is essential since it exposes the site for binding LF and EF. Domain 2 (residues 250 - 487) possesses a chymotrypsin-sensitive region critical in the translocation process and has a β-barrel core containing a large flexible loop which has been implicated in membrane pore formation (Petosa *et al.*, 1997). At low pH, the heptamer undergoes conformational alteration and converts from a prepore to a pore and forms cation-selective channels in artificial and cell membranes (Finkelstein, 1994; Milne and Collier,

1993). Domain 3 (residues 488 - 594) is the smallest of the four domains, plays an important role in oligomerization of cleaved PA (PA_{63}), and has a hydrophobic patch that is thought to be implicated in protein-protein interaction (Petosa *et al.,* 1997). Domain 4 (residues 595 - 735) is the macrophage membrane receptor-binding domain. Flick-Smith *et al.,* (2002) cloned and expressed overlapping regions of PA and found that domain 4 contains the dominant immunogenicity protective epitopes of PA. Monoclonal antibodies directed to this region can block this binding (Little *et al.,* 1996).

Proteolytic Activation of PA and the Binding of PA with LF and EF

Cell intoxication is thought to occur when full-length PA (PA_{83}) binds to the macrophage cell surface receptor *via* domain 4, which contains the host cell receptor binding site (Brossier *et al.,* 2004). When bound to a cell surface receptor PA_{83} is cleaved to PA_{63} by furan-like proteases (Klimpel *et al.,* 1992; Leppla, 1995) releasing PA_{20} which unmasks the LF and EF binding sites on the resulting membrane bound PA_{63}. This binding site is capable of binding LF or EF. Chuahan and Bhatnagar (2002) identified four amino acid residues of PA essential for the binding of LF to PA. The removal of PA_{20} causes several residues to become exposed, resulting in the creation of a large hydrophobic surface on the rest of domain 1b which is involved in the binding of PA to LF and EF competitively (Miller, Elliot, and Collier, 1999). On binding to the host cell receptor, the N-terminal amino acids (1 to 167) of domain 1, which contains a furin protease cleavage site (Klimpel *et al.,* 1992) are cleaved off exposing the LF and EF binding site and the adjacent domain 3. Domains 2 and 3 then form part of a heptameric pore on the macrophage cell surface. The LF or EF then binds to its PA receptor site and the entire toxin complex undergoes receptor-mediated endocytosis in the cell.

Requirement of Acidic pH for Membrane Pore Formation

After proteolytic activation, ring shaped heptamers of PA_{63} are formed (Milne *et al.,* 1994) which are able to form ion-conductive pores in cell membranes with receptors. Each prepore binds a total of three EF and/or LF peptides competitively, and after endocytosis and transport of the complex to an acidic, vesicular compartment, it undergoes membrane insertion and mediates translocation of EF/LF to the macrophage cytosol where it exerts its cytotoxic effect (Friedlander, 1986; Collier, 1999). Friedlander (1986) found that mouse peritoneal macrophages were killed within 1 hr. of exposure to the LT (PA + LF). Neither protein component alone showed any toxic activity. Cells could be completely protected from the LT by pretreatment with agents such as amines (NH$_4$Cl, chloroquine) and the ionophore monensin, which dissipate intracellular proton gradients and raises the pH of intracellular vesicles. This protection was reversible by reduction of the intravesicular pH. Antitoxin added after preincubation with amines was unable to protect cells subsequently exposed to low pH treatment. These results indicate that the anthrax lethal toxin requires passage through an acidic endocytic vesicle in order to exert its toxic effect within the cytosol.

PA$_{63}$ prepared from PA$_{83}$ by trypsin nicking was subjected to pH values of 4.0 to 9.0 by Miller, Elliot, and Collier (1999) and a pH of 5.0 was found to be required for conversion of PA$_{63}$ to a membrane pore forming entity. They also found that when maintained at pH 8.0, this "prepore" dissociated to monomeric subunits upon treatment with SDS at room temperature, but treatment at pH 7.0 caused it to convert to an SDS-resistant pore-like form. Transition of domain 2 (D2L2), was evidenced by (1) occlusion of a chymotrypsin site within D2L2 and (2) excimer formation by pyrine groups linked to N306C within this loop. The pore-like form retained the capacity to bind anthrax toxin A moieties and cell surface receptors but was unable to form pores in membranes or mediate translocation which require a low pH. A mutant PA$_{63}$ in which D2L2 was deleted was inactive in pore formation and translocation, but like the prepore, was capable of forming heptamers that converted to an SDS-resistant form under acidic conditions.

In its activated 63 ka form, the PA component of anthrax toxin (PA + LF + EF) forms a heptameric prepore, which converts to a pore in endosomal membranes at low pH and mediates translocation of the toxins enzymatic components (LF and EF) to the cytosol. Once assembled into the membrane-penetrating β-barrel, the hydrolytic residues face the aqueous internal surface of the pore, and the hydrophobic residues form the exterior surface of the barrel. Benson *et al.,* (1998) tested the proposal that the prepore-to pore conversion involves a conformational rearrangement of a disordered amphipathetic loop (D2L2); residues 302 – 325), in which loops from the 7 protomers combine to form a transmembrane 14-stranded β-barrel. Individual cys substitutions were generated in 24 consecutive residues of the D2L2 loop and used to form channels in artificial bilayers with each mutant, and changes in channel conductance determined after adding the thiol-reactive, bilayer-impermeant reagent methanethiosulfonate ethyltrimethylammonium (MTS-ET) to the *trans* compartment. Native PA contains no cysteines. If the cysteine of interest lined the ion-conducting pathway, derivatization with MTS-ET was expected to introduce a positive charge within the cation-selective channel and thereby reduce channel conductance. Results yielded alternating reduction and the absence of reduction of conductance in consecutive residues over two stretches (residues 302 – 311 and 316 – 325. This pattern is consistent with alternating polar and apolar residues of the two stretches projecting into the pore lumen and into the bilayer, respectively. Residues connecting these two stretches (residues 312 – 315) were responsive to MTS-ET, consistent with this being in a turn region. Single channels formed by selected mutants inhibited multiple conductance step changes in response to MTS-ET, consistent with an oligomeric pore. These results presented strong evidence confirming the model of conversion of the prepore to a 14-stranded β-barrel pore and further enhanced the translocation concept of anthrax toxin.

Electron microscopy has revealed that PA_{63} forms ring-shaped heptamers containing a central cavity (Milne *et al.*, 1994). Miller, Elliot, and Collier (1999) documented that the initial PA_{63} heptamer is a "prepore" which when maintained at pH 8.0 dissociated to monomeric subunits on treatment with sodium dodecyl sulfate (SDS) whereas treatment at pH <7.0 caused it to convert to an SDS-resistant pore-like form. The pore-like form retained the capacity to bind LF and EF and cell surface receptors, but was unable to form pores in membranes or mediate translocation in L6 cells. Domain 2 of PA, which forms the main body of the heptamer contains a disordered amphipathic loop (D2L2) (Benson *et al.*, 1998). Premature conversion of the prepore to pore *in vitro* involves an irreversible step that leaves the oligomer unable to insert into cellular membranes or to generate a transmembrane β-barrel. These observations suggest that the ability for membrane insertion by the heptamer form of PA is dependant on the D2L2 domain remaining in its "poised" prepore state before membrane binding.

The fourth domain of PA (PA-D4) is responsible for initial binding of the anthrax toxins LT and ET to the cellular receptors A synthetic gene encoding PA-D4 was prepared and expressed as a fusion protein by Krishnanchettiar, Sen, and Caffrey (2003). The rPA-D4 was purified to near homogeneity and was found to bind to Hela cells, indicating retention of functionality.

Blaustein *et al.*, (1989) demonstrated that trypsin-cleaved PA_{83} is capable of forming cation-selective channels in planar phospholipid bilayer membranes and that this activity is confined to the PA_{63} fragment; PA_{20}, LF and EF lacked channel-forming activity.

Acidification of the endosomal vesicles is thought to induce a conformational change in the PA hexamer, exposing previously buried hydrophobic regions that are necessary for the hexamer to interact with the cells lipid bilayer membrane to result in pore formation (Defrise-Quertain *et al.*, 1989). Both LF and EF were found by Kochi *et al.*, (1994) to insert into model phospholipid lipid vesicle bilayers in a pH-dependant manner. Low pH (pH 6.0 and below) induced membrane insertion of both proteins. The proteolytically cleaved form of PA_{83}, PA_{63}, undergoes a major change in hydrophobicity at low pH, allowing the protein to insert into lipid bilayers and forming ion-conducting channels; with channel formation not occurring with uncleaved PA (Koehler and Collier, 1991; Blaustein *et al.*, 1989). Kochi *et al.*, (1994) however found that uncleaved PA_{83} was able to interact with the lipid vesicles and insert at acid pH values with the optimum pH for insertion at pH 5.0. PA_{63} exhibited a similar pH dependence of membrane insertion to the uncleaved PA_{83} (maximum at pH 5.0) and similar membrane destabilization. However, PA_{83}- and PA_{63} induced destabilization of the vesicles was less efficient than LF or EF, since 5 – 10 times more protein was required to produce a similar level of membrane destabilization.

Inter-alpha-inhibitor protein (IαIP) functions as an intracellular serine protease inhibitor in human plasma. One potential target for IαIP is furin, a cell-associated serine endopeptidase essential for the proteolytic activation of PA and formation of LT. IαIP was found by Opal *et al.*, (2005) to block furin activity *in vitro* and to provide significant protection against cytotoxicity for murine peritoneal macrophages exposed to up to 500 ng/ml of LT. A monoclonal antibody (mAb), that specifically blocked the enzymatic activity of IαIP was found to eliminate its protective effect against LT-induced cytotoxicity. IαIP (30 mg/kg body weight) administered to BALB/c mice 1 hr. prior to an *i.v.* challenge with LT resulted in 71% survival after 7 days compared to no survivors among control mice.

Translocation of LF and EF

The N-terminus of LF and EF initial enters the mouth of the beta-barrel and is bound to domain 1 of the heptameric pore (Fig. **1**). The carboxy terminus then proceeds to pass through the pore (Krantz, Finkelstein, and Collier, 2006). The lumen of the β-barrel is ~15A wide and is able to accommodate secondary structural proteins only as large as an α helix (Krantz, 2006). Translation thus requires the peptide to unfold. Krantz *et al.,* (2005) found that during conversion of the heptameric precursor to the pore, the seven phenylanaline-427 residues converge within the lumen of the pore, generating a radially symmetric heptad of solvent-exposed aromatic rings. This configuration is envisioned as resulting in a narrow aromatic "iris" or ring of rings within the lumen of the pore. This aromatic iris or disk creates a structural bottleneck, requiring that the LF and EF peptides unfold for translocation into the cell. The φ-clamp is thought to grasp hydrophobic sequences as they unfold from the molten peptide globule to propel the translocating chain through the channel. Positively charged segments flanking the hydrophobic segments are repulsed by the disk. In addition, it is also thought to effectively maintain the proton gradient by acting as a hydrophobic seal. The higher concentration of protons in the *cis*-side (mouth) of the pore, inserted into the membrane of a mature endosome (pH ~5.5), is required to acid-destabilize and unfold LF and EF. This *cis* end of the pore is kept separate from the lower concentration of protons *trans* to the φ-clamp which facilitates translocation of LF and EF into the cytosol (Fig. **2**) that has a pH of ~pH 7.3 (Krantz, Finkelstein, and Collier, 2006).

The φ–clamp has been proposed to be a hydrophobic ratchet (repulsing positive charges) and is based on the strategy that the LF or EF polypeptides have anionic, cationic, hydrophobic, and hydrophilic, functionalities distributed across their sequences (Krantz, Finkelstein, and Collier, 2006).

The destabilization energy required to unfold the tertiary structure of LF and EF is thought to originate partly from the acidic pH of mature endosomes which causes their N-terminal domains to become molten globules (Krantz *et al.*, 2004).

Because the φ-clamp forms a narrow iris (Fig. **1**) it can effectively grasp the translocating polypeptide LF or EF chains as they unwind so as to direct them through the channel. This occurs as a result of the φ–clamp site creating an environment that minimizes the hydrophilic core of the unfolding LF or EF peptides. The PA_{63} heptameric pore is therefore envisioned to function as a Brownian ratchet, enabling the unwound linearized leading segment of a translocating protein to move through the channel (Fig. **2**) and the trailing part of the protein to more readily unfold (Krantz *et al.*, 2005).

A polypeptide chain such as LF or EF has a significant amount of Brownian thermal energy available which cannot do useful work, such as driving the translocating peptide chain through the pore unidirectionally. However, a transmembrane chemical gradient (△pH) can bias the Brownian fluctuation toward unidirectional translocation. Protonation or deprotonization of the acidic side chains comprising the translocating polypeptide can occur on either the *cis* or the *trans* side of the membrane and decrease or increase respectively the energy barriers for entry of the translocating polypeptide into the pore (Krantz, Finkeltein, and Collier, 2006).

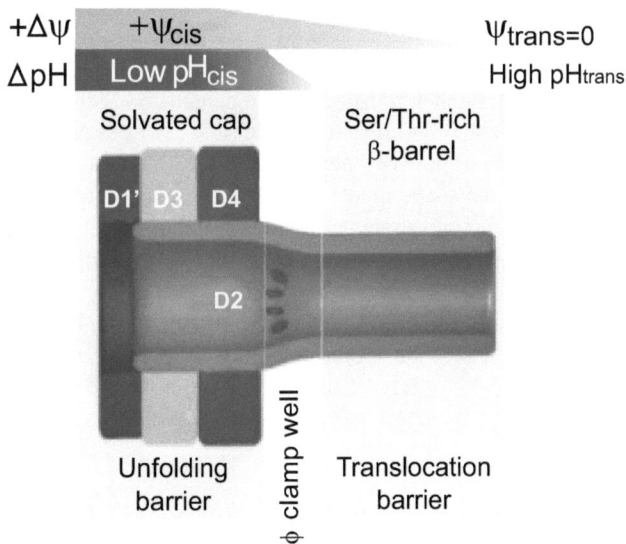

Figure 1: Pictorial visualization of the β-barrel and its translocation function. PA_{63} domains D1, D3, and D4 comprise the hydrophilic cap, while domain D2 constitutes the hydrophobic lumen of the pore with the restricting disc depicted in the φ-clamp well. The unfolding barrier is on the low pH *cis* side of the membrane while the translocation barrier is on the high pH *trans* side of the membrane entering into the cellular cytosol. From Krantz, Finkelstein, and Collier (2006) with permission.

Krantz, Finkelstein, and Collier (2006) found that translocation of LF, through PA$_{63}$ channels was markedly stimulated by a proton gradient ($\triangle$pH); that is, when the pH of the *trans* side (pH *trans*) of the lipid bilayer was greater than that of the *cis* side (pH *cis*) The LF and EF polypeptide chains are assumed to consist of segments of negatively and positively charged residues. Protonation at the *cis* side of the pore is thought to impart a net positive charge to these segments which is

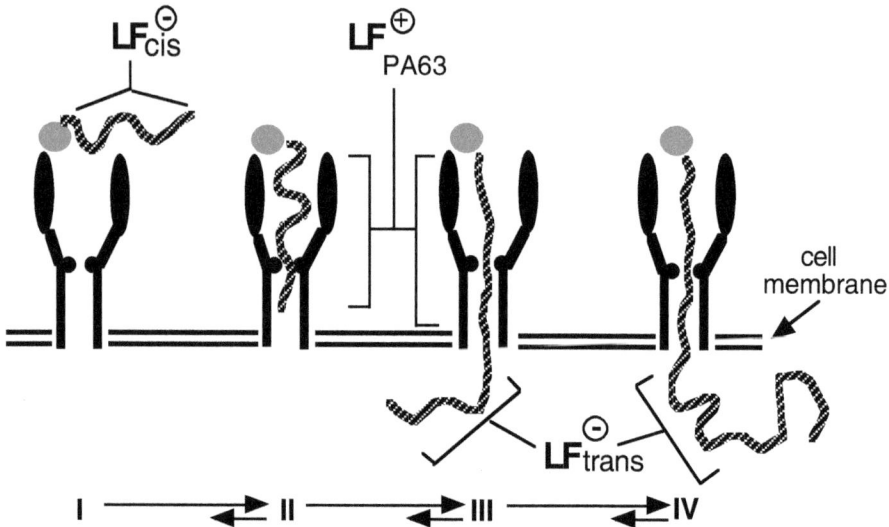

Figure 2: A tandem Brownian ratchet translocation mechanism. Stage I: Initially, the amino terminus of LF is outside the channel and negatively charged residues are in a deprotonated LF$^{(-)}$ *cis* state. Stage II: A conductance-blocked intermediate forms when a hydrophobic, aromatic and/or cationic sequences binds the ϕ-clamp site, which is the narrow constriction on the *cis* side of the extended β barrel that effectively maintains the ΔpH. Negatively charged residues in the translocating chain that have entered the channel are protonated making the stretch net–positive, LF$^{(+)}$ PA63; this chain is then compatible with the cation-selective channel. Stage III: The translocating chain proceeds through the ϕ-clamp site as the hydrophilic sequence enters the β barrel and exits to the *trans*-side solution. Translocation is favored over retro-translocation to the *cis* side under a + ΔpH. Stage IV: Successive iterations of this ΔpH driven, charge-state ratchet lead to processive translocation. Redrawn from Krantz, Finkelstein, and Collier (2006) with permission.

compatible with the cation-specificity of the pore. As the positively charged segment of LF or EF exits the channel it is then deprotonated, resulting in a net negative charge to the specific segment. Translocation therefore consists of the following process where the *trans* pH is higher than the *cis* pH (Krantz *et al.*, 2006).

$$\text{LF}_{\text{cis}}^{\ominus} \xrightarrow[\longleftarrow]{\text{low pH}_{\text{cis}}} \text{LF}_{\text{PA63}}^{\oplus} \xleftarrow{\text{pH}_{\text{trans}} < \text{pH}_{\text{cis}}} \text{LF}_{\text{trans}}^{\ominus}$$

The net negative charges of the deproteinated segments prevents reverse translocation. Thus the charge-state ratchet acts processively upon individual segments of the translocating polypeptide rich in acidic residues. The channel therefore functions as a proton/protein symporter (Krantz *et al.*, 2006) where the pH-driven charge-state ratchet results from electrostatic repulsion between the channel and anionic charges. Brossier and Mock (2001) diagrammatically presented the mechanistic aspects of the translocation process resulting in the LT and ET entering into the macrophage cytosol (Fig. **3**).

Cultural, Environmental, and Genetic Factors Controlling Toxin Production

The specific nutritional environment has been found critical for obtaining elevated yields of PA, LF, and LT from broth cultures of *B. anthracis*. Ristroph and Ivins (1983) developed a completely defined superior medium derived from the Casamino acids medium of Haines, Klein, and Lincoln (1965) designated R medium. The quantities of individual amino acids in 3.6g of yeast extract were substituted for 3.6g of Casamino acids, the concentration of glucose was increased from 0.2 to 0.25%, KH_2PO_4 was omitted, and the concentration of K_2HPO_4 was increased from 0.088 to 0.3% to provide increased buffering and activated charcoal was omitted. The completely defined R medium contained 27 individual nutrient components including 0.8% $NaHCO_3$ and was adjusted to pH 8.0. prior to filter sterilization. After 16 hrs. of incubation at 37 °C the pH dropped to 7.2 - 7.4. Use of the R medium resulted in a two-fold increase in LF with the Sterne and V770-NP1-R strains of *B. anthraci*s and a 5-fold increase with the Vollum 1B strain.

Two environmental factors, bicarbonate and temperature have been found to control the three toxin genes (*pag, cya, lef*). At 28 °C, expression levels for all three genes were four- to six-times lower than those observed at 37 °C (Sirard, Mock, and Fouet, 1994). These results explain the observation of Metchnikoff in 1884 that frogs and lizards were not normally susceptible to anthrax. However, if these reptiles were kept at 35 to 37 °C and then infected with a virulent strain they were killed (Magasanik and Coons, 1984).

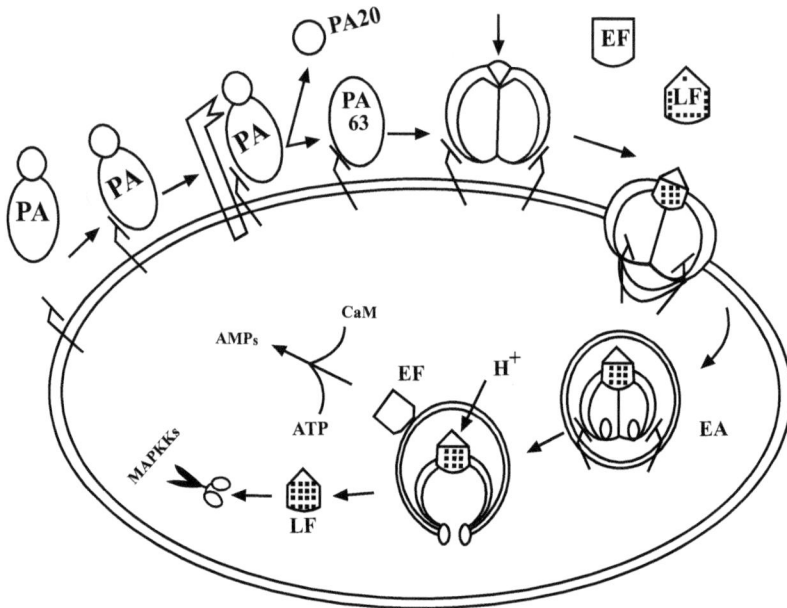

Figure 3: Mechanistic cellular model of action of anthrax toxins. Cellular toxicity can be divided into 6 stages. (1) Protective antigen (PA_{83}) binds to its specific cell membrane receptor. (2) PA_{83} is then proteolytically cleaved to PA_{63} releasing PA_{20} and exposing the binding site for EF an LF. (3) The PA_{63} peptide then oligomerizes to ring-shaped heptameric prepores. (4) EF and LF then bind to the heptameric prepore and endocytosis occurs with endosomal acidification. (5) At acid pH the prepore converts to a pore resulting in translocation of LF/EF into the cytosol. (6) The LF, as a zinc protease, then cleaves the amino terminus of mitogen-activated protein kinases MAPKKs), thereby disrupting signal pathways. From Brossier and Mock (2001) with permission.

The structural genes *cya, lef,* and *pagA* that encode the proteins EF, LF, and PA respectively are located non-contiguously within a 30-kb region of the pXO1 plasmid. The genes required for capsule synthesis, *capB, capC, capA*, and the gene encoding the depolymerization of the D-glutamic acid capsule *dcp* are located contiguously in the same direction of transcription on the plasmid pXO2 (Uchida *et al.,* 1993b).

Two genes have been identified as *trans*-acting regulators of gene transcription *acpA* located on pXO2, activates expression of *capB* on pXO2, and *atxA* which is located on pXO1 and activates expression of *capB* (Guignot *et al.,* 1997; Uchida *et al.,* 1997); and the toxin genes *cya, lef,* and *pagA* (Uchida *et al.,* 1993; Koehler, Dai, and Kaufman-Yarbray, 1994; Dai *et al.,* 1995). *atxA* is required for CO_2/bicarbonate-enhanced transcription of toxin genes in cells growing in laboratory media and also appears to regulate toxin gene expression during infection. An *atxA* –null mutant was found to be avirulent in a mouse model for anthrax and the antibody response to the three toxin proteins was deceased notably in mice infected with the *atxA*-null mutant (Dai *et al.*, 1995). Hoffmaster and Koehler (1999) found that an *atxA*-null mutant exhibited phenotypes unrelated to toxin and capsule synthesis. An *atxA*-null mutant was found to grow poorly on minimal media and sporulated more efficiently than the parent strain. In addition, the *atxA*-activated *pag* gene was found to be co-transcribed with a 300-bp gene, *pagR*, located downstream of *pagA*. *pagR* was also found to control expression of certain CO_2/*atx*-activated transcriptional fusions on pXO1 that do not correspond to the toxin genes.

Involvement of PA in the Virulence of *B. anthracis*

Ivins and Welkos (1986) cloned the *pagA* gene that encodes PA into *B. subtilis*. Two clones, designated PA1 and PA2, were identified which produced PA in liquid culture at levels of 20.5 to 41.9 µg/ml. Addition of LF or EF resulted in the formation with recombinant PA (rPA) of LT and ET respectively. Immunization (*i.m.*, twice) with the live recombinant strain of *B. subtilis* protected guinea pigs from lethal challenge with virulent *B. anthracis* spores, and the immunization partially or completely protected rats from *i.v.* challenge with LT.

Pezard, Berche, and Mock (1991) genetically engineered EF⁻ and LF⁻ derivatives of the Sterne strain (pX02⁻). LF⁻ and PA⁻ mutant strains were not lethal to Swiss mice, even at high spore doses of *s.c.* infections as high as 10^9 spores, whereas the LF⁺ EF⁻ mutant induced lethal infection with an LD_{50} spore dose of 10^7. These results clearly indicated that LF in combination with PA is a virulence factor required for lethality. Skin edema formation was observed with an LF⁻ EF⁺ mutant, which produced only the combination of PA and EF. However, EF⁻ and LF⁻ mutants were significantly less efficient at inducing, respectively lethality and edema than was the parental Sterne strain. These results suggested that the three toxin components may act synergistically *in vivo* to cause lethality and edema formation.

Kobiler *et al.*, (2006) demonstrated that the detection and level of PA in infected host sera can serve as a reliable diagnostic marker of active anthrax infection.

Two animal inhalation models utilizing rabbits and guinea pigs infected with Vollum spores were studied. With both animal models qualitative and quantitative correlations between levels of bacteremia and PA concentrations in the sera of sick animals were obtained. The detection of PA in the blood serum with an ELISA assay was coincidental with the occurrence of bacteremia. However, experience with bacteremic patients has indicated that administration of antibiotics to anthrax patients that have reached the bacteremic state is usually unable to prevent death.

Moayeri, Wiggins, and Leppla (2007) injected mice *i.v.* with PA_{83} and PA_{63} and observed that both wild-type PA and a receptor-binding-defective mutant PA were cleaved to PA_{63} at the furin cleavage site, independent of their ability to bind to cells. PA cleavage was detected in blood serum as early as 5 min. after PA injection. This suggested a PA-acting protease activity in the blood. After injection, PA_{83} was rapidly cleaved in mice but not rats. Cleavage occurred *via* a leupeptin-calcium-sensitive protease. Cleavage did not require binding to PA receptors on cell membranes. Little (2004) separated the exotoxin components PA, LF, and EF from 24 isolates of *B. anthracis* of diverse global origins by isoelectric focusing gel electrophoresis and detected the individual proteins by Western blot analysis with monoclonal antibodies. Only two isoforms for each were observed for PA and EF. Four isoforms were identified for LF.

Production of PA by vegetative cells of *B. anthracis* reaches a peak at the end of the exponential growth phase (Leppla, 1988). PA has however been found to be associated with spores. Cote *et al.,* (2005) found that PA-specific mRNA was detected in spores by RT-PCR within 15 min. of exposure to spore germinants. PA protein was detected by immunomagnetic elecrochemiluminescence (ECL) on spores within 1 hr. of exposure to a germination medium and was rapidly released into the supernatant. PA was not detected in ungerminated spores by RNA analysis, ECL, or spore-based anti-PA ELISA. Interestingly, PA was detected on nongerminated spores by immunoelectron microscopy, which may reflect the ability of PA to bind to the surface of newly formed spores in growth media. Anti-PA IgG was found to retard germination of spores *in vitro*. These authors concluded that anti-PA IgG-mediated anti-spore activities may play a role in protection during the early stages of an anthrax infection.

THE LETHAL FACTOR (LF)

Mechanism of Action of the LF and its Effect on Macrophages

PA-dependant antibody response to LF induced by immunization with live *B. anthracis* was used by Brossier *et al.,* (2000) to follow the *in vivo* interaction of LF and PA. Point mutations and deletions were introduced into the genes encoding LF, EF, and PA functional domains. The synergistic effects of the LT and ET were found to result solely from their enzymatic activities, suggesting that *in vivo* these toxins presumably act together. The PA-dependant antibody response to LF induced by immunization of OH1 outbred mice with live *B. anthracis* was used to follow the *in vivo* interaction of LF and PA. The binding of LF to PA *in vivo* was found necessary for a strong antibody response against LF and PA, whereas neither LFS activity nor binding of LT to the cell surface was required.

It is now well recognized that the LT of *B. anthracis* is the major factor resulting in shock and death from systemic anthrax (Friedander, 1986; Hanna *et al.*, 1993) succumbed in a manner that closely resembles the natural systemic infection (Hanna *et al.*, 1993). Early experiments administering antibiotics to *B. anthracis* infected guinea pigs at different stages of infection revealed the "principle of no return" whereby once the infection reached a certain stage, the animal was irreversibly doomed, even after antibiotic destruction of the organism (Lincoln and Fish, 1970; Stephen, 1986). This effect is clearly recognized as being due to the effects of the LT produced during systemic anthrax infections (Hanna, 1999). The concept that anthrax is a toxigenic condition is supported by studies that immunization of animals against LT protects them from challenge by virulent *B. anthracis* (Turnbull, 1992) and that LT deficient *B. anthracis* strains are attenuated 1000-fold (Cataldi, Labrugere, and Mock, 1990).

Major insight regarding the role of LT occurred when Friedlander (1986) discovered that macrophages were uniquely killed by LT. Hanna, Acosta, and Collier (1993) depleted mice of macrophages using a sequence of SiO_2 injections which are selectively toxic to macrophages. The macrophage depleted mice were found to be resistant to LT, with 100% survival compared to <10% survival of control animals following a LT challenge.

LF is a highly specific zinc metallo protease (Mock and Fouet, 2001) that cleaves members of the mitogen-activated protein kinase (MAPKK) family near to their amino terminus, leading to inhibition of one or more signaling pathways. LT also suppresses the communication between innate and adaptive immune responses. In addition, LT acts directly on T and B lymphocytes, blocking antigen receptor-dependant proliferation, cytokine production and Ig production. In this manner LT establishes a broad-based attack on host immunity, thus providing *B. anthracis* with a variety of mechanisms for circumventing protective host responses (Xu and Frucht, 2007).

Results with mice have implicated hypoxia-induced tissue injury (Moayeri *et al.,* 2003) and genetic factors (Moayeri *et al.*, 2004) in L-mediated lethality rather than induction of proinflammatory cytokines as suggested earlier (Mock and Fouet, 2001). Pannifer *et al.,* (2001) described the crystal structure of LF and its complex with the N terminus of MAPKK-2. LF comprises four domains: domain I binds the membrane-translocating component of anthrax toxin, PA; domain II, III, and IV together create a long deep grove that holds the 16-residue N-terminal tail of MAPKK-2 before cleavage. Domain II resembles the ADP-ribosylating toxin from *B. cereus*, but the active site has been mutated and recruited to augment substrate recognition. Domain III is inserted into domain II, and appears to have arisen from a repeated duplication of a structural element related to the zinc metalloprotease family, and contains the catalytic protease center and it also resembles domain I. The authors concluded that the crystal structure reveals a protein that has evolved through a process of gene duplication, mutation and fusion, into an enzyme with high and unusual specificity.

Vitale *et al.,* (1999) showed that LF cleaves the MAP kinase kinases (MAPKKs) Mek1 and Mek2 *in vitro* an *in vivo*, hydrolyzing a Pro8-Ile9 and a Pro10-Arg11 peptide bond in Mek1 and MeK2, respectively. The removal of the amino terminus of MAPKKs eliminates the "docking site" involved in the specific interaction with MAPKs and interferes with the phospho-activation of the MAPKs ERK1 and ERK2, which become phosphorylated in cultured macrophages following toxin challenge.

Friedlander (1986) was the first to document that exposure to low pH by LT is required for macrophage toxicity. Mouse peritoneal macrophages were killed within 1 hr. of exposure to the toxin. Cells could be completely protected from exposure to LT by pretreatment with the lysomotropic amines NH_4Cl and chloroquine, or with the ionophore monessin which dissipate intracellular proton gradients and raise the pH of intracellular vesicles. These results strongly suggest that LT exerts its effect within the cytosol after passage through an acidic intracellular vesicle or compartment. This protection was reversible and could be overcome by lowering the intravesicular pH. Antitoxin added after preincubation of LT treated macrophages with amines was unable to protect cells subsequently exposed to a low pH of 4.5. These results indicated that after pretreatment with NH_4Cl, the toxin is not accessible to antitoxin and supports the concept that LT after acidification is present in an intracellular location before expressing its toxic effect on the cell.

Groups of Fischer 344 rats were injected *i.v.* by Ivins, Ristroph, and Nelson (1989) with filter sterilized supernatants from with *B. anthracis* Vollum 1B and Stern strains containing crude anthrax toxins. Times to death of rats given identical toxin preparations varied directly with the weights of the rats. In contrast to earlier reports, the data indicated that rat weight must be taken into account during *in vivo* assays of anthrax LT activity. This same relationship may also apply to other mammalian species.

The oxidative burst is one of several microbicidal response mechanisms macrophages possess. Certain of the resulting reactive oxygen species generated are toxic to macrophages, which must therefore closely regulate their production and/or restrict their production to subcellular locales where their effects can be tolerated. Major symptoms and death from systemic *B. anthracis* infections are mediated by the action of the organisms LT on host macrophages. Although LT is fully capable of entering most cell types, toxicity is limited to macrophages, a major producer of reactive oxygen species. Hanna *et al.,* (1994) found that murine macrophages treated with high levels of LT released large amounts of superoxide anion beginning at about 1 hr., which correlated with the onset of cytolysis. Cytolysis could be blocked with various exogenous antioxidants or with N-acetyl-L-cysteine and methionine which promote the production of the intracellular antioxidant glutathione. Mutant marine macrophage lines deficient in production of reactive oxygen intermediates were relatively insensitive to the lytic effects of

LT whereas a line with increased oxidative burst potential exhibited elevated sensitivity. In addition cultured blood monocyte-derived macrophages from a patient with chronic Granulomatous Disease, in which the phagocytic oxidative burst is disabled, were totally resistant to the LT of *B. anthracis*, in contrast to control monocytes when BALB/c mice were pretreated with N-acetyl-cysteine or mepacrine (an oxygen scavenger).

The mitogen-activated protein kinase (MAPK) signal transduction pathway, is an evolutionarily conserved pathway that controls cell proliferation and differentiation. In response to extracellular signals, MAPK is phosphorylated and activated by MAPK kinases 1 and 2 (MAPKK1 and MAPKK2). Duesbery *et al.*, (1998) found that human, frog, and mouse MAPKK1 are all substrates for the protease activity of LF. In addition, LF was found to cleave MAPKK2 between residues 9 and 10. MAPKK1 mutants in which either Pro^5 or Pro^7 was replaced by Ala were resistant to cleavage by LF, indicating that the proline residues constitute important components of the cleavage site.

Duesbery and Vande Woude (1999) found that anthrax LT inhibited progesterone-induced meiotic maturation of frog oocytes by preventing the phosphorylation and activation of mitogen-activated protein kinase (MAPK). Similarly, LT prevented the activation of MAPK in serum stimulated, ras-transformed NIH3T3 cells. *In vitro* studies indicated that LF proteolytically modified the NH_2-terminus of both MAPK1 and 2, rendering them inactive and hence incapable of activating MAPK.

Popov *et al.*, (2002a) found that human peripheral monocytes (PBMC) were resistant to lysis by the LT of *B. anthracis*, in contrast to mouse macrophages, confirming earlier reports. Infection of PBMC with *B. anthracis* spores led to the appearance of large numbers of cells stained positively for apoptosis, with a reduced capacity to eliminate spores and vegetative cells, particularly when the PBMC were additionally stressed by removal of serum.

The Fas/FasL system is considered the best characterized of the known apoptosis pathways (Askenazi and Dixit 1998). Popov *et al.*, (2002b) therefore examined the possibility that Fas mediates LT-induced apoptosis. Anti-Fas-antibodies applied to RAW264.7 macrophages treated with LT greatly reduced the number of otherwise resulting dead cells, and also resulted in the number of apoptotic cells increasing. In addition, with murine RAW264.7 macrophage cells, lytic amounts of LT triggered intracellular signaling events typical of programmed cell death (apoptosis), including changes in membrane permeability, loss of mitochondrial membrane potential, and DNA fragmentation. The cells were protected from LT by specific inhibition of caspase. Phagocytosis activity of the macrophages was inhibited by sublethal concentrations of LT. Infection of cells with anthrax Sterne spores impaired their bactericidal activity which was reversed by the LT inhibitor bestatin. These observations indicated a substantial reduction in the apoptotic process. The authors suggested that apoptosis rather than direct lysis is biologically relevant to intracellular activity of infected macrophages.

Gutting *et al.,* (2005) undertook studies to determine if several different macrophage lines can influence the activity of LT when cultured with an LT-producing strain of *B. anthracis*. Macrophages were co-cultured with the LT^+ Vollum strain and assayed for macrophage cell death by morphology, trypan blue staining, neutral red activity, and lactic dehydrogenase activity in the culture medium. Following the addition of spores, 50% mortality of macrophages occurred with macrophage line RAW264.7 and 5774A.1 at 7.5 and 10 hrs. respectively. By 15 hrs, both RAW264.7 and 5774A.1 macrophages exhibited 100% motality. In contrast, macrophage cells of line IC-21 under identical culture conditions, remained 98% viable and active throughout the study (>24 hrs.). the mechanism of macrophage death appeared to involve LT because the *B. anthracis*-induced cytotoxicity was dose-dependently reversed by the addition of PA antibody to the macrophage culture media.

The mechanism of IC-21 macrophage resistance is thought to be due to the inability of IC-21 macrophages to generate reactive oxygen intermediates in response to LT. During the respiratory burst, macrophages can produce large concentrations of microbicidal superoxide anion (O_2^-), hydroxyl radical ($^\cdot OH$), hydrogen peroxide (H_2O_2) and hypochlorate ($^\cdot OCl$), through the NADH oxidase complex (Ismail *et al.,* 2002). IC-21 macrophages lack the ability to produce reactive oxygen species associated with the respiratory burst (Scott, James, and Sher, 1985) and are resistant to the LT-induced cytotoxicity, suggesting that the respiratory burst may be involved in macrophage sensitivity (Hanna *et al.*, 1994). In addition, BALB/C from which the RAW364.7 and J77A.1 macrophage lines are derived are more sensitive to lethal infection by *B. anthracis* than are C57/B16 mice from which the IC-21 macrophage line is derived (Moayari *et al.,* 2003; Welkos, Keener, and Gibbs, 1986; Welkos, *et al.,* 1989).

The sudden bursting of macrophages with simultaneous release of intracellular mediators of shock was thought to contribute to the "sudden death" observed with anthrax. Support for the involvement of cytokines in systemic shock was provided by the observation that mice immunized passively with neutralizing antisera against tumor necrosis factor α (TNFα) and inerleukin-1B (IL-1B) were protected against LT (Hanna, 1998).

Mouse peritoneal macrophages and certain macrophage-like cell lines are sensitive to anthrax lethal toxin (LT). It causes over-production of certain cytokines such as IL-1β and TNF-α (Pellizzari *et al.*, 1999). Studies have indicated that LF acts as a Zn^{++} dependant endopeptidase that cleaves the amino end of mitogen-activated protein kinase kinase 1 (MAPKK1) which contains the mitogen-activated protein kinase (MAPK) binding site (Duesbury, *et al.,* 1998; Popov *et al.*, 2002b; Vitale *et al.*, 2000; Vitale *et al.*, 1998). LF therefore prevents the association of MAPK1 with its substrate and inhibits the MAPK signal transduction pathway (Ahuja, Kumar, and Bhatnagar, 2001). Kim *et al.,* (2003) cloned the complete LF (*lef*) gene (without the signal peptide) for over expression of LF. LF was purified for use as a protease. To facilitate the function and folding during purification and immobilized enzyme activity as a GST fusion protein, a 5-glycine linker sequence (GGT-GGT-GGT-GGT-GGA) was introduced in-frame between GT and multiple cloning sites. Recombinant lethal factor (GSST-LF) was purified on a glutathione (GSH)-Sepharose 6B column) resin column. The GST tag was then removed with thrombin and the thrombin removed with benzamidine-Sepharose resulting in pure recombinant LF (rLF). The proteolytic activity of LF was determined from the cleavage of the full-length recombinant GST-MEK which is a specific cellular substrate of LF. Cleavage of the substrate (GST-MEK) produced GST containing a short N-terminal MEK peptide of 28 kDa and a resultant MEK of 44 44 kDA. Results confirmed that rLF is a Zn^{++}-dependant metalloendopeptidase rather than a metalloaminopeptidase.

Gozes *et al.,* (2006) utilized the classic Miles leakage assay (Miles and Miles, 1952) to quantify vascular leakage in mice injected *i.d.* with LT. The Miles assay uses *i.v.* injection of Evans blue dye (which binds to serum albumin), 30 min prior to *i.d.* injection of LT as a tracer to assay for macromolecular leakage from peripheral blood vessels. Quantification of the extent of leakage was assessed by extraction of skin regions surrounding *i.d.* injection sites and A_{620} of extracts determined. In some inbred mouse strains, but not in others, leakage was induced as early as 15 to 25 min. The leakage could be inhibited by ketotifen, an inhibitor of mast cell degranulation, but not by azelastine, a histamine receptor1 antagonist, or by ketanserin, a seratonin 5-HT2A receptor antagonist.

B. anthracis lethal toxin (LT) causes vascular collapse and high lethality in BALB/cJ mice, intermediate lethality in C57BL/6J mice, and no lethality in DBA/2J mice. Moayeri *et al.,* (2005) found that adrenalectomized (ADX) mice of all three strains had increased susceptibilities to LT. The increased susceptibility of ADX-DBA/2J mice was not accompanied by changes in their macrophage sensitivity or cytokine response to LT. BALB/cJ mice showed a five-old increase in serum corticosterone, while DBA/2J mice showed no change in serum corticosteroid levels in response to LT injection. These results indicated that susceptibility to anthrax LT in mice depends on a fine balance of endocrine functions easily perturbed.

Mice that have been depleted of macrophages are markedly more susceptible to aerosol and parenteral challenge with the spores of the Ames strain of *B. anthracis* (Cote *et al.,* 2004; Cote, van Rooijen, and Welkos, 2006). This observation suggests that macrophages are not required for *in vivo* germination of *B. anthracis* spores and that macrophages play a significant protective role in anthrax pathogenesis. Resistance of mice to Ames spore infection can be increased by supplementing the animal's native macrophage population with exogenous RAW264.7 wild type murine macrophages injected *i.p.,* immediately prior to *B. anthracis* spore challenge (Cote *et al.,* 2004; Cote, van Rooijen, and Welkos, 2006).

Anthrax toxins have been found to mediate immune evasion of the bacterium by interfering with both innate and adaptive immune responses. Injection of a sublethal dose of LT was found to disrupt the adaptive immune response by inhibiting the functions of dendritic cells (Agrawal *et al.*, 2003) and *in vitro* exposure to LT and ET inhibits the activation of mouse T cells (Paccani *et al.*, 2005). Comer *et al.,* (2005a) isolated CD4+ T cells from mice and observed that LT and ET inhibited T-cell proliferation in a dose-dependant manner. At the highest concentrations of LT (1.0 µg/ml of PA and 0.2 µg/ml of LF), and ET (2.5 µg/ml of PA and 0.625 µg/ml of EF), activation of T cells was completely inhibited. In the presence of a 1,000-fold-lower concentration of the toxins, T lymphocytes responded normally to stimulation with anti-CD3 and anti-CD28 antibodies. Both LT and ET notably inhibited secretion of IL-2 by the CD4$^+$ T cells. T cells isolated from LT injected mice and stimulated with anti-CD3 and anti-CD28 antibodies underwent a significant decrease in phosphorylation of the extracellular signal-regulated kinase 1 and 2 (ERK1/2), the stress-activated kinase p38, the activating transcription factor 2 (ATF-2), the serine/threonine kinase AKT, and the glycogen synthase kinase 3α and –β (GSK-3α and β). Treatment of mice with ET did not affect the phosphorylation of these kinases following T-cell stimulation. However, phosphorylation of the stress activated factor c-Jun NH$_2$-terminal kinase (JNK) was inhibited by treatment of mice with TT but not by treatment with LT. These observations clearly indicated that CD4$^+$ T cell activation is inhibited by LT or RT and that sublethal doses of LT or ET injected mice blocks the ability of T lymphocytes to respond to stimulation through the T-cell receptor. Data also indicated that LT directly disrupts the activation of T lymphocytes by blocking cellular signaling.

Cote *et al.,* (2008) used murine macrophages RAW264.7 and a mutant cell line RSD that is PA-receptor negative to test the role of toxin-targeting of macrophages. When mice were *i.p.* supplemented with wild type RAW264.7 or mutant R3D macrophages (~1 x 10^7) and then exposed to a lethal *i.p.* challenge with heat activated Ames spores the RSD macrophages resulted in 8% survival whereas the wild type RAW264.7 macrophages resulted in only 3% survival and controls (without supplementary macrophages) 0% survival. These results indicate that the LT targeting of macrophages is an important virulence mechanism for *B. anthracis* and that macrophages play a dual role in the pathogenesis of and protection against anthrax.

The statin family of HMG-coenzyme A reductase inhibitors have powerful anti-inflammatory effects independent of their cholesterol-lowering properties, which have been ascribed to modulation of Rho family GTPase activity. The Rho GTPases regulate vesicular trafficking, cell adhesion and migration, production of reactive oxygen species, and cell survival and proliferation (Benard, and Bokoch, 2002; Bokoch, 2005; Jaffe, and Hall, 2005). The only known cytoplasmic targets of LF are mitogen-activated protein kinase (MAPK) kinases, also known as the MAPK/ERK kinases (MKKs or MMEKs, respectfully) (Duesbery *et al.*, 1998; Pelizari *et al.*, 1999; Vitale *et al.*, 2000; Vitale *et al.*, 1998). MKKs in turn phosphorylate the MAPK family members p38, ERK, and ERK2 (p42/44), an JNK/SAPK, which are essential for the regulation of the immune response and cell survival. LF cleaves MKKs at their NH_2 termini, resulting in removal of the MAPK-binding "D" domain and preventing activation of appropriate downstream MAPK family members (Tanoue *et al.,* 2000; Xia *et al.*, 1998). Decathelineau and Bokoch (2009) found that statins delay LT-induced death and MKK cleavage in RAW macrophages and that statin-mediated effects on LT action are due to disruption of Rho GTPases. The Rho GTPase-inactivating toxin, toxin B, did not significantly affect LT binding or internalization, suggesting that the Rho GTPases regulate trafficking and/or localization of LT once internalized.

Ribot *et al.,* (2006) investigated the effects of LT on freshly isolated nonhuman primate (macaque) alveolar macrophages (AM). When AM and J774A.1 macrophages were exposed to 0.1 µg/ml of LT for 4 hrs., less than 10% of the J774A.1 cells were viable, compared to 98% survival for similarly treated AM which underwent cleavage of MEK1. These results demonstrated that LT is not cytolytic for AM compared to 5774Al.1 macrophages. Treatment of AM with LT alone had no effect on the secretion of inerleukin-1β (IL-1β), tumor necrosis factor alpha (TNF-α), IL-6, or IL-8. Addition of LPS to AM elicited production of TNF-α, IL-1β, IL-6, and IL-8. Indicating that the AM were capable of producing these cytokines. Simultaneous treatment of AM with LT and LPS resulted in greatly reduced levels of all four cytokines compared to levels produced when AM were treated with LPS alone. These results suggest that the consequence of MEK1 cleavage for AM is impaired induction of proinflammatory cytokine genes partially regulated by signaling in this pathway. When AM were treated for 16 hrs. with 0.1 μg of LT and then infected with *B. anthracis* spores for 4.5 hrs, bactericidal activity was substantially reduced. The authors concluded that the primary function of LT in anthrax pathogenesis may be to facilitate bacterial survival and outgrowth from AM but not AM death. The inability of AM pretreated with LF to completely eradicate the infecting bacteria suggest that intracellular secreted LF and EF help overcome the lethal consequences of phagocytosis of *B. anthracis* spores and resulting vegetative cells following germination.

Moayeri *et al.*, (2006) hypothesized that the ATP-activated macrophage P2X7 receptors implicated in nucleotide-mediated macrophage lysis could play a role in LT-mediated cytolysis and discovered that a potent P2X7 antagonist, oxidized ATP (O-ATP), protects macrophages against LT. Other P2X7 receptor antagonists, however, had no effect on LT function, while oxidized nucleotides, O-ADP, O-GTP, and O-ITP, which did not act as receptor ligands, provided protection. Cleavage of the LT substrates, the mitogen-activated protein kinases, was inhibited by O-ATP in RAW274.6 macrophages and CHO cells. The binding of the PA component of LT cells and the enzymatic proteolytic ability of the LF component of LT were unaffected by O-ATP. Instead, O-ATP inhibited formation of the sodium dodecyl sulfate-resistant PA oligomer which occurs in acidified endosomes, but did not prevent cell surface PA oligomerization. O-ATP also protected cells from ET. In addition, O-ATP, when injected into BALB/cJ mice, were completely protected against the lethality of injected LT.

Multiple microbial components trigger the formation of an inflammasome complex that contains binding domain (NOD)-like receptors (NCRs), caspase-1 and in some cases the scaffolding protein ASC. The NRL protein Nalp1b has been linked to LT-mediated cytolysis of murine macrophages (Boyden and Dietrich, 2006; Roberts *et al.*, 1998). Nalp1b belongs to the NRL family of intracellular proteins able to recognize conserved pathogen-associated motifs including LPS (Petrilli *et al.*, 2007; Sirard *et al.*, 2007). Nour *et al.*, (2009) demonstrated that in unstimulated J774A.1 murine macrophages, caspase-1 and Nalp1b are membrane associated and are part of ~200- and ~800-kDa complexes, respectively, while the capase-1 substrate interleukin-18 was cytosolic. LT treatment of these cells resulted in caspase-1 recruitment to the Nalp1-containing complex, concurrent with processing of cytosolic caspase-1 substrates. The authors also demonstrated that Nalp1 and caspases-1 are able to interact with each other. Caspase-1 associated inflammasome components included the caspase-1 substrate-enolase, proinflammatory caspase-11 and Nalp1b. These observations indicated that LT triggers the formation of a membrane-associated inflammasome complex in murine macrophages, resulting in cleavage of cytosolic caspase-1 substrates and cell death. Genetic mapping linked the LT susceptibility of murine macrophages to the NCR protein Nalp1b. Human NLR proteins closely related to Nalp1b, such as NALP3, and NALP1 have been shown to stimulate the innate immune response *via* caspase-1 activation. In turn, caspase-1 controls the LT-mediated cytolysis of murine macrophages: caspase-1 deficient macrophages are highly resistant to LT-mediated cytolysis (Boyden and Dietrich, 2006). In addition, caspase-1 inhibitors block LT killing of susceptible murine macrophages (Squires, Muchlbauer, and Brojatsch, 2007), proteosome inhibitors prevent cytolysis of LT-treated macrophages by blocking caspase-1 (Muchlbauer *et al.,* 2007; Squires, Muchlbauer, and Brojatsch, 2007), and Gadis, Avramova, and Chmielewski, (2007) found a derivative of catechol, N-oleoyldopamine (OLDA) to be a potent inhibitor of LF (Fig. **4A**) having a Ki for LF of 3.0 µM and a 50% inhibition concentration (IC$_{50}$) in a cellular assay of 15 µM. This compound was found to specifically inhibit LF and to protect cells. Subsequent studies identified an even more potent inhibitor of LF that resulted from the addition of an OH group in the para position of OLDA that had a Ki of 1.7 mM and an IC value of 13 µM (Fig. **4D**). A noncompetitive mode of inhibition for both catechol derivatives indicated that OLDA and substrate may co-occupy LF, producing an inactive enzyme-substrate inhibitor complex. Consistent with the fact that catechols are known to bind Zn^{++}, the authors found they resulted in the loss of LF inhibition.

Autophagy

Autophagy is an evolutionarily conserved intracellular process whereby cells break down long-lived proteins and organelles. The process is morphologically characterized by the formation of many large autophagic vacuoles in the cytoplasm. Autophagy is activated during periods of physiological stress such as starvation as a means to sustain viability (Martinet *et al.,* 2006). In addition, autophagy is also implicated as a protective cellular response for the elimination of infectious agents

(Schmid and Munz, 2007). The process of autophagy begins with the formation of an isolation membrane or phagophor followed by sequestration of organelles or invading bacterial cells to form an autophagosome which subsequently fuses with a lysosome to form an autolysosome which degrades the contained organelles or bacterial cell and releases the products into the cytoplasm. LF exerts its toxic effect through the disruption of the mitogen-activated protein kinase kinase (MAPKK) pathway, which is essential in mounting an efficient and rapid innate immune response against an invading pathogen (Bodart *et al.*, 2002). Tan *et al.*, (2009) were the first to describe LT-induced autophagy of murine macrophages. When RAW 264.7 murine macrophages were treated with LT an increase in acridine orange staining of acidic-vacuoles (AVO) occurred which was dependant on the LT dose. Increased staining of AVO is a typical feature observed in cells undergoing autophagy. In addition, when human promyelocytic leukemia HL-60 cells that had differentiated into macrophage-like cells were treated first with 3-methyladenine (3MA) to inhibit autophagy, accelerated cell death occurred when cells were then treated with LT compared to controls (not treated with 3 MA). These observations suggest that autophagy may function as a defense mechanism against LT intoxication.

Protein Expression Patterns of Macrophages Treated with LT

Comer *et al.*, (2005b) performed Genechip analyses on RNA from RAW264.7 murine macrophages following treatment with LT. Expression of several mitogen-activated protein kinase-regulatory genes was affected within 1.5 hrs. post-exposure to LT. By 3.0 hrs., the expression of 103 genes was altered, including host involved in intracellular signaling, energy production, and protein metabolism. Most of the up-regulated genes encoded inflammatory and signal transduction proteins as well as several transcription factors. Among the 62 down-regulated genes, 35% were involved in protein metabolism, including genes that coded for 15 ribosomal subunits. Several genes involved in the electron transport chain were also down-regulated, including ATP synthase epsilon chain, cytochrome c oxidase subunit VIIa polypeptide 2-like, NADH dehydrogenase (ubiquinone), alpha subcomplexes 2 and 7, ubiquinone-cytochrome c reductase subunit, and ubiquinol-cytochrome *c* reductase complex 11-kDa protein, which indicated a decrease in energy production in intoxicated cells.

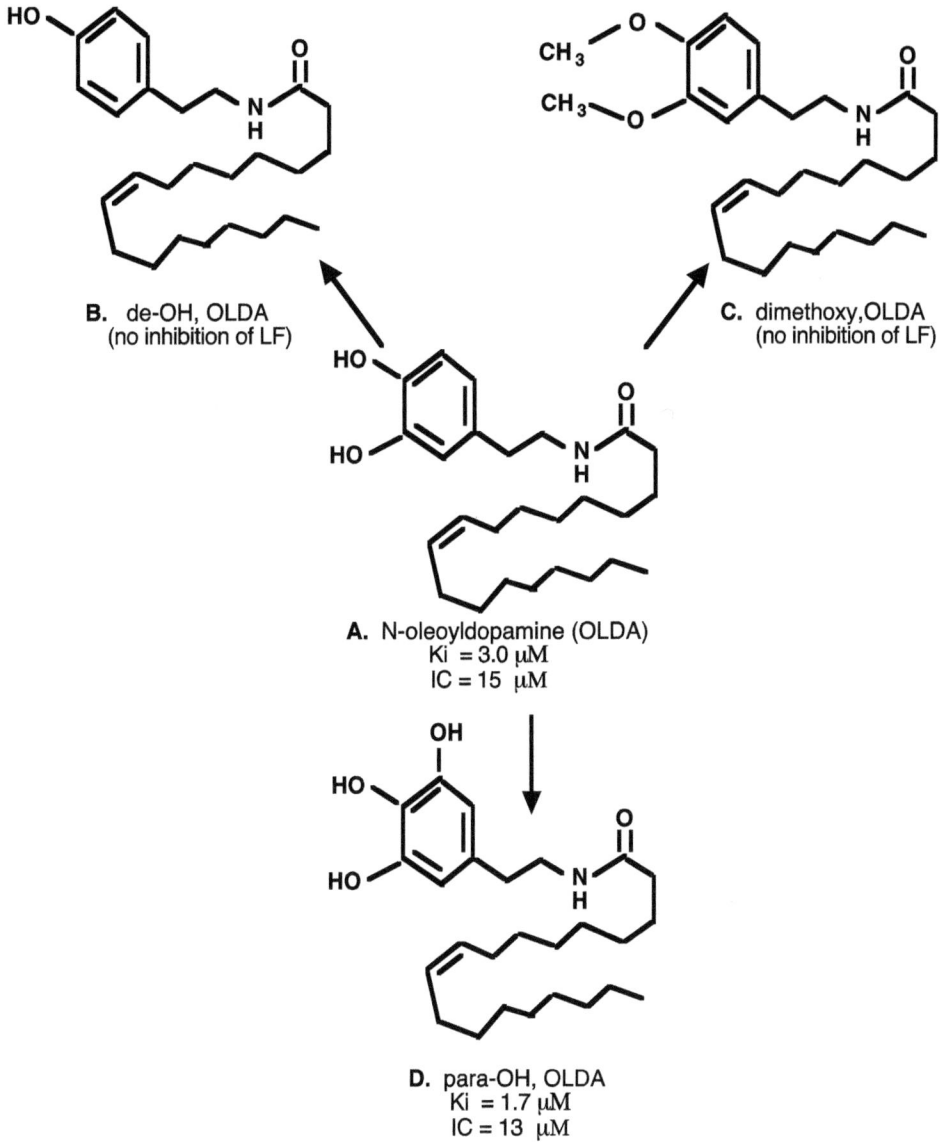

Figure 4: Inhibition of LF by catechol derivatives. Redrawn in part from Gaddis, Avramova, and Chmielewski (2007) with permission.

Comer *et al.,* (2005b) also found that G-protein signaling 16(Rgs16) was upregulated in LT-treated macrophages Rgs proteins inhibit signal transduction by increasing the GTPase activity of G protein α subunits, thereby driving them into their inactive GDP-bound form. Rgs16 specifically inhibits Ga_{i3} and subsequently antagonizes RoA signaling (Johnson *et al.,* 2003) which perturbs several intracellular signaling processes, including regulation of the cytoskeleton (Paterson *et al.*, 1990; Ridley and Hall, 1992), phospholipid metabolism (Bae *et al.*, 1998; Ren *et al.*, 1996; Zhang *et al.*, 1993), and cell migration (Fukata *et al.*, 1999). Rgs16 also preferentially attenuates the activation of p38 (Zhang *et al.*, 1999), which mediates the cleavage of MKK3 and MKK6 (Vitale *et al.*, 2000). LT also up-regulated prostoglandin 2 receptor, subtype 4 (EP_4). EP4 stimulates cyclic AMP (cAMP) formation.

Bergman *et al.*, (2005) reported on transcriptional changes within macrophages infected with *B. anthracis* (Sterne 34F2 strain) and the identification of several hundred host genes *via* microarray analysis that were differentially expressed during intracellular infection. The transcriptional response to anthrax LT in activated macrophages indicated that many of the host inflammatory response pathways are down-regulated.

Jung *et al.*, (2008) treated murine macrophages with LT and then resolved the resulting cellular proteins with two dimensional polyacrylamide gel electrophoresis and generated peptide mass-spectroscopic fingerprints after trypsin digestion. Among the differentially expressed proteins, cleaved-activated protein kinase kinase, which acts as a negative element in the signal transduction pathway and the formation of glucose-6-posphate dehydrogenase (G6PD) which plays a role in the protection of cells from hyperproduction of active oxygen were up-regulated in the LT treated macrophages. Mitogen-activated protein kinase kinase (MAPKK2 n20) was notably reduced at 9 min and disappeared after 18 min following treatment with LT. This observation implies that LF mediated cleavage of MPKK was initiated after LF translocation and that both G6PD and 6-phospho-gluconic dehydrogenase 6PGDH increase, with G6LPD more greatly increased. G6PD was up-regulated in as little as 6 min in LF treated macrophages. G6PD is the first and rate-limiting enzyme of the pentose phosphate pathway, which controls NADPH concentration *via* glucose metabolism. 6PGDH is a downstream enzyme of G6PD which produces NADH. These up-regulated enzymes play a crucial role in protection from redox-stress-induced apoptosis. In this conjunction, Hanna *et al.*, (1994) reported that murine macrophages treated with high levels of LT produced reactive oxygen intermediates. Jung *et al.*, (2008) also noted alteration in the regulation of two additional proteins. One, designated

PP1c was down-regulated and was proposed as the phosphorylated species of PP1c, which in the presence of LT is dephosphorylated and up-regulated. PP1 is a major eukaryotic serine/threonine phosphatase that regulates a variety of cellular functions through the interaction of its subunit PP1c (Cohen, 2002). Phosphorylated PP1c, which occurs in threonine residues, is inactivated; while dephosphorylation results in active PP1c. Dephosphorylated PP1c is responsible for dephosphorylation of Bad which leads to cell death (Garcia *et al.*, 2003). Bad is a pro-apoptotic member of the Bcl2 family of proteins that consists of crucial cell death regulators. The authors proposed that LT toxicity may to some extent be a result of an increased dephosphorylation of Bad by dephosphorylated (active) PPic.

Sapra *et al.*, (2006) analyzed the proteomics response of two murine macrophage cell lines (5774.1A and RAW264.7) following exposure to LT. A set of proteins in each cell line was consistently upregulated when the two macrophage cell lines were treated with LT. The upregulated proteins included those involved in energy metabolism, cytoskeleton structure, and stress response. A subset of five proteins (ATP synthase β subunit, β-actin, Hsp70, vimentin, and an Hsp60 homolog) was identified that were commonly upregulated in both cell lines.

Factors Influencing the Activity of the LT on Macrophages

Calcium

Removal of calcium from the culture medium of macrophages has been found to protect the cells from the action of LT (Bhatnagar *et al.*, 1989). Calcium removal had a minimal effect on proteolytic conversion of PA to PA_{63} and had no effect on LT binding to macrophages. Cells to which LT had been bound in the presence of calcium were protected when transferred to calcium depleted culture medium, indicating a role for calcium at a postbinding stage.

Vacuolar ATPase Proton Pump

LF requires passage through acidic vesicles in order to exert its effect within the cytosol. The vacuolar ATPase proton pump has been found to be required for the cytotoxicity of the LT from *B. anthracis*. Ménard *et al.*, (1996) found that bafilomycins A1, B1, C, and D and concanamycin A, all known selective inhibitors of the vacuolar ATPase proton pumps, are powerful inhibitors of LT toxicity towards macrophages. Bafilomycin A1 and concanamycin A were the most powerful inhibitors of LT induced toxicity. Both substances were found to prevent the acidification of intracellular acidic compartments, and that within 15 min. of their addition, macrophages no longer possessed transmembrane pH gradients. These inhibitors were found to be fully active long after (up to 120 min.) LF addition to macrophages, suggesting that LF enters the cytosol after having reached a late endosomal compartment.

Proteasome Activity

The proteasome is a multicatalytic protease that is responsible for the major extra lysosomal degradation of cellular proteins. The 20S proteasome is a large (~700 kDa) complex composed of four stacked rings, each containing seven subunits. The 20S proteasome forms the catalytic core of the larger 26S proteasome, which is responsible for the ATP-dependant degradation of proteins designated for destruction (Tanaka and Tsurumi, 1997) The subunits of the inner rings are associated with the proteolytic activities of the complex.

Tang and Leppla (1999) showed that proteasome inhibitors such as lactacystin and acetyl-Leu-Leu-norleucinal, and MG132, efficiently blocked LT toxicity, whereas other protease inhibitors did not. The inhibitors did not interfere with the proteolytic cleavage of MEK1 in LT treated cells, indicating that they do not directly block the proteolytic activity of LF. However, the proteasome inhibitors did prevent ATP depletion, an early effect of LT. These results suggest that the proteasome mediates a cytotoxic process initiated by LF in the cytosol which possibly involves degradation of unidentified molecules essential for macrophage homeostasis.

Association of LT with the Capsule of B. Anthracis

Ezzel *et al.,* (2009) characterized the LT in blood plasma from *B. anthracis* aerosol spore infected Guinea Pigs, African Green Monkeys, and Rabbits. In all cases, during the terminal phase of infection only the protease-activated 63-dalton form of PA_{63} and the residual 20-kDa fragment were detected in the plasma. No uncut PA with a molecular mass of 83-kDa as detected in plasma from toxemic animals during the terminal stage of infection. PA_{63} was largely associated with LF, forming LT. The antiphagocytic poly-γ-D-glutamic acid (γ-DPGA) capsule released from *B. anthracis* bacilli was found to be physically associated with LT in animal plasma. A portion of these LT/γ-DPGA complexes retained F protease activity. The authors concluded that the *in vivo* LT complexes differ from *in vitro*-produced LT and that the inclusion of γ-DPGA analysis when examining the effects of LT on specific immune cells *in vitro* may reveal novel and important roles for γ-DPGA in anthrax pathogenesis.

THE EDEMA FACTOR (EF)

Mechanism of Action of EF

Intracellular cyclic AMP (cAMP) is known to regulate production of cytokines such as interleukin-6 (IL-6) that modulate edema formation. The edema factor (EF) is encoded by the *cya* gene on the pXO1 plasmid. The edema toxin (EF) has been found to influence cytokine production by enhancement of intracellular cAMP in human monocytes and to increase IL-6 production (Hoover *et al.*, 1994)) and to inhibit the phagocytic activity of human neutrophils (O'Brien *et al.*, (1985)), thereby increasing host susceptibility to infection. EF has been found to be a Ca^{++}-calmodulin-dependent adenylate cyclase (Mock and Fouet, 2001) that catalyzes the synthesis of cyclic AMP (cAMP) thereby increasing intracellular levels of cyclic AMP, which is responsible for typical edema in patients with cutaneous anthrax. The cAMP level of Chinese hamster cells was found to increase about 200-fold above normal when exposed to both PA + EF (Leppla. 1982, 1984).

The cAMP response generated by the edema toxin (ET) was analyzed by Kumar, Ahuja, and Bhatnagar, (2002)s in a variety of cells including CHO, macrophage-like RAW264.7, human neutrophils, and human lymphocytes. Observations indicated that after EF reaches the cell cytosol, a rapid influx of calcium is triggered in the host cell that has a critical role in determining cAMP response of the affected cells. Agents that blocked the uptake of calcium also inhibited ET-induced accumulation of cAMP in the host cells. The authors concluded that by causing the accumulation cAMP, a potent inhibitor of immune cell function, ET may be poisoning the immune system so as to facilitate the survival of invading bacteria in host cells. The nonpermeable calcium chelator EGTA abolished the calcium influx in EF treated cells and attenuated the cAMP response. LA3+ which competes with calcium for several calcium channels without being transported across the plasma membrane, completely prevented the entry of calcium and also the ensuing cAMP response. Nifedipine, a dihdroproline antagonist of L-type calcium channels, effectively blocked calcium influx in toxin treated CHO cells and reduced cAMP accumulation six-fold. Flunarizem, a T-type calcium channel antagonist yielded similar results with CHO cells. EF administered to lymphocytes was also found to require calcium influx for reducing cellular toxicity.

Neutrophils are considered to constitute the first line of defense against bacterial infections. These phagocytic cells are able to migrate to the site of infections, and defects in neutrophil chemotaxis compromise the innate immune response. Chemotaxis is accompanied by shape changes that are mediatcd by rapid assembly and disassembly of actin filaments. Anthrax LT has been shown to impair neutrophil chemotaxis. And chemokinesis by reducing the formation of actin filaments (During *et al.*, 2005). ET alone was found by Szarowicz *et al.*, (2009) to significantly inhibit chemotaxis of neutrophils and that together with LT the effects were additive, resulting in paralysis of neutrophil chemotaxis which was associated with a notable decrease in actin synthesis.

A striking effect of ET at lethal doses in mice is adrenal necrosis. Firovid *et al.*, (2007) showed that doses of ET (10 μg) that produce no overt signs of illness in mice cause substantial adrenal lesions. These lesions were found not to be associated with reduced corticosterone production; instead, ET treated mice have increased corticosterone production.

Interference with glucocorticoid (GC) production *via* adrenalecomy, the inhibition of GC function by the GC receptor antagonist RU-486, or the administration of excess GC's have all been found to enhance sensitivity to LT in mice (Moayeri *et al.*, 2005). DBA/2J mice are normally resistant to a single dose of LT at doses that are 100% lethal in many other murine strains. Firoved *et al.*, (2007) found that an *i.v.* dose of 5 μg of ET administered simultaneously with 100 μg of LT administered *i.v.* resulted in 90% lethality whereas controls (ET alone or LT alone) resulted in no more than 20% lethality. These observations clearly indicate that with DBA/1J mice the *B. anthracis* ET enhances the lethality of LT synergistically so as to overcome the natural resistance of his murine strain to LT.

Dal Molin *et al.,* (2008) made use of fluorescence resonance energy transfer (FRET)-based single cell imaging to monitor the spacially-and temporally-resolved increase of intracellular cAMP levels as a result of ET action. Two different cell lines were used: human bronchial epithelial (HBE) cells, and intestinal epithelial cells (human colony adenosarcoma cells (Caco-2)). ET was found to be poorly active in Caco-2 cells, while with HBE cells an increase in cAMP cytosolic concentration resulted. In addition, ET induced a cAMP concentration gradient which decreased from the nucleus to the cell periphery.

Yeager *et al.,* (2009) found that ET significantly suppressed human macrophage phagocytosis of Ames spores. Cytoskeletal changes such as decreased cell spreading and lowered F-actin content were also observed with toxin-treated macrophages. ET also altered protein levels and activity of protein kinase A (PKA) activity and exchange protein activity by cAMP (Epac), a cAMP binding molecule. These observations suggest that ET weakens the host innate immune response by increasing the cAMP levels, which then signal *via* PKA and Epac to prevent macrophage phagocytosis and interfere with cytoskeletal remodeling.

Maldonado-Arocho and Bradley (2009) demonstrated that ET induces a maturation state in human dendritic cells (DCs) similar to that induced by lipopolysaccharide (LPS). ET resulted in downregulation of DC-SIGN, a marker of immature DCs, and upregulation of DC maturation markers CD83 and CD86. Maturation of DCs by ET is accompanied by an increased ability to migrate toward the lymph-node-homing chemokine macrophage inflammatory protein 3β. Interestingly, while co-treatment with LT did not affect the ET-induced decease in DC-SIGN expression, it did dampen the increase in both CD83 and CD86 expression. Anthrax ET alters cytokine profiles during a *B. anthracis* infection, with a decrease in tumor necrosis factor α and IL-12 secretion and an increase in IL-10 secretion (Cleret *et al.,* 2006; Tournier *et al.,* 2005). These findings revealed a mechanism by which ET impairs normal innate immune function and may explain the reported adjuvant effect of ET.

Comer *et al.,* (2006) used GeneChip analysis to examine global transcriptional profiles of ET-treated RAW 264.7 murine macrophage-like cells and identified 71 and 259 genes whose expression was significantly altered by ET at 3 and 6 hrs., respectively. The genes with up-regulated expression in macrophages in response to ET-treatment were known to be involved in inflammatory responses, regulation of apoptosis, adhesion, immune cell activation, and transcriptional regulation. ET also inhibited tumor necrosis factor alpha (TNF-α) production and crippled the phagocytic ability of macrophages.

The sera of rabbits 48 hrs. after subcutaneous infection with spores of a lethal strain of *B. anthracis* was recently found to contain LF and EF in a ratio of 5:1 in all five rabbits, despite their different concentrations (Dal Molin *et al.,* 2008). This ratio is similar to that obtained in bacterial cultures (Leppla, 1988; Sirard, Mock, and Fouet, 1994) and indicates that there is no differential regulation of the LF and EF genes in *B. anthracis* in *in vitro* culture and in rabbits undergoing anthrax infection and that EF and LF have the same stability in rabbit serum. PA_{83}, the full length inactive form of PA was not found in any blood sample. The cleaved activated PA63 form was always found as a 60 kDa protein and as a doublet of ~50 kDa This indicates that in rabbits with anthrax, there are one or more proteases that cleave PA at different sites. Cleavage between the N-terminal PA20 domain and PA_{63} has been reported by Ezzell and Abshire (1992) to occur in a variety of animal species, while a cleavage product of 50 kDa from PA was only recently was reported in murine and human serum (Goldman *et al.,* 2008).

EVIDENCE FOR SUPPLEMENTARY FACTORS OF VIRULENCE

Genomic Factors

Drysdale *et al.,* (2005) showed that capsule synthesis by *B. anthracis* is absolutely required for lethality in Balb/c mice, which is considered the most resistant murine mouse strain to anthrax. Heninger *et al.,* (2006) however, utilizing a Sterne strain transduced to pXO2$^+$ found that mutants deficient in EF, LF, both EF and LF, or PA resulted in LD_{50} values and mean time to death for mice similar to those of the parent strain administered spores by inhalation and *i.v.* delivery of vegetative cells. These observations suggested that genomic factors may be responsible for lethal infections by *B. anthracis* in Balb/c mice. Nonetheless, histopathological examination of tissues revealed subtle but distinct differences in infections by the parent compared to some toxin mutants, suggesting that the host response is affected by toxin proteins synthesized during infection.

Multiple Inflammatory Mediators

Tessier *et al.,* (2007) evaluated the direct effect of edema toxin (ET) on several cell types *in vitro* and assessed the possibility that mediators of vascular leakage, such as histamine which contributes to edema in rabbits administered ET intradermally. The screening of several drugs by intradermal treatment prior to injection of ET demonstrated reduced ET-induced vascular leakage with a cyclo-oxygenase inhibitor (indomethacin), agents that interfere with histamine (pyrilamine or cromolyn), or a neurokinin antagonist (spantide). Systemic administration of indomethacin or celecoxib (cyclooygenase inhibitors), pyrilamine, aprepitant (a neurokinin 1 receptor antagonist), or indomethacin with pyrilamine significantly reduced vascular leakage associated with ET. These results were interpreted to indicate that ET stimulates the production/release of multiple inflammatory mediators, specifically neurokinins, prostanoids, and histamine. The authors concluded that these mediators increase vascular permeability, and interventions directed at these mediators may benefit hosts infected with *B. cereus*.

Virulence of Toxin Mutants of *B. anthracis*

To determine the mouse virulence of non-toxigenic variants of virulent strains, Welkos, Vietri, and Gibbs (1993) isolated pXO1- derivatives of the Vollum 1B strain and the more "vaccine resistant" Ames strain which carried pXO2 from either the Ames or the Vollum B1 strain. The 50% lethal dose (LD_{50}) values of the derivatives of both strains which carried the Ames pXO2 plasmid were not significantly different from the LD_{50} values of the $pXO1^+$ $pxO2^+$ strains (and were lower than those of $pXO1^+$ $pXO2^-$ strains). $pXO2^+$ derivatives of strain UM23-1 ($pXO1^-$) were less virulent than the comparable $pXO1^-$ Ames and $pXO1^-$ Vollum 1B strain, emphasizing a role of chromosomal loci in virulence of the latter two strains. Non-toxigenic isolates ($pXO1^-$) which carried the Ames pXO2 plasmid were more virulent for CBAA/J mice than those with the Vollum 1B pXO2 plasmid, and the differences were dependant on the mouse strain. The pXO1-pXO2$^+$ strains multiplied and achieved high cell numbers systemically.

REFERENCES

Agrawal, A., Lingappa, J., Leppla, S., Agrawal, S., Jabbar, A., Quinn, C., Pulendran, B. (2003). Impairment of dendritic cells and adaptive immunity by anthrax lethal toxin. *Nature.* 434:329-334.

Ahuja, N., Kumar, P., Bhatnagar, R. (2001). Rapid purification of recombinant and anthrax-protective antigen under nondenaturing conditions. *Biochem. Biophys. Res. Commun.* 286:6-11.

Askenazi, A., Dixit, V. (1998). Death receptors: signaling and modulation. Science. 281:1305-1308.

Bae, C., Min, D., Flemming, I., Exton, J. (1998). Determination of interaction sites on the small G protein RhoA for phoshpolipase D. *J. Biol. Chem.* 273:11596-11604.

Bergman, N., Passalacqua, K., Gaspard, R., Shetron-Rama, L., Quackenbush, J., Hanna, P. (2005). Murine macrophage transcriptional responses to *Bacillus anthracis* infection and intoxication. *Infect. Immun.* 73:1069-1080.

Benson,E., Huynh, P., Finkelstein, A., Collier, R. (1998). Identification of residues lining the anthrax protective antigen channel. Biochem. 37:3941-3948.

Bhatnagar, R., Singh, Y., Leppla, S., Friedlander, A. (1989). Calcium is required for the expression of anthrax lethal toxin activity in the macrophage-like cell line J774A.1. Infect. Immun. 57:2107-2114.

Blaustein, R., Koehler, T., Collier, R., Finkelstein, A. (1989). Anthrax toxin: channel- forming activity of protective antigen in planar phospholipids bilayers. Proc. Natl. Acad. Sci. U.S.A. 86:2209-2213.

Bokoch, G. (2005). Regulation of innate immunity by Rho GTPases. Trends Cell Biol. 15:163-171.

Benard, V., Bokoch. G. (2002). Assay of Cdc42, Rac, and Rho GTPase activation by affinity methods. Methods Enzymol. 345:349–359.

Bodart, J., Chora, A., Liang, X., Duesbury, N. (2002). Anthrax, MEK and cancer. Cell cycle. 1:10-15.

Boyden, E. Dietrch, W. (2006). Nalp1b controls mouse macrophage susceptibility of anthrax lethal toxin. Nat. Genet. 38:240-244.

Bradley, K., Mogridge, J., Mourez, M., Collier, R., Young, J. (2001). Identification of the cellular receptor fo anthrax toxin. Nature. 414:225-229.

Brossier F., Weber-Levy, M., Mock, M., Sirar, J. (2000). Role of toxin functional domains in anthrax pathogenesis. Infect. Immun. 68:1781-1786.

Brossier, F. Mock, M. (2001). Toxins of *Bacillus anthracis.* Toxicon. 39:1747-1755.

Cataldi, A., Labruyere, E., Mock, M. (1990). Construction and characterization of a protective antigen-deficient *Bacillus anthracis* strain. Molec. Microbiol. 4:1111-1117.

Chauhan, V., Bhatnagar, R. (2002). Identification of amino acid residues of anthrax protective antigen involved in binding with lethal factor. Infect. Immun. 70:4477- 4484.

Cleret, A., Quesnel-Hellman, A., Mathieu, J., Vidal, D., Tournier, J. (2006). Resident CD11c$^+$ lung cells are impaired by anthrax toxins after spore infection. J. Infect. Dis. 194:86-94.

Cohen, P. (2002). Protein phosphatase 1-targeted in many directions. J. Cell Sci. 115:241-256.

Collier, R. (1999). Mechanism of membrane translocation by anthrax toxin: insertion and pore formation by protective antigen. J. Appl. Microbiol. 87:283.

Comer, J., Copra, A., Peterson, J., Konig, R. (2005a). Direct inhibition of T- lymphocyte activation by anthrax toxins *in vivo*. Infect. Immun. 73:8275-8281.

Comer, J., Galindo, C., Copra, A., Peterson, J. (2005b). GeneChip analyses of global transcriptional responses of murine macrophages to the lethal toxin of *Bacillus anthracis*. Infect. Immun. 73:1879-1885.

Comer, J., Galindo, C., Zhang, F, Wenglikowski, A., Buch, K., Garner, H., Person, J., Chopra, A. (2006). Murine macrophage transcriptional function responses to *Bacillus anthracis* edema toxin. Microb. Path. 41:96-110.

Cote, C., Rea, K., Norris, S., Van Rooijen, N., Welkos. S. (2004). The use of a model of *in vivo* macrophage depletion to study the role of macrophages during infection with *Bacillus anthracis* spores. Microb. Pathog. 37:169–175.

Cote, C., Rossi, C., Kang, A., Morrow, P., Lee, J., Welkos, S. (2005). The detection of protective antigen (PA) associated with spores of *Bacillus anthracis* and the effects of anti-PA antibodies on spore germination on spore macrophage interactions. Microbial. Path. 38:209-225.

Cote, C., Van Rooijen, N., Welkos1, S. (2006). Roles of macrophages and neutrophils in The early host response to *Bacillus anthracis* spores in a mouse model of infection. Infect. Immun. 74:469-480.

Cote, C., diMezzo, T., Banks, D., France, B., Bradley, K., Welkos, S. (2008). Early interactions between fully virulent *Bacillus anthracis* and macrophages that influence the balance between spore clearance and development of lethal infection. Microbes and Infect. 10:613-619.

Dai, Z., Sirard, J., Mock, M., Kohler, T. (1995). The *atxA* gene product activates transcription of the anthrax toxin genes and is essential for virulence. Molec. Microbiol. 16:1171-1181.

Dal Molin, F., Zornettqa, I., Puhar, A., Tonello, F., Zaccolo, M., Montecucco, C. (2008). cAMP imaging of cells treated with pertussis toxin, cholera toxin, and anthrax edema toxin. *Biochem, Biophys. Res. Commun.* 376:429-433.

Decathelineau, A., Bokoch, G. (2009). Inactivation of Rho GTPases by s atins attenuates anthrax lethal toxin activity. Infect. Immun. 77:348-359.

Defrise-Quertain, F., Cabiaux, V., Vandenbranden, M., Wattiez, R., Falmagne, P., Ruysschaert, J., (1989). pH-dependant bilayer destabilization and fusion of phospho-lipidic large unilamellar vesicles induced by diphtheria toxin and its fragments A and B. *Biochem.* 28:3406-3412.

Drysdale, M., Heninger, S., Hutt, J., Chen, Y., Lyons, C., Koehler, T. (2005). Capsule synthesis by *Bacillus anthracis* is required for dissemination in murine inhalation anthrax. EMBO J. 24:221-227.

Duesbery, N., Vande Woude, G. (1999). Anthrax lethal factor causes proteolytic inactivation of mitogen-activated protein kinase kinase. *J. Appl. Microbiol.* 87:289-293.

Duesbery, N, Webb, C, Leppla, S. Gordon, V., Klimpel, K., Copeland., Ahn, N., Oskarsson, M., Fukasawa, K., Paull, K., Vande Woude, G. (1998). Proteolytic inactivation of MAP-kinase-kinase by anthrax lethal factor. Science. 280:734-737.

During, R., Li, W., Hao, B., Koenig, J., Stephens, D., Quinn, C., Southwick, F. (2005). Anthrax lethal toxin paralyzes neutrophil actin-based motility. J. Infect. Dis. 192:837- 845.

Ezzel, J., Abshire, T., Panchal, R., Chabot, D., Bavari, S., Lefel, E., Purcell, B., Friedlander, A., Ribot, W. (2009). Association of *Bacillus anthracis* capsule with lethal toxin during experimental infection. Infect. Immun. 77:749-755.

Firoved, A., Moayeri, M., Wiggins, J., Shen, Y., Tang, W., Lepla, S. 2007. Anthrax edema toxin sensitizes DBA/2J mice to lethal toxin. *Infect. Immun.* 75:2120-2125.

Finkelstein, A. (1994). The channel formed in planar lipid bilayers by the protective antigen component of anthrax toxin. Toxicol. 87:29-41.

Flick-Smith, H., Walker, N., Gibson, P., Bullifent, H., Hayward, S., Miller, J., Titball, R, Williamson, D. (2002). A recombinant carboxy-terminal domain of the protective antigen of *Bacillus anthracis* protects mice against anthrax infection. Infect. Immun. 70:1653-1656.

Friedlander A. (1986). Macrophages are sensitive to anthrax lethal toxin through an acid-dependant process. *J. Biol. Chem.* 261:7123-7126.

Friedlander, A. (1996). Characterization of the lethal factor binding and cell receptor binding domains of protective antigen of *Bacillus anthracis* using mono- clonal anti bodies. Microbiol. 12:707-715.

Fukata, Y., Osiro, N., Kinosita, N., Kawano, Y., Matsuoka, Y., Bennett, V., Matsuura, Kaibuci, K. (1999). Phosphorylation of adducin by Rhokinase plays a crucial role in cell motility. J. Cell Biol. 145:347-361.

Gaddis, B., Avramova, L., Chmielewski, J. (2007). Inhibitors of anthrax lethal factor. Bioorg. Medic. Chem. Lett. 17:4575-4578.

Garcia A., Cayla, X., Guergnon, J., Dessauge, F., Hospital, V., Rebollo, M., Fleischer, A., Rebollo, A. (2003). Serone/threonine protein phosphatases PP1 and PP2A are key players in apoptosis. Biochimmie. 85:721-726.

Gozes, Y., Moayeri, M., Wiggins, J., Leppla, S. (2006). Anthrax lethal toxin induces ketotifen-sensitive intradermal vascular leakage in certain inbred mice. Infect. Immun. 74:1266-1272.

Guignot, J., Mock, M., Fouet, A. (1997).AtxA activates the transcription of genes harbored by both *Bacillus anthracis* virulence plasmids. FEMS Microbiol. Lett. 147:203-207.

Gutting, B, Gaske, K., Schilling, A, Slaterbeck, A, Sobota, L., Mackie, R., Buhr, T. (2005). Differential susceptibility of macrophage cell lines to *Bacillus anthracis*—Vollum 1B. Tox. *in Vitro*:19:221-229.

Haines, R., Klein, B., and Lincoln, R. (1965). Quantitative assay for crude anthrax toxins. J. Bacteriol. 89:74-83.

Hanna, P., (1998). Anthrax pathogenesis and host response. Curr. Top. Microbiol. Immunol. 225:13-35.

Hanna, P. (1999). Lethal toxin actions and their consequences. J. Appl. Microbiol. 87:285-287.

Hanna, P., Acosta, D., Collier, R. (1993). On the role of macrophages in anthrax. Proc. Natl. Acad. Sci. USA. 90:10198-10201.

Hanna, P., Kruskal, B., Ezekowitz, R. Boom, B, Collier, R. (1994). Role of macrophage oxidative burst in the action of anthrax lethal toxin. Molec. Med. 1:7-18.

Heninger, S., Drysdale, M., Lovchik, J., Hutt, J., Lipscomb, M., Koehler, T., Lyons, C. (2006). Toxin-deficient mutants of *Bacillus anthracis* are lethal in a murine model of pulmonary anthrax. *Infect. Immun.* 74:6067-6074.

Hoffmaster, A., Koehler, T. (1999). Control of gene expression in *Bacillus anthracis. J.* Appl. Microbiol. 87:279-281.

Hoover, D., Friedlander, A, Rogers, L., Yoon, I., Warren, R., Cross, A. (1994). Anthrax edematoxin differentially regulates lipopolysccharide-induced monocyte production of tumor necrosis factor alpha and interleukin-6 b y increasing intracellular cyclic AMP. Infect. Immun. 62:4432-4439.

Ismai, N., Olano, J., Feng, H., Walker, D. H. (2002). Current status of immune mechanisms of killing intracellular microorganisms. FEMS Microbiol. Lett. 207:111- 120.

Ivins, B., Ristroph, J., Nelson, G. (1989). Influence of body weight on response of Fischer 344 rats to anthrax lethal toxin. Appl. Environ. Microbiol. 55:2098-2100.

Ivins, B., Welkos, S. (198)6. Cloning and expression of the *Bacillus anthracis* protective antigen gene in *Bacillus subtilis*. Infect. Immun. 54:537-542.

Jaff, A., Hall, A. (2005). Rho GTPases: biochemistry and biology. Ann. Rev. Cell Dev. Biol. 21:247-269.

Johnson, E., Seasholtz, T., Waheed, A., Kreutz, B.,Suzuki, N., Kozasa, T., Jones, T., Bown, H., Druey, K., (2003). RGS16 inhibits signaling through the G alpha 13-rho axis. Nat. Cell Biol. 5:1095-1103.

Jung, K., Seo, G., Yoon, J., Park, K., Kim, J., Kim, S., Oh, K., Lee, J., Chai, Y. (2008). Protein expression pattern of murine macrophages treated with anthrax lethal toxin. *Biochem. Biophys. Acta.* 1784. 1501-1506.

Kim, J., Kim, Y., Ko, B., Chae, Y., Yoon, M. (2003). Production and proteolytic assay of lethal factor from *Bacillus anthracis*. Protein Expr. Purif. 30:293-300.

Klimpel, K., Molloy, S., Thomas, G., Leffla, S. (1992). Anthrax toxin protective antigen is acivated by a cell surface protease with the sequence specificity and cataltic properties of furin. *Proc. Natl. Acad. Sci. U.S.A.* 89:10277-10281.

Kobiler, D., Weiss, S., Levy, H., Fisher, M., Mechaly, A., Pass, A., Alboum, Z. (2006). Protective antigen as a correlative marker for anthrax in animal models. *Infect. Immun.* 74:5971-5876.

Kochi, S., Martin, I., Sciavo, XG., Mock, M., Cabiaux, V. (1994). The effects of pH on the Interaction of anthrax toxin letha and edema factors with phospholipid vesicles. Biochem. 33:2604-2609.

Koehler, T., Collier, R. (1991). Anthrax toxin: low-pH induced hydrophobicity and channel formation in liposomes. *Mol. Microbiol.* 5:1501-1506.

Koehler, T., Dai, Z., Kaufman-Yarbray, M. (1994). Regulation of the *Bacillus anthracis* Protective Antigen Gene: CO_2 and a *trans*-acting element activate transcription from one of Two Promoters. J. Bacteriol. 176:586-595.

Krantz, B, Trivedi, A., Cunningham, A., Christensen, K., Collier, S. (2004). Acid- induced unfolding of the amino-terminal domains of the lethal and edema factors of anthrax toxin. J. Mol. Biol. 344:739-756.

Krantz, B., Melnyk, R., Zhang, S, Juris, S., Lacy, D, Wu, Z., Finkelsein, A., Collier, R. (2005). A phenylalanine clamp catalyzes protein translocation through the anthrax toxin pore. Science. 309:777-781.

Krantz, B., Finkelstein, A., Collier, R. (2006). Protein translocation through the anthrax toxin transmembrane pore is driven by a proton gradient. J. Mol. Biol. 355:968-979.

Krishnanchettiar, S., Sen, J., Caffrey, M. (2003). Expression and purification of the *Bacillus anthracis* protective antigen domain 4. Protein Expr. & Purif. 27:325-330.

Kumar, P., Ahuja, N., Bhatnagar, R. (2002). Anthrax edema toxin requires influx of calcium for inducing cyclic AMP toxicity in target cells. Infect. Immun. 70:4997-5007.

Lacy, D., Collier, R. (2002). Structure and functionality of anthrax toxin. Current topics in Microbiol. Immunol. 271:61-85.

Leppla, S. (1982). Anthrax toxin edema factor: A bacterial adenylate cyclase that increases AMP concentrations in eukaryotic cells. Proc. Natl. Acad. Sci. USA. 79:3162- 3166.

Leppla, S. (1984). *Bacillus anthracis* calmodulin-dependant adenylate cyclase: chemical and enzymatic properties and interactions with eukaryotic cells. Adv. Cyclic Nucleotide Protein Phosphorylation Res. 17:189-198.

Leppla, S. (1988). Production and purification of anthrax toxin. Meth. Enzymol. 165:103-116.

Leppla, S. (1995). Anthrax toxins. In J. Moss, B. Iglewski, M., Vaughn, A. Tu, (eds.) Bacterial toxins and virulence factors in disease. *Handbook of Natural Toxins*. Vol 8: pp. 543-572.

Lincoln, R., Fish, D. (1970). Anthrax toxin. In T. C. Monte, S. Kadis, and S. J. Ajl (eds.) Microbial Toxins III ed. Pp. 361-414. Academic Press, New York. Little, S. (2004). Western blot analysis of the exotoxin components from *Bacillus anthracis* separated by isoelectric focusing gel electrophoresis. Biochem. Biophys. Res. Commun. 317:294-300.

Little, S., Knudson, G. (1986). Comparative efficacy of *Bacillus anthracis* live spore vaccine and protective antigen vaccine against anthrax in the guinea pig. Infect. Immun. 52:509-512.

Little, S., Novak, J., Lowe, J., Leppla, S., Singh, Y., Klimpel, K., Lidgerding, B., Friedlander, A. (1996). Characterization of lethal factor binding and cell receptor binding domains of protective antigen of *Bacillus anthracis* using monoclonal antibodies. Microbiol. 142:707–715.

Liu, S., Leppla, S. (2003). Cell surface tumor endothelium marker 8 cytoplasmic tail- independent anthrax toxin binding, proteolytic processing, oligomer formation, and internalization. J. Biol Chem. 278:5227-5234.

Maldonado-Arocho, F., Bradley, K. (2009). Anthrax edema toxin induces maturation of dendritic cells and enhances chemotaxis towards macrophage inflammatory protein 3b. Infect. Immun. 77:2036-2042.

Martinet, W., de Meyer, G., Andries, L., Herman, A, Kockx, M. (2006). *In situ* detection of starvation-induced autophagyJ. Histochem. Cytochem. 54:85-96.

Ménard, A., Altendorf, K., Brevens, D., Mock, M., Montecucuo, C. (1996). The vacuolar ATPase proton pump is required for the cytotoxicity *of Bacillus anthracis* lethal toxin. FEBS Lett. 386:161-164.

Miles, A., Miles, E. (1952). Vascular reactions to histamine, histamine-liberator and leukotaxine in the skin of guinea-pigs. J. Physiol. 118:228–257.

Miller, C., Elliot, J., Collier, R. (1999). b Anthrax protective antigen: prepore-to pore- conversion. Biochem. 38:10432-10441.

Milne, J., Collier, R. (1993). pH-dependant permeabilization of the plasma membrane of mammalian cells by anthrax protective antigen. J. Mol. Microbiol. 10:647-653.

Milne, J., Furlong, D., Hanna, P., Wall, J., Collier, R. (1994). Anthrax protective antigen forms oligomers during intoxication of mammalian cells. J. Biol. Chem. 269:20607- 20612.

Moayari, M, Haines, D., Young, H., Leppla, S. (2003). *Bacillus anthracis* lethal toxin induces TNF-α-independent hypoxia-mediated toxicity in mice. J. Clin. Investig. 112:670-682.

Moayeri, M., Martinez, N., Wiggins, J., Young, H., Leppla S. (2004). Mouse susceptibility to anthrax lethal toxin is influenced by genetic factors in addition to those controlling macrophage sensitivity. Infect. Immun. 72:4439-4447.

Moayeri, M., Webser, J., Wiggins, J., Leppla, S., Sternberg, E. (2005). Endocrine Perturbation increases susceptibility of mice to anthrax lethal toxin. Infect. Immun. 73:4238-4244.

Moayeri, M., Wickliffe, K., Wiggins, J., Leppla, S. (2006). Oxidized ATP protection against anthrax lethal toxin. Infect. Immun. 74:3707-3714.

Moayeri, M., Wiggins, J., Leppla, S. (2007). Anthrax protective antigen cleavage and clearance from the blood of mice and rats. Infect. Immun. 75:5175-5184.

Mock, M., Fouet, A. (2001). Anthrax. Ann. Rev. Microbiol. 55:647-671.

Muchlbauer, S., Evring, T., Bonucelli, G., Squires, R., Ashton, A., Porcelli, S., Lisanti, M., Brojatsch. (2007). Anthax lethal toxin kills macrophages in a strain-specific manner by apoptosis or caspase-1-mediated necrosis. Cell cycle. 6:758-766.

Nour, A., Yeung, Y., Santambrogio, L., Boyden, E., Stanley, E., Brojatsch, J. (2009). Anthrax lethal toxin triggers the formation of membrane-associated inflammasome complex in murine macrophages. Infect. Immun. 77:1262-1271.

O'Brien, J., Friedlander, A., Dreier, T., Ezzell, J., Leppla, S. (1985). Effects of anthrax toxin components on human neutrophils. Infect. Immun. 47:306-310.

Opal, S., Artenstein, A., Cristofaro, P., Jhung, J., Palady, J., Parejo, N., Lim, Y. (2005). Intra-alpha-inhibitor proteins ar e ndogenous furin inhibitors and provide protection against experimental anthrax intoxication. *Infect. Immun.* 73:5101-5105.

Paccani, S., Tonello, F., Gittoni, R., Natale, M., Muraro, L., D'Elios, M. Tang, W., Montecucco, C., Baldari, C. (2005). Anthrax toxins sppress T lymphocyte activation by disrupting antigen receptor signaling. J. Exp. Med. 201:325-331.

Pannifer, A., Wong, T., Schwarzenbacher, R., Renatus, M., Petosa, C., Blenkowska, J., Lacy, D., Collier, R., Park, S., Lepla, S., Hanna, P., Liddington, R. (2001). Crystal structure of the anthrax lethal factor. Nature. 414:229-233.

Paterson, H., Self, A., Garrett, M., Just, I., Aktories, K., Hall, A. (1990). Microinjection of Recombinant P21rho induces rapid changes in cell morphology. J. Cell Biol. 111:1001-1007.

Pellizzari, R., Guidi-Rontani, C., Vitale, G., Mock, M., Montecucco, C. (1999). Anthrax lethal factor cleaves MKK3 in macrophages and inhibits the LPS/IFNα-induced release of NO and TNFa. FEBS Lett. 462:199-204.

Petosa, C., Collier, R., Klimpel, K., Leppla, S, Liddington, R. (1997). Crystal structure of the anthrax toxin protective antigen. Nature. 385:833-838.

Petrilli, V., Dosert, C., Murave, D., T schopp, T. (2007). The inflammasome: a danger sensing complex triggering innate immunity. Curr. Opin. Immunol. 19:615-622.

Pezard, C., Berche, P., Mock, M. (1991). Contribution of individual toxin components to virulence of *Bacillus anthracis. Infect. Immun.* 59:3472-3477.

Popov, S., Villasmil, R., Bernardi, J., Green, E., Cardwell, J., Popova, T., Wu, A., Alibek, D., Bailey, C., Alibek, K. (2002a). Effect of *Bacillus anthracis* lethal toxin on human peripheral blood mononuclear cells. FEMS Lett. 527:211-215.

Popov, S., Villasmil, R., Bernari, J., Grene, E., Cardwell, J., Wu, A., Alibek, D., Baile, C., Alibec, K. (2002b). Lethal toxin of *Bacillus anthracis* causes apoptosis of macrophages. Biochem. Biophys. Res. Commun. 283:349-355.

Ren, X., Bokoch, G., Traynor-Kaplan, A., Jenkins, G., Anderson, R., Schwartz, M. (1996). Phsical association of the small GTPase ho with a 68-kda phosphotidylinositol 4- phosphae 5-kinase in Swiss 3T3 cells. Mol. Biol. Cell. 7:435-442.

Ribot, W., Panchal, R., Brittingham, K., Ruthel, G., Kenny, T., Lane, D., Curry, B., Hoover, T., Friedlander, A., Bavari, S. (2006). Anthrax lethal toxin impairs innate immune functions of alveolar macrophages and facilitates *Bacillus anthracis* survival. Infect. Immun. 74:5029-5034.

Ridley, A., Hall A. (1992). The small GTP-binding protein rho regulates the assembly of Focal adhesions and actin stress fibers in response to growth factors. Cell. 70:389-399.

Ristroph, J., Ivins, B. (1983). Elaboration of *Bacillus anthracis* antigens in a new, defined culture medium. Infect. Immun. 39:483-486.

Roberts, J., Watters, J., Ballard, J., Dietrich, W. (1988). Ltx1, a mouse locus hat influences the susceptibility of macrophages to cytolysis caused by intoxication with *Bacillus anthracis* lethal factor, maps to chromosome 11. Mol. Microbiol. 29:581-591.

Sapra, R., Gaucher, S., Lachmann, J., Buffleben, G., Chirica, G., Comer, J., Peterson, J., Chopra, A., Singh, A. (2006). Proteomic analyses of murine macrophages treated with *Bacillus anthracis* lethal toxin. Microb. Path. 41:157-167.

Schmid, D., Munz, C. (2007). Innate and adaptive immunity through autophagy. Immunity. 27:11-21.

Scobie, H., Rainey, G., Bradley, K., Young J. (2003). Human capillary morphogenesis protein 2 functions as an anthrax toxin receptor. Proc. Natl. Acad. Sci. USA. 100:5170-5174.

Scott, P., James, S., Sher, A. (1985). The respiratory burst is not required for killing of Intracellular and extracellular parasites by a lymphokine-activated macrophage cell line Eur. J. Immunol. 15:553-558.

Sirard, J., Mock, M., Fouet, A. (1994). The three *Bacillus anthracis* toxin genes are coordinately regulated by bicarbonate and temperature. J. Bacteriol. 176:5188-5192.

Sirard, J., Vignal, C., Dessein, R., Chamaillard, M. (2007). Nod-like receptors: cytosolic watchdogs for immunity against pathogens. PLos Pathog. 3:c152.

Squires, R., Muchlbauer, S., Brojatsch, J. (2007). Proteosomes control caspase-1 activation in anthrax lethal toxin-mediated cell killing. J. Biol. Chem. 282:34260-34267.

Stanley, L., Smith, H. (1961). Purification of factor I and recognition of a third factor of the anthrax toxin. J. Gen. Microbiol. 26:49-66.

Stephen, J. (1986). Anthrax toxin. In F. Dorner and J. Drew (eds), Pharmacology of Bacterial Toxins. Pp. 381-395.

Pergamon Press, Oxford, England. Szarowicz, S., During, R. Li, W., Qinn, C., Tang, W., Southwick, F. (2009). *Bacillus Anthracis* edema toxin impairs neutrophil actin-based motility. Infect. Immun. 77:2455-2464.

Tan, Y., Kusuma, C., St. John, L., Vu, H., Alibek, K., Wu, A. (2009). Induction of Autophagy by anthrax lethal toxin. Biochem. Biophys. Res. Commun. 379:293-297.

Tanaka, K., Tsurumi, C. (1997). The 26S proteasome: subunits and functions. Mol. Biol. Rep. 24:3-11.

Tang, G., Lepla, S. (1999). Proteasome activity is required for anthrax lethal toxin to kill macrophages. *Infect. Immun.* 67:3055-3060.

Tanoue, T., Adachi, M., Moriguchi, T., Nishida, E. (2000). A conserved docking motif in MAP kinases common to substrates, activators, and regulators. Nat. Cell. Biol. 2:110- 116.

Tessier, J., Green, C., Padgett, D., Zhao, W., Schwartz, L. Hughes, M., Hewlett, E. 2007.) Contributions to histamine, prostanoids, and neurokinins to edema elicited by edema toxin from *Bacillus anthracis*. Infect. Immun. 75:1895-1903.

Tournier, J., Quesnel-Hellman, A., Mathieu, J., Montecucco, C., Tang, W., Mock, D., Vidal, D., Goossens, P. (2005). Anthrax edema toxin cooperates with lethal toxin to impair cytokine secretion during infection of dendritic cells. J. immun. 174:4934- 4941.

Turnbull, P. (1992). Anthrax vaccines: past, present, and future. Vaccine. 9:533- 539.

Uchida, I., Makino, S., Sasakawa, C., Yoshikawa, M., Sugimoto, C., Terakado, N. (1993). Identification of a novel gene, *dep*, associated with depolymerization of the capsular polymer in *Bacillus anthracis*. Molec. Microbiol. 9:487-496.

Uchida, I., Makino, S., Sekizaki, T., Terakado, N. (1997). Cross-talk to the genes for *Bacillus anthracis* capsule synthesis by *atxA*, the gene encoding the *trans*-activator of anthrax toxin synthesis. *Molec. Microbiol.* 23:1229-1240.

Vitale, G., Bernardi,L., Napolitani, G., Mock, M., Montecucco, C. (2000). Susceptibility of mitogen-activaed protein kinase kinase family members to proteolysis by anthrax lethal factor. Biochem. J. 352:739-745.

Vitale, G., Pellizzari, R., Recchi, C., Napolitani, G., Mock, M., Montecucco, C. (1999). Anthrax lethal factor cleaves the N-terminus of MAPKKS and induces tyrosine/threonine phosphorylation of MAPKS in cultured macrophages. J. Appl. Microbiol. 87:288.

Vitale, G., Pelizzari, R., Reccchi, C., Napolitani, G., Mock, M., Montecucco, C. (1998). Biochem. Biophys. Res. Commun. 248:706-711.

Welkos, S., Keener, T., Gibbs, P. (1986). Differences in susceptibility of inbred mice to *Bacillus anthracis*. Infect. Immun. 51:795-800.

Welkos, S., Vietri, N., Gibbs, P. (1993). Non-toxigenic derivatives of the Ames strain of *Bacillus anthracis* are fully virulent for mice: role of plasmid pXO2 and chromosome in strain-dependent virulence. Microb. Pathog. 14:381-388.

Welkos, S., Trotter, R. Becker, D, Nelson, G. (1989). Resistance to the Sterne strain of *B. anthracis*: phagocytic cell responses of resistant and susceptible mice. Microbial. Path. 7:15-35.

Xia, Y., Wu, Z., Su, B., Murray, B., Karin, M. (1998). JNKK1 organizes a MAP kinase module through specific and sequential interactions with upstream and downstream components mediated by is amino-terminal extension. *Genes Dev.* 12:3369-3381.

Xu, L, Frucht, M. (2007). *Bacillus anthracis*: a multi-faceted role for anthrax lethal toxin In thwarting host immune defenses. Int. J. Biochem. Cell Biol. 39:20-24.

Yeager, L., Chopra, A., Peterson, J. (2009). *Bacillus anthracis* edema toxin suppresses human macrophage phagocytosis and cytoskeletal modeling *via* the protein kinase A and exchange protein activated by cyclic AMP pathways. Infect. Immun. 77:2530- 243.

Zhang, J., King, W., Dillon, S., Hall, A., Feig, I., Rittenhouse, S. (1993). Activation of platelet phosphatidylinositide 3-kinase requires the small GTP-binding protein Rho. J. Biol. Chem. 268:22251-22254.

Zhang, Y., Neo, S., Han, J., Yaw, L., Lins, S. (1999). RGS16 attenuates G alpha q-dependent p38 mitogen-activated protein kinase activation by platelet-activating factor. J. Biol. chem. 274:2851-2857.

CHAPTER 5

Vaccine Development and Immunological Aspects of *Bacillus anthracis*

Abstract: A number of attenuated, partially attenuated, and fully virulent strains of *B. anthracis* have been utilized over the years for vaccine development and animal challenge studies which are fully described. A description of early veterinary vaccine development followed by the development of human vaccines and their relative safety and incidence of adverse reactions are presented in considerable detail. Studies on the elucidation of the major factors of virulence, the lethal and edema toxins and capsule formation and their encoding by plasmids constitutes a major advance in the understanding of the genetic factors influencing anthrax and the critical antigens involved in the development of vaccines. The development of a variety of different potential vaccines are discussed including attenuated whole cell vaccines, unattenuated dead cell vaccines, purified protein based vaccines, recombinant protein based vaccines, vaccines utilizing bacterial vectors, viral vectors, and eukaryotic vectors and the efficacy of various adjuvants are presented in detail. A notably comprehensive presentation of macrophage studies associated with various animal models of anthrax is used to facilitate a clarified presentation of the biochemical factors involved in anthrax infections and their influence on the innate immune system. Detection of *B. anthracis* in environmental samples is invariably based on detection of spores. A variety of immunological methods are described for the rapid detection of *B. anthracis* spores in environmental samples.

Keywords: Major strains *of B. anthracis*; plasmid pXO1 plasmid pXO2; capsule; innate immune response; immunological detection; adjuvants; eucaryotic vaccine vectors; viral vaccine vectors; bacterial vaccine vectors, passive immunization; cytokines; chemokines; oxidative burst; complement; mice; rats; rabbits, guinea pigs; monkeys.

INTRODUCTION

Anthrax vaccines can be divided into veterinary and human. Anthrax vaccine development, particularly for livestock, originated in the late 1800's with the use of non-toxin producing strains. Most cell-free vaccine development studies have made use of the protective antigen (PA), the lethal toxin (LT), or both. Adjuvants have been found to play a major role in the delivery of all effective vaccines with a variety used over the years. This chapter deals in detail with all of these areas of vaccine development, including adverse reactions, the various antigens involved, their efficacy, and the various animal vaccine models that have been utilized.

DIFFERENCES IN VIRULENCE AMONG DIFFERENT VACCINE CHALLENGE STRAINS OF *B. anthracis*

Major Strains of *B. anthracis*

The Ames strain of *B. anthracis* was originally isolated from a diseased cow from a Texas cattle grazing field in the U.S. in 1980 and is fully virulent (pXO1$^+$ pXO2$^+$). It was forwarded to Iowa State University in Ames Iowa and was thereafter designated the "Ames" strain. It was used as a major experimental challenge strain after laboratory animal immunization prior to 2001. Its availability is presently restricted.

The British Vollum strain of *B. anthracis* was isolated from an oxford cow just before WWII. This strain was then passed through a series of infected monkeys to increase its virulence yielding substrain "M36". In 1951, strain M36 resulted in a lethal inhalation anthrax infection of a British microbiologist William Boyle. The resulting strain isolate from Boyle was then designated 1B indicating "from Boyle". The Vollum 1B strain is fully virulent but is somewhat less so than the Ames strain. These are the standard reference strains used in most animal challenge studies involving vaccine efficacy. The Sterne strain was originally isolated from South Africa and is pXO1$^+$ pXO2$^-$ and therefore produces no capsule, is considered avirulent, and is used as a major veterinary vaccine strain. The Ames strain of *B. anthracis* is representative of strains which have been designated "aberrant" or "vaccine resistant". Such strains are usually more virulent than other strains such as Vollum 1B.

The toxin-encoding plasmid pXO1 and capsule-encoding plasmid pXO2 are required for full virulence of *B. anthracis* in some animals. However, the non-toxigenic pXO1 cured derivatives of certain anthrax strains are not completely attenuated for mice, and their virulence is dependant on the bacterial strain. The strain related differences were found to be partially associated with plasmid pXO2 as demonstrated by pXO2 transductants of the attenuated vaccine strain UM23-1 cured of pXO1 (Welkos, 1991).

VACCINES

Early Vaccines

Live attenuated vaccines against anthrax for livestock in the early 1880's were among the first vaccines developed. Later, protein-based vaccines for human use were developed in the 1970s. Human cutaneous anthrax is thought to give rise to immunity because reinfections are rare and tend to be much less severe (Hodgson, 1941). Animals surviving anthrax infection have been found to be more resistant to subsequent challenge (Friedander, 2002).

The first studies on the development of veterinary vaccines for anthrax were undertaken in 1878 by the English veterinarian Burdon Sanderson. He diluted blood from infected cattle and then injected subinfectious doses (low numbers of anthrax bacilli) to susceptible cattle who acquired immunity to an otherwise lethal subsequent challenge dose of the organism.

The next major attempt to develop a veterinary vaccine was by the French veterinarian and biologist Jean-Joseph Henri Toussaint. Toussaint heated the filtered blood of anthrax-infected animals to 55° C which yielded a noninfectious fluid, presumably toxic to invading bacilli. It is quite possible that antibodies present in the filtered blood imparted passive immunity to the recipient animals, depending on the volume administered. Alternatively, the protein toxins produced by the organism would be expected to pass through the filter and result in a humoral or antibody response to the toxins.

Toussaint also treated infected blood with carbolic acid to destroy the cells of *B, anthracis* and found that it successfully protected 16 of 20 sheep from a lethal challenge of fully virulent blood. However, his vaccine preparations failed to yield consistent protection.

While Toussaint utilized a killed vaccine, Pasteur developed a live attenuated veterinary vaccine consisting of viable spores that eventually was universally accepted. The discovery by Pasteur of reducing the virulence of *B. anthracis* by growth at elevated temperature (42.5 °C) was effective in development of the first effective anthrax vaccine. Mikesell *et al.,* (1983) found that growing fully virulent strains at 42 °C cured the cells of the pXO1 plasmid (Fig. **1**), which encodes the three proteins that comprise the two anthrax toxins, the lethal factor (LF), the edema factor (EF) and the protective antigen (PA) (see chapter 4 for the interactive mechanisms involving these proteins).

The three genes that encode these toxins are harbored on plasmid pXO1, while the genes encoding the antiphagocytic capsule are encoded on plasmid pXO2. The isolation of an attenuated nonencapsulated toxigenic variant (pXO1$^+$, pXO2$^-$, Fig. **2**) presently referred to as the Sterne strain and its development as a live attenuated spore vaccine has greatly influenced the control of the disease in livestock (Sterne, 1937; Sterne, Nichol, and Lambrechts, 1942). Live vaccines have often demonstrated superior protection in experimental animals compared to non-live component vaccines (Friedlander, Welkos, and Ivins, 2002).

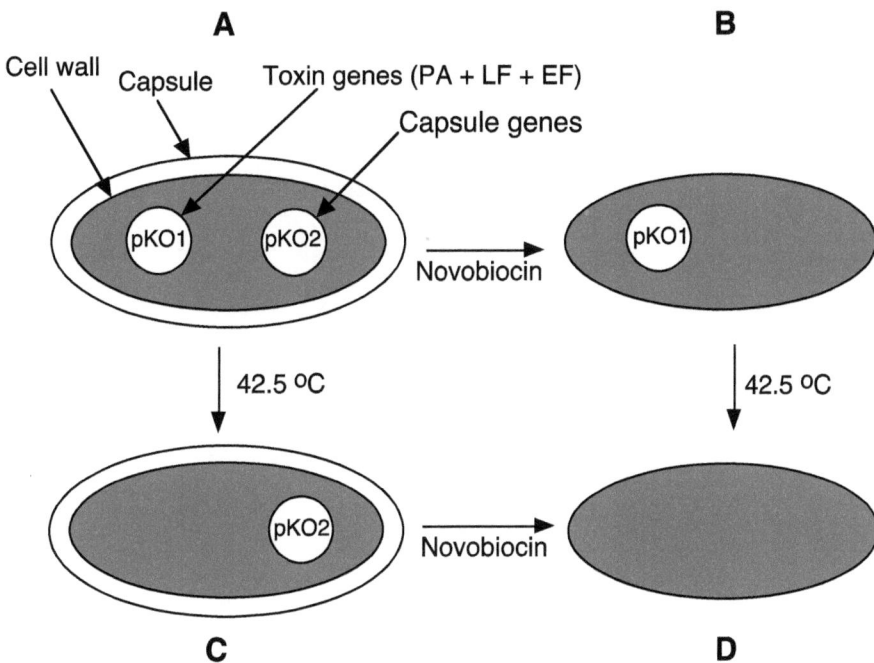

Figure 1: Diagrammatic representation of virulence plasmids of *B. anthracis* and the results of their selective curing. A: Fully virulent cell producing both the toxin complex and capsule. B: Curing of the pXO2 plasmid by growth in the presence of novobiocin results in a Sterne-type veterinary vaccine strain lacking the capsule. C: Curing of the pXO1 plasmid by growth at 42.5 °C results in a Pasteur-like veterinary vaccine strain lacking toxin production. D: Curing of both plasmids by growth in the presence of novobiocin at 37 °C followed by growth at 42.5 °C. Adapted from WHO (2008).

Veterinary Vaccines for Livestock and Other Mammals

Pasteur's Livestock Vaccine Strains

When repeatedly subcultured on agar media at an elevated temperature (42.5 °C) wild-type virulent strains gradually become avirulent which was the method used by Pasteur to obtain his famous attenuated live vaccine. Avirulent strains resulting from growth at elevated temperatures usually form capsules *in vivo* but do not produce toxins. The loss of toxin production is now recognized to be a result of plasmid curing at elevated 1 2 3 4 5 6 7 8 growth temperatures. The toxin genes are encoded on plasmid pXO1 while the capsule genes are encoded on plasmid pXO2 (see chapter 4). The Pasteur strain is pXO1⁻ pXO2⁺ and the Stern strain is pXO1⁺ pXO2⁻.

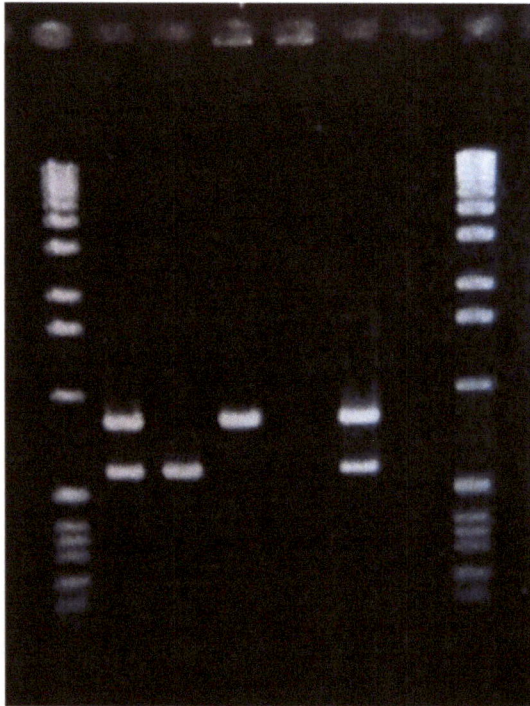

Figure 2: PCR confirmation of *B. anthracis*. Lanes 2 and 6 illustrate the presence of the protective antigen gene (*pagA*) and one or more of the genes in the capsule biosynthesis operon (*capBCAD*) representing a fully virulent strain of *B. anthracis*. Lane 3 shows the presence of a gene for one of the toxin components (PA) but no capsule gene; the strain would be similar to the Sterne veterinary vaccine strain. Lane 4 shows the presence of a capsule gene but no toxin gene; the strain would be similar to the Pasteur veterinary vaccine strain. From WHO (2008) with permission.

Pasteur used two vaccines for the subcutaneous vaccination of cattle and sheep. The first vaccine (strain type 1°) consisted of bacilli grown at 42.5 °C for 15 to 20 days. After 12 days a second inoculation of bacilli cultured at 42.5 °C for 10 to 12 days (strain type 2°). The first vaccine was notably less virulent than the second vaccine and only caused anthrax symptoms in small experimental animals, mice and young but not adult guinea pigs. In contrast, Pasteur's second vaccine was noninfectious for rabbits but infected adult guinea pigs. These two vaccines resulted in complete immunity of livestock to injection of a fully virulent strain. It is presently assumed that the first vaccine strain had a higher percent of cells cured of plasmid pXO1 than the second vaccine strain.

An additional innovation by Pasteur was the addition of saponin to his vaccines. Saponins are produced by many plants and consist of a steroid or triterpene and a sugar. All saponins are hemolytic, foam strongly when shaken with water and form oil-in water emulsions. Saponin suspends the spores uniformly and also causes a strong inflammatory reaction at the injection site. The saponin-caused inflammation is thought to allow the anthrax spores to germinate in the tissue injected, but reduces any immediate septicemia derived from the resulting vegetative cells.

In the 1930s the South African veterinarian Max Sterne found that a much lower concentration of saponin (0.5%) similarly enhanced the level of resulting immunity without severe inflammation at the injection site. Sterne then developed the "Sterne 34F2" strain of *B. anthracis* that yielded uncapsulated vegetative cells as a viable veterinary spore vaccine (Sterne, 1939a; Stern, 1939b). These bacilli had the potential to generate protective immunity without causing an infection in animals. Stern suspended 6×10^5 and 1.2×10^6 spores of the 34F2 strain in 0.5% saponin in a 50% glycerine saline solution to yield a vaccine with minimum inflammatory effects and maximum protection. Sterne noted that goats and llamas were unusually susceptible to developing anthrax from the vaccine itself. The Sterne 34F2 strain continues to be the most widely used veterinary vaccine for cattle.

In Argentina and Uruguay, Pasteur's vaccine was administered from the spring of 1904 to September, 1905 to 140,000 cattle, 50,000 sheep, and 2,000 horses. No fatalities resulted from the vaccination and almost complete eradication of anthrax occurred (Swiderski, 2004).

Present Livestock Vaccines

Present vaccines for veterinary immunization of livestock against anthrax consist of spores of attenuated strains of *B. anthracis*. These veterinary strains are classified into three groups:

1. Live attenuated vaccines, non-toxigenic and encapsulated (tox⁻ cap⁺).

2. Live attenuated vaccines, toxigenic and non-encapsulated (tox⁺ cap⁻).

3. Live attenuated vaccines, toxigenic and encapsulated (tox⁺ cap⁺).

The Pasteur vaccines belong to the first group with a plasmid content of pXO1⁻ pXO2⁺. On the basis of residual virulence, the Pasteur vaccines are classified into two groups (Turnbull, 1991) as follows:

1. Pasteur vaccine $1°$ type, pathogenic for mice and non-pathogenic for guinea pigs.

2. Pasture vaccine $2°$ type, pathogenic for guinea pigs and non-athogenic for rabbits.

The Sterne strain of *B. anthracis* has been used as a vaccine against anthrax for over 70 years. The Sterne vaccine is characterized by an elevated protective capacity and very low residual virulence and lacks the pXO2 plasmid that harbors the *capBCAD* operon encoding enzymes involved in capsule synthesis. Although the live attenuated (capsule free) spore Sterne vaccine has been highly successful in reducing the incidence of anthrax in cattle and in humans, the immunogenicity of the Sterne strain is compromised by its residual pathogenicity due to toxin production in mammals. Certain undesirable side-effects, such as fever, and sever edema are observed in vaccinated animals and death can occur in susceptible species such as goats.

The *B. anthracis* strain named "Carbosap" is presently used in Italy as a live veterinary vaccine for immunizing cattle and sheep and it is also protective for goats and horses (Chiocco and Sobrero, 1985). Since its isolation prior to 1949 it has been considered to be in the Pasteur vaccine $1°$ type group. Fasanella *et al.,* (2001) genetically characterized the Carbosap strain with a series of nine PCR primer pairs from Ramiss *et al.,* (1996) (Table **1**, chapter 6) specific for the *B. anthracis* Bad 813 chromosomal sequence and the toxin genes on the pXO1 plasmid and the *cap* genes on the pXO2 plasmid. The presence of the cap genes was confirmed. Primer pairs amplifying sequences derived from the *lef* (lethal factor)*, cya* (edema factor), and *pag* (protective antigen) genes yielded amplicons of the correct size for these genes as did the Ba813 primers. These results indicate that the Carbosap strain possesses both plasmids pXO1 and pXO2 and the toxin and capsule genes associated with these respective plasmids. *In vivo* staining of tissue from autopsied animals infected with the Carbosap strain confirmed the presence of capsules. The carbosap vaccine strain was found to be virulent for mice and guinea pigs but not virulent for rabbits. The Stern F34 vaccine strain was found to be virulent for mice and non virulent for guinea pigs and rabbits. The Pasteur SS104 strain was non-virulent for all three animal species. A fully virulent strain of *B. anthracis* was found pathogenic for all three animal species. These results indicated that the Carbosap strain possesses somewhat greater virulence for small laboratory animals compared to the Sterne F34 and Pasteur S104 strains but a lower level of virulence than a fully virulent strain. Interestingly, the plasmid profile of the Carbosap strain ($pXO1^{+}$ $pXO2^{+}$) is similar to that of several attenuated atypical Pasteur strains possessing both plasmids (Fasanella *et al.*, 2001) and is therefore considered to fall into a third or "atypical" group and can no longer be considered as a member of the Pasteur $1°$ type of veterinary vaccine strains. The attenuated virulence of the carbosap strain may be due to either a low level of expression of either PA or LF or both *in vivo* or alternatively to a low level of expression of one or more virulence factors residing on the chromosome such as hemolysins normally produced anaerobically but not aerobically by *B. anthracis* (Klichko *et al.*, 2003) which are referred to as anthrolysins or anthraysins are discussed later in this chapter. A detailed tabulation of the virulence genes present in all of these three types of veterinary strains and control strains is presented in Table **1**.

The STI vaccine strain of *B. anthracis* is an avirulent isolate (pXO1+ pXO2) used extensively in the former Soviet Union as an attenuated live spore vaccine for livestock and humans (Shlyakhov and Rubenstein, 1994).

Mice appear to be highly susceptible to replicating cells of fully virulent *B. anthracis* while somewhat resistant to the anthrax toxin. Rabbits are susceptible to infection by virulent *B. anthracis* while they are resistant to the Stern strain (pX01$^+$ pX02$^-$) and to Carbosap (pX01$^+$ pX02$^+$) attenuated vaccine strains. In contrast, mice and guinea pigs are susceptible to the Carbosap vaccine strain. The European Pharmacopoeia specifies that anthrax vaccines for veterinary use must be tested in laboratory animals. In the potency test for anthrax veterinary vaccines the test animals must be vaccinated with 1/10 of the normal recommended dose and this should protect them against a 200LD$_{50}$ challenge dose of a fully virulent strain while an unvaccinated control group should succumb to a 20LD$_{50}$ dose of the same strain. The lack of availability of the Ames strain after 2001, resulted in strain A0843 being assessed as a suitable *B. anthracis* challenge strain by Fasanella, Scasiamacchia, and Garafolo (2009). This strain was isolated from a sheep that had died in Italy in 1993 from an anthrax outbreak. The suitability of this strain was validated in male and female New Zealand rabbits by subcutaneous injection of a 2LD$_{50}$ dose (40,000 spores) which resulted in 44.7%, 70.2%, and 100% of the rabbits succumbing at 72, 96, and 120 hrs. respectively after injection of the spores. The authors proposed that the consistent performance of strain A0843 makes it suitable as a potency challenge strain in rabbits in place of the Ames strain.

Human Vaccines

Efficacy of Human Vaccines

The licensed US human anthrax received US FDA approval in 1970. The vaccine consists of a cell-free filtrate prepared from a formalin-treated culture supernatant of a nonproteolytic, toxigenic, nonencapsulated, avirulent *B. anthracis* strain (pXO1$^+$ pXO2$^-$), designated V777-NP1-RR, adsorbed to the adjuvant aluminum hydroxide. Its efficacy is primarily due to the adsorbtion of PA in the culture supernatant to aluminum hydroxide. A similar vaccine, prepared by adsorbing a sterile culture supernatant filtrate of the 32F2 Sterne strain to potassium aluminum sulfate, is licensed for use in the U.K. AVA has been recommended for individuals working with *B. anthracis* in the laboratory. A notable degree of protection (93%) was observed when the vaccine was administered to goat hair mill workers in Northeastern US during 1955 to 1958 (Friedlander *et al.*, 2002). In guinea pigs AVA does not provide effective protection against aerosol challenge (Ivins *et al.*, 1995; Fellows *et al.*, 2001) while Rabbits and nonhuman primates are fully protected by AVA (Fellows *et al.*, 2001).

Table 1: Characteristics of *B. anthracis* live veterinary vaccine strains and controls. Adapted from Fasanella *et al.,* (2011)

Characteristic	*B. anthracis* fully virulent strain FG/05/93	*B. anthracis* Sterne F34vaccine strain	*B. anthracis* Carbosap vaccine strain	*B. anthracis* Pasteur SS104 strain	*B. cereus* ATCC 14579
Ba813 Chromosomal DNA Sequence	+	+	+	+	+
lef (LF)	+	+	+	+	-
pag (PA)	+	+	+	+	-
cya (EF)	+	+	+	+	-
cap (capsule)	+	+	+	+	+
Plasmid pX01	+	+	+	+	-
Plasmid pX02	+	+	+	+	-
Mouse virulence (Albino Swiss)	+	+	+	+	-
Guinea pig virulence (Albino)	+	+	-	+	-
Rabbit virulence (New Zealand)	+	+	-	-	-

Protection of animals vaccinated with live spore vaccines is recognized to be superior and more prolonged than is protection with protein vaccines. Ezzell and Absire (1988) compared the sera from guinea pigs vaccinated with a live veterinary anthrax spore vaccine (Sterne strains) with the sera from guinea pigs vaccinated with the MDPH human vaccine which consists of culture protein of *B. anthracis* adsorbed by aluminum hydroxide. Animals vaccinated with the MDPH human vaccine had two-to four-fold higher titers to PA than did animals vaccinated with the spores of the *B. anthracis* Sterne strain. Sera from the animals vaccinated with the spore vaccine recognized two major *B. anthracis* vegetative cell-associated proteins that were either not recognized or poorly recognized by sera from animals that received the human vaccine. These proteins, were termed extractable antigens 1(EA1) and 2 (EA2). The EA1 protein appeared to be coded by the chromosomal DNA, whereas the EA2 protein was only detected in strains that possessed the pXO1 plasmid. Both of the extractable antigen proteins were serologically distinct from the components of ET and LT. In guinea pigs vaccinated with the MDPH human vaccine, PA was the primary antigen that was recognized, as was also the case with antisera from MDPH-vaccinated humans.

Vaccine trials with partially purified EA1 demonstrated that it does not elicit protective antibody against anthrax. In addition, animals vaccinated with sterile gamma-irradiated cell walls had significant antibody titers to the N-acetylglucosamine-galactose polysaccharide of *B. anthrac*is but were not protected against the virulent Ames strain. These results reflect the fact that antibody to PA plays a central role in protection against anthrax and that other antigens may be of little consequence. These results also agree with the observation that live vaccines composed of *B. anthracis* strains which lack either both plasmids or just the pXO1 plasmid, and which therefore do not produce PA, are not protective (Ivins *et al.*, 1986a, Uchida *et al.*, 1986). In contrast, *B. subtilis* clones possessing the PA encoding gene *pag*, when used as a live vaccine, provided protection against virulent *B. anthracis* spores in guinea pigs and to LT in rats (Ivins *et al.*, 1986b).

In the former Soviet Union, a skin test utilizing "Anthraxin" first licensed for use in 1962, has become widely used for retrospective diagnosis of human and animal anthrax and for vaccine evaluation (Shylakhov, Rubinstein, and Novikov, 1997). This is a commercially produced heat-stable protein-polysaccharide-nucleic acid complex without capsular or toxigenic material, derived from the edematous fluid of animals injected with the vaccine ST1-1 or Zenkowski strain of *B. anthracis*. It is sterilized by autoclaving. The test involves intradermal injection of 0.1 ml of Anthraxin. A positive test is defined as erythema of $\geq$ 8 mm with induration persisting for 48 hrs. This delayed-type hypersensitivity is seen as reflecting anthrax cell mediated immunity and was reportedly able to diagnose anthrax retrospectively some 31 years after primary infection in up to 75% of cases (Shylakhov, Rubinstein, and Novikov, 1997).

Shlyakhov, Rubinstein, and Novikov (1997) subjected seven groups (2,596 human subjects) to vaccination with a live anthrax vaccine (HLAV) by three different routes (scarification, subcutaneous, and aerosol). The vaccines were tested for anthrax cell-mediated immunity using the "Anthraxin" skin test at 7, 15, 30, 90, 180, and 365 days following vaccination. The kinetic pattern of immunogenic response showed a significant five-phase response curve: phase I (2-6 days post-vaccination) showed a slow increase in positive Anthraxin skin reactions. Phase II (7-15 days post-vaccination), showed an exponential rise to a maximum at day 15. Phase II (16-30 days post-vaccination) exhibited a decrease to day 30. Phase IV (31-90 days post-vaccination) led to a relative restoration of the positive skin reactions. During phase V (91-365 days post-vaccination) there was a continuous decrease in positive Anthraxin skin reactions. The loss of the skin test reaction on day 30 is considered a characteristic feature of post-vaccination anthrax cell-mediated immunity. The authors concluded that this may be due to a blockade of macrophage by lethal anthrax toxin released by the multiplying vaccine cells.

The active component of the licensed human anthrax vaccine (BioThrax^{TM} or AVA) is the PA. Zhou *et al.,* (2008) isolated and analyzed the PA-specific antibody repertoire from an AVA vaccinated individual. All PA-specific antibodies were found to have undergone somatic hypermutation and class switch recombination, both signs of affinity maturation. All variable heavy (V_H) chain genes and variable light (V_L) chain genes were found to be mutated compared to their germline gene of origin. Mutations were rarely observed in the constant light or heavy chain regions, indicating that the mutations observed most likely arose during somatic hypermutation. The majority of the V_H and V_L mutations were associated with the RGYW/WRCY sequence motifs which are recognized as mutation host spots (Wagner and Neuberger, 1996). Although the antigenic epitopes recognized by the antibody response were distributed throughout the PA monomer, the majority of antibodies were found to recognize determinants located on the amino-terminal (PA_{20}) sub-domain of PA. This could potentially have the effect of blocking the proteolytic cleavage of PA_{83} to its active PA_{63} form.

In Russia, the viable *B. anthracis* strain STI-1 has been used extensively for human anthrax vaccination for more than four decades (Stepanov *et al.*, 1996). However, it has been found to result in low immunogenicity. In addition, general and local responses occur both after primary and revaccination with the frequency of adverse effects increasing with the number of vaccinations. Studies have shown that with experimentally immunized animals, the spores of *B. anthracis* can persist for long periods in lymph nodes, spleen, and bone marrow without germinating (Kolesov, Mikhailov, Presnov, 1962; Ipatenko *et al.*, 1990). Stepanov *et al.*, (1996) undertook studies on the effects of specific humoral immunity caused by antianthrax γ-globulin and its fractions (IgG, IgA, and IgM) on the initial stages of spore germination. Total γ-globulin derived from vaccination of animals with the STI-1 strain was found to decrease spore germination two-fold, whereas the immunoglobulins isolated from it, except for IgG, did not exhibit such an effect. Opsonization activity of the anthrax immune serum was found to be largely determined by IgM which influences the germinating spore, thus providing the initiation of phagocytosis. IgM stimulated phagocytosis only with regard to germinating spores and did not affect vegetative cells. The rate of uptake of inert particles (India ink) or spores (dormant or activated) by lymphoid macrophages was determined in nonvaccinated and vaccinated white mice. The uptake rates were found to be similar both for vaccinated and nonvaccinated animals which means that spores operate as inert particles. A considerable difference however was observed with activated spores. In this case, the uptake rate in non-immunized animals decreased more than 5-fold during the first 2 hrs, which is indicative of infection development. In the immunized animals, spore germination by activated spores was blocked and the uptake rate did not change. The authors proposed that the immunity in anthrax has both an antitoxic but also an antisporal character and that vaccines should contain the entire complex of anthrax antigens consisting of spore, the three toxin components, and the proteins encoded by the *capA, capB,* and *capC* genes of the pXO2 plasmid. This approach might be of value, particularly if the toxin proteins were modified so as to eliminate toxicity but still retain protective antigenicity.

The seroconversion rates and geometric mean concentrations (GMC) of IgG anti-PA for stored sera from U.S. military personnel (compliant with the AVA immunization schedule) with the U.S. licensed anthrax vaccine AVA were reported for 246 individuals by Singer *et al.,* (2008). Seroconversion was defined as a >4-fold rise of serum anti-PA IgG as determined using an ELISA assay. Pre-dose samples for the 4th and 6th doses were drawn up to 56 days prior to such respective vaccinations. Post-dose samples were drawn 14 – 42 days after vaccination. After the third dose, the increase in overall GMC was from 1.83 to

59.9 µg/ml (33-fold) and the sero conversion rate was 85.3%. After the fourth dose, the overall increase in the GMC was from 24.61 to 157.4 µg/ml and the seroconversion rate was 67.9%. After the sixth dose, the GMC increased from 92.4 to 277 µg/ml and the seroconversion rate was 45.5%.

Synthetic oligonucleotides (ODN) containing immunostimulatory CpG motifs have been found to boost the immune response to co administered antigen, including AVA (Jones *et al.*, 1999; Klinman, 2004; Klinman *et al.*, 2004; Verthelyi *et al.*, 2002). CpG ODN induces the functional maturation of professional antigen-presenting cells and trigger the production of immunostimulatory cytokines and chemokines (Ballas, Rasmussen and Krieg, 1996; Halpern, M., Kurlander, and Pisetsky, 1996, Klinman, 2004, Klinman *et al.*, 1996; Krieg *et al.*, 1995). Xie *et al.*, (2005) examined whether CpG ODN 1555 (GCT-AGA-CGT-TAG-CGT) adsorbed onto cationic polyactide-co-gycolide (PLG) microparticles resulting in (CpG ODN-PLG) would increase the speed and magnitude of protective anti-PA immunity induced by coadministering CpG ODN-PLG with AVA. One group of A/J mice were vaccinated with AVA (200 µl) only and a second group received CpG ODN-PLG simultaneously with AVA. Coadministration of CpG ODN with AVA increased the IgG anti-PA titer by five-fold above that of AVA alone 14 days after immunization and protected mice against a lethal spore challenge 7 days after vaccination. When CpG ODN-PLG was coadministered with AVA the titer was increased ~300-fold above that of AVA administered alone. Mice immunized with AVA alone were highly susceptible to infection when challenged with a lethal dose of anthrax spores 14 days after immunization (79% mortality). Interestingly, >80% of the mice immunized with CpG ODN-PLG plus AVA survived challenge at this post-immunization period. PLG or ODN-PLG co-administered with AVA increased the IgG anti-PA titer ~two-fold above that of AVA alone. ODN coadministered with AVA yielded the same IgG level as AVA alone 14 days after administration.

Fellows *et al.*, (2002) demonstrated that the golden Syrian hamster is not an appropriate model for investigating human anthrax vaccine efficacy. Groups (10 animals/group) of hamsters were vaccinated *i.m.* with the AVA vaccine at either 4 weeks or at 4 and 8 weeks and were then challenged *s.c.* at 10 weeks with spores of various fully virulent *B. anthracis* isolates (Ames, Zimbabwe, Namibia,

Vollum 1B). Although ELISA and toxin neutralization assays exhibited high titers, none of the AVA-vaccinated hamsters were protected from challenge or demonstrated a significantly extended time to death compared to that of control animals.

The potency test for the anthrax vaccine currently licensed for human use in the USA (Anthrax Vaccine Adsorbed, AVA) involves the protection of actively immunized guinea pigs from a lethal challenge with a virulent strain of *B. anthracis*. Pombo *et al.,* (2004) developed a potential alternative anti-PA-ELISA assay that is nonlethal and utilizes mice vaccinated with the AVA vaccine or vaccines under development. The assay can presumably be used in assessing consistency of manufactured batches of a given vaccine which is a reported variable with the AVA vaccine. The parameters of specificity, linearity, accuracy, precision, detection limit, and quantification limit were determined for validation of the ELISA assay. Recombinant PA (rPA) was used for coating of the wells and an anti-PA preparation was obtained by immunizing CD-1 mice with rPA combined with aluminum hydroxide. After addition of polyclonal murine antibody and washing the wells, goat anti-mouse IgG conjugated to alkaline phosphatase was added and then incubated for 2 hrs. at room temperature. After rinsing the wells, p-nitrophenyl phosphate was added for color development at room temperature. The reader should note that the temperature of the ELISA assay was not controlled which could constitute an uncontrolled seasonal variable. In addition, no attempt was made to correlate IgG responses to PA in the murine model to IgG responses to PA in the guinea pig model or to guinea pig or mouse survival on challenge with respect to variable PA levels and variable levels of resulting IgG. The reader should also note that ELISA assays for anti-Pa IgG have been found not to reflect a direct correlation with the level of immune protection compared to toxin neutralization titers (Reuveny *et al.,* 2001; Weiss *et al.*, 2006, Reason *et al.*, 2009). However, the authors did document and validate the described parameters of the murine ELISA which was the stated goal of the study.

Antibiotic Administration in Conjunction with AVA

Anthrax Vaccine Adsorbed (AVA) is ineffective when delivered after anthrax spore challenge as proliferative bacteria reach toxic levels before protective immunity develops. Klinman and Tross (2009) examined the utility of administering a single dose of the long-acting lipopeptide antibiotic Dalbavancin in conjunction with CpG-adjuvant AVA. CpG oligonucleotides as adjuvants interact with Toll-like receptor 9 expressed by B cells and plasmacytoidal dendritic cells, improving antigen presentation and triggering the production of Th1 and pro-inflammatory chemokines and cytokines resulting in accelerated and enhanced protection Klinman and Tross (2009). A/J mice were immunized *i.p.* with 10 μL of AVA plus 20 μg of CpG ODN (GCT-AGA-CGT-TAG-CGT and TCA-ACG-TTG-A) and were also injected with 150 μg of dalbavancin. Mice were challenged *i.p.* With 30 LD_{50} of the Stern strain (pXO1$^+$ pXO2$^-$) which is lethal to A/J mice (*i.p.* LD_{50} = 1.1 x 10^3 spores). Dalbavancin administered from 0 to 15 days prior to challenge protected all mice but failed to protect if administered 30 or more days prior to challenge. When Dalbavancin was administered from 0 to 24 hrs. after challenge, all mice (10 to 18 per group) survived. When delivered 3 days after challenge, nearly half the mice succumbed to infection. Dalbavancin was ineffective if administered four or more days post-challenge due to irreversible toxemia and systemic infection. The CpG-adjuvant AVA vaccine induced partial protection within five days of administration and complete protection by day 10. Mice vaccinated less than five days prior to challenge succumbed from infection. When mice were administered both the CpG-adjuvant AVA vaccine and Dalbavancin on the same day of challenge and after one day after challenge continuous protection resulted.

The Safety of the B. anthracis AVA Vaccine and Adverse Reactions in Humans

The safety of anthrax vaccination of humans has become controversial in recent years. AVA is made from an avirulent, nonencapsulated strain of *B. anthracis*. The primary series is administered at 2, 4, and 6 weeks, followed by injections given at 6, 12 and 18 months. Subsequent booster doses are recommended annually.

Brachman *et al.,* (1962) conducted the only randomized, placebo-controlled trial of the efficacy of a protective antigen anthrax vaccine including the recording of immediate onset of adverse events. The vaccine involved was not AVA but was an earlier formulation fro the R1-NP strain of *B. anthracis*. The study performed in 1955 involved four mills in eastern U.S. that processed raw goat hair, which was commonly contaminated with anthrax spores. The worker population eligible for the study included 1,249 men and women with no history of prior anthrax infection. Study participants received subcutaneous inoculations of 0.5 c.c of either vaccine or placebo (0.1% alum). Significant local adverse reactions were observed in about 2 to 15% of immunized persons after he first through the fourth inoculations and in about 40% of immunized persons after the seventh inoculation. The most common local reactions-erythema, pruritis, and a small area of induration-were mild and disappeared within 24 to 48 hrs. Overall, local reactions of any type, from mild to severe, were observed following 35% of inoculations. Severe edema-producing reactions occurred following 2.8% of vaccinations. Systemic reactions were observed in two of the vaccine recipients and consisted of "some malaise of 24 hrs. duration".

Pittman *et al.,* (2001) analyzed U.S. Army Medical research data involving 25 years (1973-1999) of AVA use to protect personnel at Fort Detrick, Maryland. The total number of personnel involved was 1583 with 10,722 AVA doses administered. Females reported a higher incidence of injection site and adverse systemic events than males. Local reactions of some type were reported for 3.6% of the 10,722 doses. Erythema and/or induration, the most common local reactions, were reported for 3.2% of the doses administered. Systemic reactions of any sort were reported for 101 of the 10,722 doses (0.94%) and were more frequent with women than men. Overall, the most common systemic reactions were: headache (0.4%), malaise (0.4%), myalgia (0.3%), nausea (0.1%), and dizziness (0.1%). Data indicated a difference in incidence of local reactions between AVA lots. After an initial large local reaction, most individuals had no reaction to a subsequent dose of AVA.

In 1999, 2.4 million U.S. Armed forces personnel were ordered to receive the anthrax vaccine. Adverse reactions were found to occur at an incidence of 0.02 to 0.1% or higher. In some military locals the incidence has been as high as 7% (Nicolson, Nass, and Nicolson, 2000). Unusual chronic effects have included vomiting, diarrhea, polyarthralgias fever, splenic tenderness, cognitive problems, polymyalgia, weakness, and numbness. A severe allergic reaction has also been noted resulting in a loss of epidermis and the lining of the GI tract. More severe symptoms have also been reported among numerous anthrax vaccine recipients including seizures in addition to respiratory distress and a variety of pulmonary illnesses.

Niu *et al.,* (2009) reviewed in detail the adverse events (AEs) following human anthrax vaccination from 1990 to 2007. In 2002, the Institute of Medicine's report (IOM, 2002) of AEs after AVA vaccination found that AEs were similar in type and rate of occurrence to AEs from other adult vaccines. The Anthrax Vaccine Expert Committee (AVEC) in 2002 concluded that there was not a "high frequency or unusual pattern" of serious AEs, but did identify certain rare events such as aggravation of spondyloarthropathy (joint inflammation), anaphylactoid reaction (severe systemic hypersensitive allergic reaction), arthritis, bronchiolitis obliterans (constrictive bronchiolitis) organizing pneumonia, as possibly or probably being caused by the AVA vaccine.

More recently the U.S. FDA summarized reports to the Vaccine Adverse Event Reporting System (VAERS) received through August 15, 2005. This report identified certain serious injection site and allergic reactions as being likely related to AVA. It also identified other conditions: arthritis, systemic lupus erythematosus (an autoimmune inflammatory disease of connective tissue), multisymptom illness, atrial fibrillation, pemphigus vulgaris (an autoimmune blistering disease that affects the skin and mucous membranes), diabetes mellitus type 1, optic neuritis, Guillain-Barré Syndrome (GBS, an autoimmune disorder affecting the peripheral nervous system) and facial palsy without establishment of any causal relationship to AVA.

During March 1, 1998 to May 7, 2007, approximately 6 million AVA doses were administered to US. Military personnel. As of January 16, 2007, there were 4,753 reports filed of adverse reactions (AEs) following vaccination with AVA of which 4,273 (90%) were non-serious, 455 (9.6%) serious, and 25 deaths (0.5%), the majority of which involved serious maladies prior to AVA vaccination.

The U.S. FDA's review of VAERS AVA reports through August 15, 2005 found no causal relationship between death and other serious AEs (other than some injection site reactions and some reports of allergic reactions) and the administration of AVA, but recommended some conditions (*e.g.*, systemic lupus erythematosus, optic neuritis, and arthritis) for further study.

Serious AEs such as multiple sclerosis, transverse myelitis, amylotrophic lateral sclerosis (ALS), rheumatoid arthritis, polyarteritis nodosa, and psoriasis were below the background rate in the general population (Niu *et al.,* (1999)). Reviews of serious AEs and death reports did not show a distinctive pattern indicative of a causal relationship to AVA vaccination. However, like many vaccines, AVA can cause local injection site and allergic reactions.

Pittman *et al.,* (2002) compared the use of fewer AVA doses administered *i.m.* or *s.c.* with the recommended schedule and route. A total of 173 volunteers divided into 7 groups were given AVA once *i.m.* or *s.c.*, two doses, two or four weeks apart, *i.m.* or *s.c.* or six *s.c.* doses at 0, 2, 4 weeks and 6, 12, and 18 months (control group), licensed schedule route). *i.m.* administration of AVA resulted in fewer injection site reactions than *s.c.* administration. Following the first *s.c.* dose of AVA, females had a significantly higher incidence of injection site reactions such as erythema, induration, and subcutaneous nodules than males which had been previously reported. The incidence of reactions decreased with a larger dose interval between the first two doses. The peak anti-PA IgG antibody response of subjects given two doses of AVA four weeks apart *i.m.* or *s.c.* was comparable to that of subjects who received three doses of AVA at two-week intervals.

Vaccines from Toxin Deficient and Other Mutants of *B. anthracis*

Mutants of toxigenic, nonencapsulated strains of *B. anthracis* dependant on aromatic amino acids for growth designated Aro⁻ have been isolated with the use of the tetracycline transposon Tn916 (Ivins *et al.,* 1990a). They were notably attenuated for mice and guinea pigs and protected these highly susceptible species. The Aro⁻ mutants grew much slower than the wild type, which may account for their reduced lethality. However, the presence of the tetracycline resistance determinant prevented their use since their reversion to the Aro⁺TCˢ phenotype has prevented their application as live vaccines. In addition, concern regarding side effects of live vaccines has prevented their acceptance as human vaccines in Western countries.

Derivatives of an attenuated strain of *B. anthracis* deficient in each of the three toxin proteins have been constructed. The genes for PA (*pag*), EF (*cya*), and LF (*lef*) were inactivated by a large internal deletion and the insertion of an antibiotic-resistance cassette (Erm^r or Kan^r) using recombinant technology (Pezard, Berche, and Mock, 1991; Pezard, Duflot, and Mock, 1993) and the constructs separately introduced into *B. anthracis* Sterne on a shuttle plasmid. An EF$^-$ strain and an LF$^-$ strain were then isolated after homologous recombination with the resident toxin-encoding plasmid pXO1 and loss of the introduced plasmid resulting in replacement of the wild-type gene on pXO1 with the corresponding deleted copy harboring the antibiotic-resistance cassette. Spores from these mutants and from a previously constructed PA- mutant were used to inoculate mice, and the lethality and local edema monitored. Skin edema was observed with an EF$^+$ LF$^-$ PA$^+$ mutant. LF$^-$ PA$^-$ mutants were not lethal even at high inocula (1 x 10^9 cfu) whereas an EF$^-$ LP$^+$ PA$^+$ mutant induced lethal infections. These results indicated that LF in combination with PA is a key virulence factor required for lethality. In addition, the parental strain producing all three proteins (PA, EF, and LF) was more lethal and caused more edema than the mutants producing the individual toxins. This indicated that the toxins are synergistic *in vivo*. In a later study, Pezard *et al.,* (1995) found that high titers of antibody to PA were observed after immunization of mice with all strains producing PA, while titers of antibodies to EF and LF were weak in animals immunized with strains producing only EF or LF. In contrast, immunization with strains producing either PA + EF or PA + LF resulted in increased antibody response to EF or LF respectively. The extent of antibody response to LF was found to be dependant on the presence of PA. The binding of LF to PA is therefore necessary for good antibody response to LF. These results confirmed the role of PA as a major protective antigen in the humoral response and also indicated a significant contribution of LF and EF to immunoprotection.

Cohen *et al.,* (2000) found that a single *s.c.* immunization of guinea pigs with 5 x 10^7 spores of a nontoxigenic and nonencapsulated strain of *B. anthracis* expressing rPA chromosomally induced high titers of neutralizing anti-PA antibodies. The immune response was long lasting (at least 12 months) and provided 100% protection against a lethal spore challenge (30LD$_{50}$) of the virulent

Vollum strain. In contrast, vaccination with a similar number of vegetative cells of the same strain resulted in 75% protection. The result clearly indicated a direct correlation between the level of expression of PA and the potency of the vaccine, and suggested that some *B. anthracis* spore-associated antigens may contribute significantly to protective immunity.

Pezard *et al.,* (1995) examined the antibody response to the *in vivo* production of PA, LF, and EF in Swiss mice immunized *s.c.* with spores (1×10^6 to 1×10^8) of *B. anthracis* strains producing these proteins. The parental Sterne 7702 strain along with single and double mutant strains were used. Single mutant strains produced EF-LF, PA-LF, and PA-EF. Only PA-LF was still lethal. Double mutant strains produced only PA, LF, and EF respectively. All PA-producing strains elicited high levels of specific antibodies against PA. In mice immunized with strains producing solely EF, LF, or EF plus LF, the immunological response to these proteins was weak with low specific antibody titers. In contrast, when LF or EF was produced by strains which also produced PA, a significant increase in the response against LFS and EF was observed. These results suggest that the increased immunogenicity of LF and EF originates from a protein complementation mechanism. Thus, the ability of PA to bind to cell surface receptors and facilitate the internalization and intracellular processing of EF or LF may be a prerequisite for the higher antibody response to these two proteins when bound to PA. The authors speculated that when EF and LF are produced by strains deficient in PA production, these two toxin components may not have the opportunity to interact with the appropriate effector cells.

Singh, Ivins, and Leppla (1998) described the immunological activity of purified PA, unable to bind LF or EF, derived from a recombinant strain of *B. subtilis*. The immunological efficacy of the purified mutant PA alone and in combination with LF and EF to protect against anthrax challenge was assessed. Both mutant and native PA preparations elicited high anti-PA titers in Hartly guinea pigs. Mutant PA, administered alone and in combination with LF and EF, completely protected the guinea pigs from *B. anthracis* spore challenge compared to nonimmunized controls.

Barnard and Friedlander (1999) assessed the efficacy of several live, recombinant anthrax vaccines given in a single-dose regimen with Hartley guinea pigs. These live vaccines were created by transforming ΔANR and ΔSterne, two nontoxigenic strains of *B. anthracis* with four different recombinant plasmids that expressed PA to various degrees. This allowed assessment of the effect of the chromosomal background of the strain, as well as the amount of PA produced, on protective efficacy of the immunizing infection. There were no significant strain-related effects on PA production *in vitro*, plasmid stability *in vivo*, survival of the immunizing strain in the host, or protective efficacy of the immunizing infection. The protective efficacy of the live, recombinant anthrax vaccine strains correlated with the ELISA determined anti-PA antibody titers they elicited *in vivo* and the level of PA they produced *in vitro*.

Brossier *et al.,* (1999a, 2000) presented genetic evidence that the functional domains on PA and the metalloprotease activity of LF and the adenylcyclase activity of EF are required for lethality and edema formation respectively. Brossier *et al.,* (2000) found that a non-encapsulated Sterne strain producing wild-type PA and mutationally inactive LF and EF was highly effective in the immunization of mice. This strain has been suggested as a possible candidate for live vaccination of humans.

Alani-Grinstein *et al.,* (2005) documented that an attenuated non toxigenic nonencapsulated *B. anthracis* oral spore vaccine expressed high levels of a rPA mutant as a protective antigen provided protection against a lethal *B. anthracis* challenge to guinea pigs. Protection was conferred when the vaccine was administered in the spore form, which unlike the vegetative cells, survived passage through the GI tract. A comparison of immunization of unirradiated spores with immunization of γ-irradiated spores demonstrated that germination and *de novo* synthesis of PA were prerequisites for an immune protective response.

Brossier, Mock, and Sirard (1999b) produced a genetically detoxified LT strain derived from the Sterne strain of *B. anthracis* that was devoid of lethal effects and was as protective as the parental Sterne strain against experimental anthrax. A mutated *lef* gene harbored on a suicide plasmid was introduced by conjugation into a recipient Sterne strain harboring a knockout mutation of the *lef* gene derived from a deletion. Crossover resulted in integration of the mutated gene into pXO1 of the recipient resulting in the strain RPL686. Strain RPL686 has a single base-pair mutation whereby histidine at the catalytic site of the LF is replaced by an alanine. The RPL686 mutant strain produced a full length stable LF protein devoid of lethality and was fully immunogenic and protective for guinea pigs challenged with the Pasteur II strain which is lethal for guinea pigs. An RPL686 Δ*cya* derivative of the RPL686 mutant strain devoid of the *cya* gene (encoding EF) was similarly protective. However, a single base-pair mutation can be expected to undergo reversion to the wild-type and the authors suggested that a double mutant would therefore ensure against such and be more suitable.

Whole Cell Nonviable Vaccines

Currently used acellular human vaccines against anthrax consist of PA. However, in experimental animals such vaccines are less protective than live attenuated strains. The administration of a vaccine consisting 10^8 formaldehyde-inactivated spores (FIS) of *B. anthracis* to PA (PA-FIS) was found to elicit total protection against subcutaneous challenge with spores of virulent *B. anthracis* strains in mice and guinea pigs (Brossier, Levy, and Mock, 2002). The inactivated spores carried point mutations within the catalytic regions of both LF and EF. Immunization with FIS alone was sufficient to protect mice partially, and guinea pigs totally. Guinea pigs receiving PA alone with aluminum hydroxide were poorly protected (22% survival). In contrast, inclusion of FI in the vaccine (1 x 10^7 spores or more) enhanced protection (100% survival) against lethal subcutaneous challenge. Immunization with spores alone, provided only partial protection of guinea pigs (25% survivors). With mice, PA alone protected 50% and FIS alone protected 33%. In contrast, 100% of mice immunized with PA plus FIS survived the subcutaneous lethal challenge. The combination of PA plus FIS therefore provided full protection under conditions where PA or FIS alone failed to do so. Interestingly, the PA-neutralizing activities of the two forms of the vaccine (PA and PA-FIS) were similar, suggesting that the higher level of protection conferred by the combination vaccine was not due to anti-PA antibodies alone but also to the contribution of antibodies directed against individual spore antigens. Kudva *et al.,* (2005) subsequently identified spore surface proteins for which AVA induced immune sera contained antibodies.

Protein Based Anthrax Vaccines

Vaccines Based on PA

PA is considered the most important antigen in protein based vaccines for anthrax in which adjuvants have a critical function. In the presence of adjuvant (aluminum potassium sulfate) also known as alum or alhydrogel, PA binds to the alhydrogel *via* electrostatic forces making it stable against proteolytic degradation. Cattle administered two sequential doses of PA using alum as adjuvant elicited similar protection as a single dose of a live spore vaccine (Jackson, Wright, and Armstrong, 1957). It is now recognized that PA must bind to LF to yield the lethal toxin (LT). The presence of antibody against PA can be considered to prevent the *in vivo* formation of the LT.

Several strains of *B. anthracis* have been reported to cause fatal infection in immunized guinea pigs (Auerbach and Wright, 1955; Ward *et al.*, 1965). Little and Knudsen (1986) immunized guinea pigs *i.m.* with either a PA vaccine or a live Sterne strain spore vaccine. Challenge doses consisted of 1,000 spores injected *i.m.* Previously reported results were confirmed with 9 of 27 challenge strains of *B. anthracis* being PA vaccine resistant. In contrast, guinea pigs immunized with the live Sterne strain spore vaccine were fully protected against these nine isolates. These results indicate that antibodies to toxin components may not be sufficient to provide protection against all strains of *B. anthracis* and that other antigens may play a role in active immunity. It follows that the efficacy of anthrax vaccines should be tested by using PA vaccine-resistant isolates to ensure protection against all possible challenge strains.

Turnbull *et al.,* (1986) used a competitive ELISA assay for detection of antibodies in human and guinea pig serum to PA and LF. Human vaccination schedule with an acellular vaccine induced predictable and lasting antibody titers to PA and LF when present in the vaccine. Live spore vaccines administered to guinea pigs in a single dose conferred superior protection than the human vaccines, although they elicited significantly lower anti-PA and anti-LF titers at the time of challenge. The authors concluded that substantial anti-PA and anti-LF titers may not, therefore, indicate solid protective immunity against anthrax infection. The reader should note that ELISA assays for determining the levels of anti-PA and anti-LF were subsequently found not to reflect a direct correlation with the level of immune protection compared to neutralization titers (Reuveny *et al.,* 2001; Weiss *et al.,* 2006, Reason *et al.,* 2009).

The current UK human anthrax vaccine has been in use since 1963. The vaccine consists of a protein precipitate from the supernatant fluid of cultures of the Sterne strain of *B. anthracis.* The major immunogen is PA. Baille, Fowler, and Turnbull (1999) compared the IgG antibody to PA with that from individuals infected with anthrax to that of healthy individuals immunized with the UK licensed anthrax vaccine. The predominant antibody subclass was IgG1. In addition, IgG3 was found in convalescent serum while vaccines produced IgG2, IgG3, and IgG4 subclasses. The significance of the IgG4 subclass is obscure. It has been proposed that the production of the IgG4 subclass antibodies is indicative of a Th2 immune response (Rodriquez *et al.,* 1996).

New Zealand white rabbits were immunized twice (four weeks apart) with varying doses of the licensed anthrax vaccine AVA by Pitt *et al.,* (1999). Rabbits were then challenged with a lethal dose of aerosolized spores of the Ames strain 10 weeks from day 0. Results indicated that both the amount of anti-PA IgG and toxin antibody titers present in sera at the time of peak antibody response were significant predictors of survival.

Oral vaccines are thought to be capable of stimulating the mucosal immune system to produce local IgA responses in addition to systemic responses so as to induce superior protection against mucosal infection. Zegers *et al.,* (1999) transformed a strain of *Lactobacillus casei* with a plasmid (pB-PA3.1) harboring the *pag* gene that encodes the protective antigen PA and the *Pldh* promoter of the lactic dehydrogenase gene of *L. casei.* The copy number of the plasmid was 30-50. The level of PA expression was highest at the mid-log phase of growth. PA expression decreased at the end of the log phase of growth and was hardly detected after 36 hr of broth incubation. Mice were immunized orally with 5 x 10^{10} or 1 x 10^9 live cells of the *L. casei* transformant or the *L. casei* wild type intranasally with 2 x 10^9 or 4 x 10^7 live cells. Oral or intranasal immunizations were administered on three consecutive days with a booster given at days 28, 29, and 30. A second booster was given at days 63, 64, and 65. In addition, one group of mice received *I.P.* a soluble fraction of a cell lysate made from the *L. casei* transformant B-PA3.1 at days 0 and 28 admixed with specol as adjuvant. This cell lysate produced high levels of IgG to PA. PA-specific antibody responses after oral and nasal immunization with the *L. casei* transformant were not detected. Mice immunized orally or intranasally produced lower levels of antibody responses to *L. casei* compared to the *i.p.* injection of the cell lysate.

Reuveny *et al.,* (2001) undertook studies to define markers for protective immunity to anthrax. Guinea pigs were immunized by different methods and the immune response and corresponding extent of protection against a lethal challenge with anthrax spores determined. Active immunization was performed by a single injection using one of two methods: (1) vaccination with increasing amounts of PA and (2) vaccination with constant amounts of PA that had been thermally inactivated for increasing periods. In both studies a direct correlation between survival and toxin neutralizing-antibody titer was found. Neutralizing-antibody titers were based on the ability to prevent PA/LF-induced mortality of J774A.1 cells exposed to PA + LF at final concentrations of 5 and 2 mg/ml respectively. In addition, in a complementary study involving passive transfer of PA hyperimmune sera to naive animals, a similar correlation between neutralizing-antibody titers and protection was found. In all three immunization studies, neutralization titers of at east 300 were sufficient to confer protection against an otherwise lethal $40LD_{50}$ dose of virulent anthrax spores of the Vollum strain. Such consistency in the correlation of protective immunity with anti-PA antibody titers was not observed for antibody titers determined with an ELISA assay. These results indicated that the neutralizing titer to PA can be used as a surrogate marker for protection.

Gaur *et al.,* (2002) investigated the immune response of mice following inoculation intranasally, subcutaneously, and through the skin on days 0, 15, and 28 with purified PA. Intranasal and subcutaneous immunization with PA resulted in high IgG ELISA titers. The predominant subclass in each group was IgG1. High titers of IgA were observed only in intranasally inoculated mice. In a cytotoxicity assay, these sera protected J774A.1 macrophage cells from lethal toxin challenge. These results suggest that non-invasive nasal immunization may be useful in improving vaccination strategies against anthrax.

The currently licensed human vaccine in the U.S. is produced by growing the V770-NPI-R strain of *B. anthracis* (pXO1$^+$ pXO2$^-$) in the presence of bicarbonate under microaerophilic conditions and adsorbing the filter sterilized culture supernatant to aluminum oxyhydroxide adjuvant. The protective component is PA. This vaccine strain is considered to have several limitations, including a sporogenic and fully toxic genotype. Production of vaccines from this strain results in lot-to-lot variability due to inconsistent PA production levels, inclusion of undefined PA proteolytic degradation products, and inclusion of other bacterial products such as LF and EF These lot-to-lot variables in turn influence the intensity of adverse events resulting from vaccination with such preparations. A significant body of work has therefore been recently developed involving the production of PA free of these contaminants.

To eliminate the undesirable and variable components of the vaccine an aviruluent non toxigenic strain designated ΔStern-1 having the phenotype (pX01$^-$ pXO2$^-$) was selected as a host for PA expression and a recombinant plasmid pPA102 encoding PA was electrotransformed into the host and subsequently an asporogenic variant was selected and designated ΔStern-1-1(pPA102)CR4 with a resulting phenotype of PA$^+$::Kan^r::LF$^-$::EAF$^-$::CAP$^-$.

PA in alhydrogel has been found to induce a high serum IgG titer in mice and the predominant antibody subclass is IgG1. When plasmid DNA encoding PA was used by Williamson *et al.,* (1999) to immunize mice, a low level of serum IgG was detected which was slightly increased by boosting with plasmid DNA. However, when mice immunized with plasmid DNA were later boosted with recombinant PA (rPA), a significant and rapid three-fold increase in titer occurred. In contrast, control animals not immunized with plasmid DNA and primed with rPA developed a 3-fold lower titer.

A genomic analysis of the *B. anthracis* virulence plasmid pXO1, aimed at identifying potential protein based vaccine candidates and virulence related genes was undertaken by Ariel *et al.,* (2002). Eleven putative proteins (See chapter 6, Table **1** for primers) were chosen as targets for functional genomic studies. DNA products of selected open reading frames (ORF's) were transcribed and directly translated *in vitro* and their immunogenetics were assessed on the basis of their anti-*B. anthracis* antisera. Among the eleven ORFs selected for analysis, nine were successfully expressed as full-length polypeptides and of these, three were found to be antigenic, and to have immunogenic potential, and to be similar to that of PA.

Marcus *et al.,* (2004) developed a guinea pig model in which a primary immunization with threshold levels of PA induced a long-term T-cell immunological memory response without inducing detectable anti-PA antibodies. A vaccination of primed animals with the same threshold PA levels was effective for memory activation, yielding a robust and rapid secondary response. Challenge with a lethal dose ($40LD_{50,}$ 2000 spores) of spores after the booster vaccinations indicated that the animals were not protected at days 2, 4, and 6 postboosting. Protection was achieved only from the eighth day postboosting, concomitant with the detection of protective levels of neutralizing antibody titers in the circulation. The authors concluded that the booster immunizations should not be used without concomitant antimicrobial treatment in postexposure situations.

Brossier *et al.,* (2004) developed two monoclonal antibodies (mAbs) against PA, designated 7.5 and 48.3, which were able to neutralize the activities of LT and ET *in vivo* and *in vitro*. mAb 7.5 was found to bind to domain 4 of PA and prevented the binding of PA to its cell receptor. mAb 48.3 was found to bind to domain 2 and blocked the cleavage of PA_{83} to PA_{63} and PA_{20}, thereby preventing LF and TF from binding to PA. The epitope recognized by this antibody was in a region involved in the oligomerization of PA_{63}; thus mAb 48.3 did not recognize the oligomer form. mAbs 7.5 and 48.3 neutralized the activities of LT and ET in mice. In addition, an additive effect was observed between the two mAbs against PA and a MAb against LF in protecting mice against a lethal challenge by the Sterne strain of *B. anthracis*.

In the adaptive immune response, antibodies are thought to represent the molecular link between recognition of a pathogen and its elimination through phagocytosis. In both cases, Fc receptors on phagocytic membranes recognize and bind to IgG. Mabry *et al.,* (2005) reported on the prevention of toxemia in guinea pigs from inhalation of *B. anthracis* spores by passive immunization. A very-high-affinity anti-PA antibody fragment (M18) lacking the Fc region conjugated to a 40-kda polyethylene gycol (PEG) polymer increased the serum half-life dramatically beyond that of full-length IgG. Administration of the conjugate by *s.c.* injection resulted in a significant increase in survival and an improved overall mean time to death (TTD) in the guinea pig model challenged with a relatively high dose of the Vollum 1B strain.

Williamson *et al.,* (2005) immunized rhesus macaques with *B. subtilis* expressed rPA bound to hydrogel and observed a significant increase in IgG response to th rPA compared to macaques receiving the existing licensed UK vaccine AVP. Immune macaque sera from all immunized groups contained toxin-neutralizing antibody and recognized all the domains of PA. Macaques immunized with rPA recognized both the N- and C-terminal domains of the PA molecule, while those immunized with the AVP and AVA vaccines recognized the N-terminal region more extensively than the C-terminal region. All immunized macaques (5 per group) challenged by a $305LD_{50}$ aerosolized spore dose of the Ames strain of *B. anthracis* survived. Purified IgG to PA passively transferred to naive A/J mice conferred protection against aerosol challenge with a 200 LD_{50} spore dose of the Ames strain of *B. anthracis* in a dose dependant manner.

Weiss *et al.,* (2006) undertook studies to determine the correlation between immunological parameters and protection against *B. anthracis* infected New Zealand white rabbits. Full immunization and partial immunization were achieved by single and multiple injections of standard and diluted doses of a PA-based vaccine. Immunized rabbits were challenged by intranasal spore instillation with Vollum and ATCC 6605 strains of *B. anthracis*. Total ELISA anti-PA antibody titers greater than 1×10^5 conferred protection, whereas lower titers (between 10^4 and 10^5) provided partial protection but failed to predict protection. Neutralizing antibody titers between 500 and 800 provided partial protection, while titers higher than 1000 conferred protection. The data indicated that regardless of the immunization regimen or the time of challenge, neutralizing antibody titers are better predictors of protection than total anti-PA titers obtained by an ELISA assay.

Reason *et al.,* (2009) established a large panel of human PA-specific monoclonal antibodies (mAbs) derived from multiple individuals vaccinated with the approved anthrax AVA vaccine BioThrax. Although these mAbs bound PA in standard ELISA binding assays, less than 25% were capable of neutralizing LT. Although neutralizing antibodies recognize determinants throughout the PA monomer, they were found o be significantly less common among those paratopes that bind to the immunodominant amino-terminal portion of the PA molecule. These findings demonstrated that PA binding alone is not sufficient to neutralize LT and suggests that for an antibody to effectively block PA-mediated toxicity, it must bind to PA in a manner that results in disruption of one of the toxin functions.

Mohamed *et al.,* (2004) described monoclonal antibodies (mABs) that function to increase the potency of LT against murine macrophage cell lines. The postulated mechanism involves binding of a mAB to PA on the macrophage cell surface to stabilize the PA by interaction of the mAB with macrophage Fcα receptors. This is thought to result in an increase in the amount of PA bound to the cell surface, which in turn leads to enhanced killing, most likely due to increased internalization of LF. Blocking of PA-receptor binding eliminated enhanced cell death by these unique mABs. The dual interaction of PA with its macrophage membrane receptor and these mABs may lead to stabilization of the PA on the membrane surface, thereby decreasing the rate of dissociation from its membrane receptor.

Antiserum raised against whole irradiated *B. anthracis* spores has been shown to result in anti-germination activity *in vitro* (Brossier, *et al.,* 2002), implying the presence of antigens on the spore surface. Spores of *B. anthracis* are shrouded with an exosporium that loosely envelopes the outer surface of the spore coat and consists of a latticework basal layer and a covering of hair-like projections (Gerhardt and Ribi (1964)). The hair-like appendages are constructed from the immunodominant spore glycoprotein Bc1A (Steichen *et al.,* 2003; Sylestre *et al.,* 2002) and represent available vaccine targets. On the basis of these observations Cybulski *et al.,* (2008) found eleven recombinant proteins of 30 derived from the exosporium to react with *B. anthracis* spore antibodies. The antigens were derived from the basal surface of the exosporium, below the Bc1A projections. A/J mice immunized with basal surface proteins p5303 or BxpB in combination with subprotection levels of PA exhibited enhanced protection against *s.c.* challenge, while neither anti-BpB or anti-p5303 antibodies reduced the rate of spore germination *in vivo*, both caused increased spore uptake and led to a higher rate of spore destruction by phagocytic cells.

Gorse *et al.,* (2006) reported on the safety and immunogenicity of a new rPA vaccine (rPA102) derived from *B. anthracis* in comparison to AVA. 100 volunteers (18 to 40 years old) were divided into five groups and received 5, 25, 50, and 75 µg of rPA102 in aluminum hydroxide *i.m.* at 0 and 4 weeks. Eighty-one volunteers received rPA102 among the four dose groups and 19 received AVA and all were blinded to the treatment assignment. The first vaccination with rPA102 (all doses) resulted in 36 (44.4%) local reactions and 40 (49.4%) systemic reactions (mostly headaches, compared to 18 (94.7%) and 3 (15.8%) respectively for AVA. The second vaccination with rPA102 (all doses) resulted in 28 (35.4%) local reactions and 23 (29.1%) systemic reactions compared to 16 (84.2%) and 4 (21.1%) respectively for AVA. The third vaccination with rPA102 (all doses) resulted in 34 (43.6%) local reactions and 18 (23.0%) systemic reactions. No severe reactions were reported after the first vaccination. After the second vaccination, severe pain-limited motion of the vaccination arm was reported by one AVA recipient, several recipients experienced headaches resulting from 5 µg of rPA102, and severe tiredness was reported by a 50 µg rPA102 recipient. After the third vaccination, redness graded as severe at the vaccination sight was reported by a different 50 µg rPA102 recipient. The only adverse event judged to be possibly related to vaccination with rPA102 was a single case of urticarial rash occurring 9 days after the first vaccination with rPA102 and further vaccination of his participant was discontinued. There was no evidence of an increasing trend in the frequency of local reactions with increasing rpA102 dose with the first vaccination. A higher proportion of the AVA group reported local reactions than any of the rPA102 groups. No comparison was made between AVA and rPA102 vaccines following the third vaccination since the AVA group received only 2 vaccinations, however after the third vaccination with the rPA102 vaccine a dose-related increase in the frequency of symptoms (from 29.4% to 61.9%) was noted (predominantly pain/tenderness). For those reporting pain/tenderness, the median duration of symptoms was 2 days for AVA and 1 day for rPA102 recipients. After the second vaccination, it increased to three days among AVA recipients and to 2 days among rPA102 recipients. After the third rPA102 vaccination the median duration of pain was 2 days. Headaches wee reported more frequently by rPA102 recipients (all dose levels taken collectively) than by AVA recipients following the first vaccination (5.5%, AVA; 33.3%, rPA102). The frequency of headaches was not significantly different between the two vaccine type groups following the second vaccination. Two weeks following the second vaccination, the mean titers for LT neutralization activity (TNA) for the 5, 25, 50 an 75 µg rPA102 and AVA groups were 38.6, 7.4, 373.9, 515.3, and 855.2 respectively. The geometric mean ELISA concentrations for anti-PA IgG were 3.7, 11.5, 25.9, 44.1, and 176.1 µg/ml respectively. Two weeks after the third vaccination, mean TNA values for the four rPA102 groups were 134.7, 719.7, 21,16.6, 2411.4, and for ELISA were 22.9, 104.7, 196.4, and 262.6 µg/ml respectively. The TNA response after two injections of the 75 µg dose of rPNA102 was similar to that after two injections of AVA.

Gwinn *et al.,* (2006) developed a scalable process to produce and purify multi-gram quantities of a highly purified, recombinant P (rPA) from *E. coli.* The rPA protein was produced in a 50-liter fermenter and was purified using anion-exchange, hydrophobic interactive, and hydroxyapatite chromatography to achieve >99% purification. The final yield of purified rPA resulted in ~2.7 g of rPA per Kg of cell paste (~270 mg/l) of highly pure, biologically active rPA protein.

Klas *et al.,* (2008) confirmed that intranasal dry powder anthrax formulations containing rPA are able to protect rabbits against aerosol challenge nine weeks after a single immunization. The optimum dose of rPA in dry powder was experimentally determined to be 150 µg. Rabbits received a single dose of either 150 µg rPA, 150 mg rPA + 150 µg of a 10 mer-peptide representing the *B. anthracis* capsule conjugated to BSA or 150 µg of conjugate alone. All dry powder vaccines contained monophosphoryl lipid A (MPL) a toll-like receptor-4 (TLR-4) agonist to enhance the induced immune response and chitosan. Significant anti-rPA titers and anthrax lethal toxin neutralizing antibody (TNA) levels occurred with both rPA containing vaccines, although rPA-specific IgG and TNA levels were reduced in rabbits immunized with rPA + conjugate. Nine weeks after immunization, rabbits were exposed to a mean aerosol challenge dose of $278LD_{50}$ of Ames spores. The rPA vaccine resulted in 8/10 rabbits surviving, while the rPA + conjugate resulted in 6/10 rabbits surviving. The conjugate alone resulted in 0/10 survivors. These observations demonstrated that a single immunization with dry powdered rPA was capable of protecting against a lethal aerosol spore challenge nine weeks after immunization.

Vaccines Based on Hemolysins

The term "anthrolysins" or "anthralysins" has been applied to hemolysins (ANLs) produced by *B. anthracis*. Anthrolysin O (ALO) is a cholesterol-dependant hemolysin that forms pores in the membranes of human and horse RBCs resulting in RBC lysis. Cowan *et al.*, (2007) constructed and characterized a novel genetic toxoid of ALO designated Δ6am.ALO which was able to bind to RBCs but was incapable of pore-formation and hemolysis. Immunization of mice with both active and non-hemolytic forms of ALO elicited protection against lethal *i.v.* challenge with hemolytic ALO in mice but neither was protective against *B. anthracis* spores in a murine *i.p.* challenge. Nonimmunized mice lethally challenged by *i.v.* injection into the tail vein expired within 1 min. Histological examination revealed tissue damage exclusively confined to the lungs. Many small to medium sized pulmonary vessels were fully occluded by thrombi which were absent in untreated control mice. ALO and other hemolysins produced by *B. anthracis* may contribute to the pulmonary venous congestion and the thrombocytopenia observed with patients suffering from inhalation anthrax (Jernigan *et al.*, 2001) although hematocrit values were reported to be normal indicating the absence of hemolysis *in vivo*.

Protection by Capsular Antigens

Kozel *et al.*, (2004) reported that a monoclonal antibody (mAb) reactive with the poly-γ-D-glutamic acid (γDPGA) capsular polypeptide of *B. anthracis* was protective against pulmonary anthrax in mice. Kozel *et al.*, (2007) subsequently examined a library of six mAbs generated from mice immunized with γDPGA. A notable diversity in reactions of the mAbs with encapsulated cells occurred. Most of the mAbs produced a rim type reaction at the capsular edge. Some of the mAbs produced a second capsular reaction well beneath the capsular edge, suggesting complexity in capsular structure. Binding of the mAbs to soluble γDPGA differed considerably in the complexity of binding curves using a fluorescence perturbation assay. mAbs producing rim type capsule reactions typically produced the more complex binding isotherms. One mAb was markedly less protective than the remaining mAbs in a murine model of pulmonary anthrax.

The capsule of *B. anthraci*s is a polymer of γ-D-glutamic acid, greatly enhances virulence and is a poor immunogen. When a synthetic peptide consisting of nine D-glutamic acid residues (γ-Dglu$_9$) linked at the γ-COOH groups was conjugated to keyhole limpet hemocyanin (KLH), mixed with Freunds complete adjuvant and injected subcutaneously into mice, antibodies to γ-Dglu$_9$ were produced (Wang *et al.*, 2004). Antibodies to γ-Dglu$_9$ bound to the surface of encapsulated *B. anthracis cells* and mediated opsonophagocytosis by murine macrophages. The authors concluded that inclusion of an immunogenic capsular component as well as PA in new anthrax vaccines would generate immune responses to both the bacteremic and toxigenic aspects of anthrax infection and thereby increase protective efficacy.

The capsule of vegetative cells of *B. anthracis* is an essential virulence factor composed solely of poly-γ-D-glutamic acid (γDPGA) which by itself is a poor immunogen. Kubler-Kielb *et al.*, (2006) undertook studies to identify the optimal construct for clinical use of synthetic γDPGA of different lengths bound to carrier proteins of different densities. The advantages of the synthetic over the natural polypeptide are the homogeneous chain length and end groups, allowing conjugates to be accurately characterized and standardized and their chemical composition to be related to their immunogenicities. The carrier proteins employed in conjugates included bovine serum albumen, recombinant *Pseudomonas aeruginosa* exotoxin A, rPA from *B. anthracis*, and tetanus toxoid. All conjugates were immunogenic. The γDPGA chain length of 10 do 15 amino acids and the average density of 15 mol γDPGA per mol of carrier protein were found to be optimum.

Non Toxin Immunogens of B. anthracis

Gat *et al.*, (2006) identified proteins from the Vollum strain of *B. anthracis* (pxO1$^+$ pXO2$^+$) able to evoke an immune response *via* a reductive genomic-serological screen of open reading frames (ORFs). The screen included *in vitro* expression of the selected ORFs by coupled transcription and translation of linear PCR-generated DNA fragments, followed by immunoprecipitation with antisera from *B. anthracis*-injected rabbits. Among 197 selected ORFs, 161 were chromosomal and 36 were on plasmids pXO1 and pXO2, and 138 ORFs had putative functional annotations (known ORFs) and 59 had no assigned functions (unknown ORFs). A total of 129 (93%) of the known ORFs could be expressed, whereas only 38 (64%) of the unknown ORFs were successfully expressed. All 167 expressed polypeptides were subjected to immunoprecipitation with anti-*B. anthracis* antisera, which revealed 52 seroactive immunogens, only one of which was encoded by an unknown ORF. Experimental findings suggested that surface-anchored proteins and adhesins or transporters, such as cell wall hydrolases, proteins involved in iron acquisition, and amino acid and oligopeptide transporters, have great potential to be immunogenic. More than 30 novel *B. anthracis* immunoreactive virulence-related proteins were listed which could be useful in diagnosis, pathogenesis studies, and future anthrax vaccine development.

Bacillus collagen-like protein of *B. anthracis* (Bcla) is the immunodominant glycoprotein on the exosporium of *B. anthracis* spores. Brahmbhatt *et al.,* (2007a) assessed the impact of BclA on spore germination *in vitro* and *in vivo*, surface charge, and interaction with host matrix proteins by constructing a *bclA* null mutant in the Sterne 34F2 strain of *B. anthracis*. The growth and sporulation rates of the Δ*bcla* and parent strains were nearly indistinguishable, but germination of mutant spores occurred more rapidly than that of wild-type spores *in vitro* and was more complete by 60 min. In addition, the mean time to death of A/J mice inoculated *s.c.* or intranasally with mutant spores was lower than that for the wild-type spores, even though the LD_{50} dose of the two strains was similar. The Δ*bcla* spores were more markedly less water repellant than wild-type spores, and presumably as a consequence, the extracellular matrix proteins laminin and fibronectin bound significantly better to mutant than to wild-type spores. These results suggested that BclA acts as a shield to not only reduce the ease with which spores germinate but to also change the surface properties of the spore, which in turn may impede the interaction of the spore with host matrix substances.

Brahmbhatt *et al.,* (2007b) hypothesized that antibodies to the *B. anthracis* spore surface protein antigen BclA are largely responsible for the augmented immunity to anthrax of animals vaccinated with inactivated spores and PA, compared to vaccination with PA alone. To test this theory rBc1A with adjuvant was injected *i.p.* into A/J mice on day zero and again *s.c.* on day 15 using 20 mice per group. Challenge with a $10LD_{50}$ dose of spores of the Sterne strain resulted in anti-rBclA antibodies but failed to protect the mice. Similar results were obtained with rabbits. However, all 20 mice that received suboptimal amounts of recombinant PA followed by administration of rBc1A two weeks later survived spore challenge. Additionally, anti-rBclA IgG, compared to anti-PA IgG, promoted a sevenfold-greater uptake of opsonized spores by mouse macrophages and markedly decreased intramacrophage spore germination.

Sensitivity and Specificity of Polyclonal and Monoclonal Antibodies Raised Against Spores

The sensitivity and specificity of polyclonal and monoclonal antibodies raised against intact *B. anthracis* spore preparations was assessed by Longchamp and Leighton (1999) using Western blotting. Secondary antibody conjugated to horseradish peroxidase was used for visualization and detection. Polyclonal antibodies were produced in rabbits by a regimen of intact spore immunizations using disrupted spore preparations followed by vegetative cell immunization with a mixture of *B. anthracis* strains. Mouse monoclonal antibodies were produced using presumably disrupted spore preparations derived from the Vollum strain. None of the antibodies studied were completely specific in recognizing the *B. anthracis* spore surface. A polyclonal serum recognized a wide range of spore surface epitopes and demonstrated limited cross-reaction with the spore surface of *B. cereus.* Two monoclonal antibodies demonstrated more extensive cross-reaction with *B. globigii* and *B. subtilis.* These monoclonal antibodies did not react with spore surface epitopes but did react with vegetative cell epitopes of all four *Bacillus* species studied. The reader is reminded that in screening for hybridoma cell lines producing monoclonal antibodies, usually only the predominant monoclonal antibodies directed against the most abundant antigenic epitopes will be detected. The authors suggested that spore preparations, or isolated spore surface antigens, of higher purity may be required to develop monoclonal antibodies that recognize the *B. anthracis* spore surface epitopes specifically.

Surrogate Hosts for PA Production

Bacterial Vectors Encoding PA

When the PA from *B. anthracis* was cloned into *B. subtilis* strong protection was elicited with guinea pigs and mice against virulent Ames strains with 95% to 100% of the former an 70% of the latter surviving (Ivins *et al.*, 1990a,b; Welkos *et al.*, 1990).

Iacono-Connors *et al.,* (1991) found that the PA expressed by a WR recombinant vaccinia virus (WR-PA) elicited a partial protective effect against a lethal *B. anthracis* intramuscular spore challenge in guinea pigs. The WR-PA recombinant protected 60% of male mice and 50% of guinea pigs. The Connaught strain of the vaccinia virus PA recombinant (Con-PA) failed to protect mice and guinea pigs. PA purified from baculovirus recombinant tissue cultures plus adjuvant partially protected male mice and completely protected female guinea pigs from challenge. Sterne-PA, with and without adjuvant failed to significantly protect male mice from challenge and resulted in only 27.3 and 8.3% survival, respectively. However, animals immunized with Sterne-PA and a TriMix Ribi adjuvant had high anti-PA titers. High anti-PA titers do not necessarily provide complete protection from challenge. Stern-PA combined with the TrisMix Ribi adjuvant did provide complete protection for guinea pigs against a lethal Ames spore anthrax challenge in female guinea pigs. The results indicated a significant level of efficacy for recombinant PA-viruses in protecting against anthrax which in certain cases were equal or superior to that of the Stern-PA vaccine.

Gupt, Waheed, and Bhatnagar (1999) expressed the structural gene for PA as a fusion protein with 6X histidine residues in *Escherichia coli* to facilitate affinity purification in the absence of LT and ET. Expression of PA in *E. coli* under the transcriptional regulation of the T5 promoter yielded intracellular inclusion bodies consisting of insoluble protein aggregates. The inclusion bodies were solubilized in 6 M guanidine-HCl and the protein purified under denaturing conditions (8-M urea) using nickel nitrilotriacetic acid (Ni-NTA) affinity chromatography. The denatured protein was renatured by gradual removal of the urea on the NI-NTA column. The PA protein was then purified using a Mono-Q column with Fast Protein Liquid chromatography (FPLC). The yield of the recombinant PA (rPA) was 2 mg/L of culture with a 3,023-fold level of purification. The rPA plus LF was able to lyse macrophages and exhibited biological activity comparable to that of PA from *B. anthracis*.

Baillie, Moore and McBride (1998) cloned the PA operon into a heat induced prophage of the *B. subtilis* φ 105 phage. Lysogenic cells of *B. subtilis* when grown at 36 °C and then shifted to 50 °C for 5 min. and then returned to 37 °C underwent prophage induction. The level of expression of the recombinant PA (rPA) was 1.65 mg/L of PA. Only two of eight female guinea pigs vaccinated with the resulting rPA preparation survived a lethal challenge with *B. anthracis*. This was presumably due to partial proteolytic degradation of the rPA.

PA produced by a recombinant *B. subtilis* strain and its efficacy when combined with Ribi adjuvant was found to perform equally well as native PA as an immunogen for subsequent aerosolized spores of the Ames strain administered to guinea pigs (Fowler *et al.*, 1999). Similar levels of IgG to PA were produced in guinea pigs by both the recombinant and native forms of PA.

The UK anthrax vaccine was first produced in the 1950's and is essentially an alum-precipitated cell free filtrate of the non-encapsulated Sterne 34F2 strain of *B. anthracis*. Studies using the Vollum challenge strain of *B. anthraci*s have shown that the UK vaccine protected rhesus monkeys from airborne challenge (Henderson, Peacock, and Belton, 1956). Similar studies with guinea pigs have also shown that the vaccine confers good protection against this challenge strain (Broster and Hibbs, 1990). However, protection is less effective in experimental animals challenged with an aerosol of spores from other strains of *B. anthracis* (Auerbach and Wright, 1955). Anti-PA titers alone do not allow the prediction as to whether or not protection will occur, which has led to speculation that cell-mediated responses may also be involved. Since PA administration alone induces poor protection, the use of adjuvants is recognized as essential for achieving satisfactory immunological protection.

Gupta, Waheed, and Bhatnagar (1999) expressed the structural gene for PA as a fusion protein with 6X histidine residues in *Escherichia coli* to facilitate affinity purification of PA in the absence of LT and ET. Expression of PA in *E. coli* under the transcriptional regulation of the T5 promoter yielded an insoluble protein aggregate of inclusion bodies. The inclusion bodies were solubilized in 6 M guanidine-HCl and the protein purified under denaturing conditions using nitrilotriacetic acid (Ni-NTA) affinity chromatograph. The denatured protein was renatured by gradual removal of the denaturant (8 M urea) on the Ni-NTA column. The PA protein was then purified using a Mono-Q column with fast protein liquid chromatography (FPLC). The yield of the recombinant PA (rPA) was 2 mg/L of culture broth with a 3023-fold-level of purification. Antisera to native PA recognized the rPA. The rPA plus LF was able to lyse macrophages and exhibited biological activity comparable to that of PA from *B. anthracis.*

Chauhan *et al.,* (2001) reported on the constitutive expression of the PA gene in *Escherichia coli* strain DH5α where the PA gene was under the control of the phage T5 promoter. PA was constitutively expressed without degradation and no plasmid instability was observed. The resulting PA was purified by metal-chelate affinity chromatography and with batch culture scale-up was found to be produced at a harvested level of 125 mg/L. Cells of recombinant *E. coli* DH5α yielded an insoluble protein aggregate that formed intracellular inclusion bodies of 83 kDa corresponding to native PA_{83}. The growth curve of the *E. coli* DH5α recombinant strain indicated that the maximum yield of PA occurred at the end of exponential growth and was stable throughout the stationary growth phase.

Garmory *et al.,* (2003) cloned the PA gene into an auxotrophic non infectious mutant of *Salmonella* Typhimurium as a fusion with the signal sequence of the hemolysin (*hlyA*) gene of *Escherichia coli* to allow export of PA *via* the Hly export system. To stabilize the export cassette, it was integrated into the chromosome of the live *Salmonella* carrier. When this carrier culture of *S.* Typhimurium was given *i.v.* to A/J mice, they developed high levels of antibody to PA and were protected against *i.p.* challenge with a $1000LD_{50}$ spore dose of *B. anthracis* strain STI (Tox^+ Cap^-). In contrast, oral administration of the carrier culture induced only low levels of antibodies to PA and failed to protect the mice against challenge.

Galen *et al.,* (2004) fused the *clyA* gene sequence within the chromosome of *Salmonella* Typhi to the domain 4 (D4) moiety of *B. anthracis* PA. The *clyA* gene encodes a cryptic hemolysin ClyA which also constitutes an endogenous export system. This fusion was used to facilitate synthesis and export of D4 by *S.* Typhi. A high copy number plasmid was used to carry the gene fusion. Among 15 mice immunized intranasally with *S.* Typhi carrying the plasmid and exporting the protein fusion, 11 manifested a four-fold or greater increase in serum anti-PA IgG compared with only 1 of 16 mice immunized with the live vector expressing only cytoplasmic D4.

Laird *et al.,* (2004) expressed soluble recombinant PA (rPA) in relatively high amounts in the periplasm of *E. coli* harvested from shake flasks and bioreactors. The rPA protein was purified 96 to 98-fold using Q-Sepharose-HP and hydroxyapatite column chromatography. Yields of 370 mg of purified PA per liter of culture medium were obtained.

Viral Vectors Encoding PA

Adenovirus vectors are promising for use in vaccinating against potential infectious agents such as *B. anthraci*s. McConnel *et al.,* (2006) described the construction and evaluation of an adenovirus vaccine expressing domain 4 (Ad.D4) of *B. anthracis* PA. Ad.D4 elicited antibodies to PA 14 days after a single *i.m.* injection, which boosting further increased. In addition, two doses of Ad.D4 four weeks apart were sufficient to protect 67% of mice from an otherwise lethal *i.v.* LT challenge. Additionally, the release of inflammatory cytokines from vaccinated mice after LT challenge was characterized. Interleukin-1b (IL-1b) levels in mice that survived LT challenge were similar to levels in nonsurvivors while IL-6 levels were higher in survivors than nonsurvivors. These findings suggested that LT-mediated death may not be a direct result of inflammatory-cytokine release.

Yin *et al.,* (2008) utilized the truncated hepatitis B virus core (HBc) protein (aa1-144) as a carrier of a 24 amino acid sequence of the immunogenic $2\beta_2$-$2\beta_3$ loop of the PA domain. The recombinant protein HBc-N144-PA-loop2 was expressed in *Escherichia coli* and was able to form HBc-like particles confirmed by electron microscopy. In mice the chimeric protein was able to induce PA specific antibodies. In guinea pigs it was able to induce PA specific antibodies regardless of whether alum was used as an adjuvant or not and was able to partially protect guinea pigs against a $40LD_{50}$ (~30,000 spores) virulent subcutaneous anthrax spore challenge resulting in 57% survival), compared to 12.5% survival with HBc-N144 plus adjuvant as vaccine. Interestingly the use of rPA plus adjuvant resulted in 66.6% survival. The level of LT neutralizing titer was 40-100 times less with the chimeric protein (HBc-N144-PA-loop2) than that obtained with the use of rPA as vaccine. This difference was presumably due to the fact that PA is an 83 kDa protein with many more immunogenic domains or epitopes than the 24 amino acid epitope of PA comprising the chimeric protein.

Recombinant cowpea mosaic virus (rCPMV) particles displaying foreign peptide antigens on the particle surface have been found suitable for development of peptide-based vaccines. However, the purified particles can be expected to remain infectious for plants. Phelps, Dang, and Rasochova (2007) developed rCPMV particles that displayed on their surface a 25 amino acid peptide derived from PA. Intact rCPMV particles were separated from contaminating cleaved forms by anion exchange chromatography. Recovery of inactivated rCPMV particles was achieved by blending 40g of frozen leaf issue with 150 ml of cold 30 mM Tris, 0.5 M ammonium sulfate 0.2 mM phenylmethanesulfonyl fluoride, pH 9.0 and the clarified extract allowed to inactivate for 20 hr. at 22 °C. Cold 20% PEG6000 in 1 M NaCl was added to a concentration of 5% PEG6000 and 0.25 M NaCl. After 1 hr. at 4 °C the preparation was centrifuged at 15,000 x g for 30 min, diluted with ultrapure water, passed through a 0.45 micron porosity membrane filter and then purified with an FPLC system. The yield of purified noninfectious rCPMV was 0.3 g/kg of leaf tissue.

Watson *et al.,* (2004) cloned the *pag* gene into a chloroplast vector with the psbA regulatory signals to enhance translation. Crude plant extracts of *Nicotinia tabacum* cv. Petit Havana tobacco leaves contained up to 2.5 mg of full length 83 kDa PA/g of fresh leaf tissue. Chloroplast-derived PA when combined with LF lysed murine macrophages. Calculations indicated that 400 million doses of vaccine (free of contaminants) could be produced per acre, which could presumably be further enhanced 18-fold using a commercial cultivar in the field.

Chichester *et al.,* (2007) successfully expressed domain 4 of PA (PAD4) and domain 1 of LF (LFD1) as fusions to lickenase (LicKM), a thermostable enzyme form

Clostridium thermocelum, and transiently expressed these fusion forms in *Nicotinia benthamiana*. Plant produced antigens were combined and used to vaccinate mice *i.p*. All animals that received the combined LicKM-PAD4 and LicKM-LFD1 vaccine developed high antibody titers that were predominantly IgG1 and were able to neutralize the effects of LT *in vitro* on mouse macrophages.

Based on the speculation that intact PA83 would be difficult to isolate due to endogenous yeast protease activity, recombinant PA63 (rPA63) was produced by Hepler *et al.,* (2006) in a protease and glycosylation-deficient strain of *Saccharomyces cerevisiae* enhanced for over-expression of the *Saccharomyces* GA44 transcription factor. Satisfactory intracellular production of rPA63 was not achieved with the use of the GAL10 *Saccharomyces* secretion promoter. Highly purified rPA63 isolated from *S. cerevisiae* under urea denaturing conditions (to reduce multimerization) when combined with LF demonstrated reduced lethality for macrophages compared to non-denatured rPA83 purified from *Escherichia coli*. However, rabbits and rhesus monkeys immunized with rPA63 adsorped onto a commercial aluminum adjuvant mounted a vigorous response, producing IgG that was cross-reactive to rPA83. Vaccination of rabbits and monkeys with the rPA63 vaccine provided protection from a lethal inhalation dose of virulent *B. anthracis* spores that compared favorably with the human vaccine Biothrax™ with both animal species.

To further the development of an orally delivered human vaccine for mass vaccination against anthrax, *Salmonella* Typhimurium was engineered by Stokles *et al*., (2007) to express full-length PA and domains 1 and 4, or domain 4 (which are known protective domains of PA) using codon optimized PA DNA fused to *S. Typhimurium* ClyA (cytolysin export system) and under control of the osmotically controlled promoter *ompc*. This clyA export apparatus requires only the fusion of proteins to the 34-kD clyA protein for the secretion of proteins from *Salmonella*. The immunogenicity of PA1-ClyA and PA4-ClyA recombinants was assessed after intragastric dosing three times at 2-weekintervals with ~10^9 CFU of *S. Typhi* (aroaA) an attenuated aromatic amino acid-dependant strain containing a new expression plasmid. After only one or two doses of the aroa *S. Typhi* recombinant mutant, specific responses were generally low. However, on day 73, all of the mice immunized with *S. typhi* expressed PA or PA fragments 1 and 4 and developed IgG responses as detected by ELISA. Immunized mice were aerosol challenged with 10^5 CFU (200 LD_{50}) of *B. anthracis* spores derived from a $pXO1^+$ $pXO2^-$ strain to which they were susceptible. Control mice succumbed within six days. Two of eight mice inoculated with a recombinant *Salmonella* PA1+4 strain survived challenge, and five of six mice inoculated with PA1-4 survived. Thus the recombinant strain of *S. Typhi* expressing the full length PA protein as a fusion with ClyA afforded significant protection against aerosolized spore challenge. Mice immunized with recombinant PA1-4, PA1+4, or PA4 were protected against challenge, while immunization of mice with PA1 protected only two of six mice from challenge. The overall result indicated that maximum protection was obtained when the full-length PA protein (PA1-4) was used in the *S. Typhi* recombinant vaccine.

Luxembourge *et al.,* (2008) found that intramuscular (*i.m.*) delivery of a eucaryotic plasmid encoding PA using electroporation (EP) rapidly induced anti-PA IgG and toxin neutralizing antibodies within two week following a single immunization in mice, rats, and rabbits. Anti-PA IgG responses were increased by one order of magnitude in mice and by one to two orders of magnitude in rats and rabbits compared to conventional *i.m.* injection. In contrast, conventional *i.m.* injection of the PA encoding plasmid did not induce a detectable neutralizing antibody response with most of the animals.

Lui *et al.,* (2008) identified a spore coat associated protein (SCAP) in *B. anthracis*. The entire amino acid sequence of SCAP of *B. anthracis* has been found to share 99.5% identity with camelysin, a casein-cleaving metalloprotease of *B. cereus*. An *Escherichia coli* vector-bound vaccine system was used to determine the immunogenicity of SCAP. ICR mice generated detectable SCAP antibodies three weeks after intranasal immunization with intact particles of U.V.-irradiated *E. coli* vector overproducing SCAP. The adjuvant effect of a U.V.-irradiated *E. coli* vector eliminated the necessity of boosting and the use of other immunomodulators. *E. coli* components such as LPS and CPG have been commonly used as highly immunogenic adjuvants and may be responsible for the enhanced immunogenicity of the SCAP.

rPA-based vaccines have been shown to induce high-titer anti-PA responses and can protect rabbits and nonhuman primates against lethal *B. anthracis* challenge

(Ivins *et al.*, 1998; Little *et al.*, 2006). However, in some studies, protection decreased dramatically over six to12 months (Little *et al.*, 2006). Another major drawback to rPA-based vaccines is that they stimulate immunity to only a single *B. anthracis* antigen, PA.

Skoble *et al.*, (2009) therefore developed a multivalent anthrax vaccine that potently stimulated a broad immune response to rPA and other anthrax antigens. The vaccine was based on a psoralen-killed but metabolically active (KBMA) bacteria that combined the safety of a killed vaccine with the potency of a live vaccine. Psoralens form covalent monoadducts and cross-links with pyrimidine bases of DNA and RNA following exposure to long-wave-length UV light. The primary mechanism by which bacteria repair psoralen-induced DNA damage is through nucleotide excision repair (NER) which is initiated by the ABC excision repair complex a product of the UV light response (uvrA, uvrB, and uvrC) genes. Bacteria defective for NER are incapable of repairing cross-links in their DNA and are extremely sensitive to photochemical inactivation by combined treatment with psoralen and UV light (Brockstedt *et al.*, 2005; Sancar and Sancar (1988)). A series of vaccine candidate strains was constructed with sequential mutations in the *B. anthracis* Sterne 7702 strain chromosome and on the pXO1 plasmid, resulting in a strain that is nonsporogenic, NER deficient, and nontoxigenic but still antigenic for PA, LF, and EF that can be produced as a KBMA vaccine. The resulting KBMA *B. anthracis* vaccine provided protective humoral immunity against lethal *B. anthracis* challenge in mice, rabbits, and guinea pigs. Vaccination was performed with mice *i.m.*, *s.c.*, and *i.v.* Rabbits and guinea pigs were vaccinated *i.m.* Mice were challenged *s.c.*, rabbits and mice were challenged intratracheally and guinea pigs were challenged intranasally. Mice vaccinated with KBMA Sterne produced the highest levels of PA-neutralizing antibody when vaccinated *i.m.* Vaccination with KBMA Sterne induced less inflammation in muscle than adjuvanted rPA and fully protected mice against challenge with lethal *s.c.* doses of toxigenic unencapsulated Sterne 7702 spores and rabbits against challenge with lethal pneumonic doses of fully virulent Ames strain spores. Guinea pigs vaccinated with KBMA were partially protected against intranasal lethal Ames spore challenge with 50% of the animals surviving compared to 33% surviving after vaccination with AVA. In addition, ELISA assays detected antibody titers against not only PA, but also against LF, and EF, whereas vaccination with rPA does not elicit antibodies to the latter two.

Venezuelan equine encephalitis (VEE) virus has been used as a vaccine vector for a variety of immunogens. The VEE vaccine vector is composed of a self-replicating RNA (replicon) containing all of the VEE virus nonstructural genes and a multiple-cloning site in place of the VEE structural genes. Several different anthrax vaccines were constructed by Lee, Hadjipanayis, and Welks (2003) by cloning the PA gene (*pag*) from *B. anthracis* into the VEE vaccine vector. The anthrax vaccines were produced by assembling the vectors into propagation-deficient CEE replicon particles *in vitro*. The Sterne strain of *B. anthracis* (pXO1$^+$ pXO2) is lethal for A/J mice even though no capsule is produced. A/J mice inoculated *s.c.* with three doses of the 83-kda PA VEE viral vaccine were completely protected from *s.c.* challenge with heat shocked spores of the Stern strain.

Expression of PA in plants through chloroplast transformation has several advantages over bacterial expression systems. Foreign proteins have been expressed at extraordinarily high levels in transgenic chloroplasts due to the presence of ~10,000 copies of chloroplast genomes per cell. Koya *et al.*, (2005) successfully expressed PA in transgenic on *Nicotinia tabacum* var. *petite Havana* leaves. Affinity chromatography was used to partially purify the PA. Mice immunized with the resulting chloroplast-derived PA and with *B. anthracis*-derived fully purified PA, adsorbed to alhdrogel adjuvant, yielded comparable IgG immune titers of ~1:3000,000. Mice immunized with chloroplast-derived PA and challenged with a 100% lethal dose of LT (LF +PA) survived. Major advantages of plant-derived recombinant PA consists of the absence of EF and LF and ease of production, without the need for expensive fermenters. The authors estimated that even with a 50% loss during purification, one acre of transgenic plants can produce 360 million doses of functional anthrax vaccine.

Eucaryotic Vectors Encoding PA

A number of studies have indicated that eucaryotic vectors harboring DNA encoding PA$_{83}$ or PA$_{63}$ when injected into laboratory animals results in antibody responses to PA and protection against challenge with otherwise lethal doses of *B. anthracis* spores. By using eucaryotic expression vectors, the encoded antigens are fully expressed resulting in *in vivo* mammalian cell production of PA_{83} or PA_{63} and the humoral antibody systems of the vaccinated animals respond with antibody production against PA.

A eucaryotic plasmid vector encoding the immunogenic and biologically active portion of PA (PA_{63}) was constructed by Gu *et al.,* (1999) and injected intramuscularly into BALB/c mice. Spleen cells of the injected mice were stimulated to secrete interferon-γ and interleukin-4 when exposed to PA *in vitro*. Spleen cells from control unvaccinated mice showed no such response. Immunized mice also mounted a humoral immune response dominated by IgG1 anti-PA antibody production, the subclass of IgG known to confer protection against *B. anthracis* LT. Mice immunized three times with the PA DNA vaccine were protected against lethal challenge with a combination of PA plus LF (LT).

The ability of DNA vaccination to protect against a lethal challenge of anthrax toxin was evaluated by Price *et al.,* (2001). BALB/c mice were immunized *via* gene gun inoculation with eucaryotic expression vector plasmids encoding either a fragment of PA (PA_{63}) or a fragment of LF. Plasmid pCLF4 harbored the DNA encoding the N-terminal region (amino acids 10 to 254) of *B. anthracis* LF cloned into the pCI eucaryotic expression vector. Plasmid pCPA harbored the DNA encoding the biologically active portion of PA (PA_{63}) cloned into the pCI expression vector. Injection was three times at two-week intervals. Titers of antibodies to both PA and LF from mice immunized with the combination of pCPA and pCLF4 were four to five-times greater than titers from mice immunized with either gene alone. All mice immunized with pCLF4, pCPA, or the combination of both survived a lethal intravenous challenge with LT (PA plus LF) whereas all unimmunized mice did not survive. These results demonstrated that DNA-based immunization alone can provide protection against a lethal toxin challenge and that DNA immunization against the LF antigen alone provides complete protection against LT.

Reimenschneider *et al.,* (2003) injected rabbits with eucaryotic plasmid expression vectors harboring the DNA encoding *B. anthracis* PA. Groups of 10 rabbits were injected three times at four-week-intervals by gene gun inoculation. All rabbits vaccinated with the DNA vaccine developed antibody responses. After the fourth vaccination, rabbits were challenged with subcutaneous injection of a 100LD$_{50}$ dose of heat activated spores of the Ames strain of *B. anthracis*. All rabbits that received the control plasmid vaccine (no PA gene insert) died within four days of challenge while 9/10 rabbits given the PA DNA vaccine and 7/10 given the human AVA vaccine survived.

Hahn *et al.,* (2004) evaluated the efficacy of genetic vaccination with plasmid vectors encoding PA, in protecting mice from a lethal challenge with spores of *B. anthracis*. Mice were immunized *via* gene gun inoculation. Three plasmid vectors were used. The vector pSecTag P83, encoded the full-length PA protein PA$_{83}$ (83kDa) and contained a signal sequence for secretion of expressed protein. The plasmids pCMV/ER PA83 and pCM/ER PA63, encoded the full-length form of PA$_{83}$ (83 kDa) and the physiologically active form PA$_{63}$ (63kDa) respectively. All three plasmids induced PA-specific humoral immune responses consisting predominantly of IgG1 antibodies in mice. Spleen cells collected from plasmid-vaccinated mice produced PA-specific interleukin-4,interleukin-5, and interferon-γ *in vitro*. Vaccination with either pSecTag PA83 or pCMV/ER PA83 resulted in significant protection of all A/J mice. The plasmid pCMV/ER PA63 was not used for this purpose as it was found to yield lower antibody titers than the two PA83 harboring plasmids.

Herrmann *et al.,* (2006) injected rabbits with several eucaryotic expression plasmids harboring DNA sequences encoding PA. The first PA gene insert, WT-PA, encoded the full-length PA (amino acids 1-764) using the original PA signal peptide as the leader sequence. The second PA gene insert, tPA-PA, included the mature 735 amino acid version of PA (amino acids 30-764) but the first 29 amino acids of PA signal peptide were replaced with the human tissue plasminogen activator (tPA) leader sequence. The third gene insert, tPA-PA$_{63}$, coded for the 63kDa biologically active form of PA (amino acids 175-764). The PA inserts were cloned into the eucaryotic DNA immunization vector PAS801 which uses a CMC IE promoter. The highest IgG titer was obtained with the tPA-PA vector. Antisera from DNA immunized rabbits were able to protect macrophages from anthrax LT and from the toxin released by intracellularly germinated spores of *B. anthracis* (Sterne strain). Passive protection of mice against pulmonary challenge to a $50LD_{50}$ dose of *B. anthracis* spores (Sterne strain) was achieved when rabbit antiserum was administered 1 hr before or 1 hr after challenge.

Unlike some other mammals (including humans), mice are quit susceptible to infection with encapsulated, non-toxigenic strains of *B. anthracis* (Welkos *et al.*, 1993). The major virulence factor in mice appears to be the polyglutamic acid capsule (Brossier *et al.*, 2002). As a consequence, vaccination of mice with PA-based human vaccines often fails to result in protection from a challenge with fully virulent *B. anthracis* spores (Ivins *et al.*, 1990a, 1992). A/J mice are deficient in complement factor C5, which increases their susceptibility to toxigenic strains of *B. anthracis* and therefore the A/J strain used in this study by Herrmann *et al.*, (2006) was considered suitable for challenge studies by lethal anthrax spores. No correlation was found between PA-specific Ig, IgG1, or IgG2 and levels of protection from anthrax infection. The authors speculated that for the mouse strain used, only relatively low neutralizing antibody titers may be required for survival.

Rhie *et al.*, (2003) conjugated PA to the capsular poly-α-D-glutamate (PGA) yielding PA-PGA to elicit the production of antibodies specific for both bacilli and toxins. In a subsequent study Aulinger *et al.*, (2005) reported on the efficacy of a vaccine comprised of the capsular PGA conjugated to a dominant-negative inhibitory mutant of PA (DNI-PGA). DNI is a translocation-deficient mutant of PA with two mutations (Sellman, Nassi, and Collier, 2001) that co-oligomerized with wild-type PA and potently blocked the translocation processor so as to inhibit toxin action. DNI was able to assemble with PA molecules into heptamers that could still bind LF and EF. However, chimeric DNI/PA heptamers were not capable of transporting LF or EF into the cell cytosol, thus preventing cell damage by LF and EF (Sellman, Nassi, and Collier, 2001). When tested in mice, DNI alone was more immunogenic than PA, and the DNI-PGA conjugate elicited significantly higher levels of antibodies against PA and PGA than PA-PGA. The authors theorized that the two point mutations in DNI may have improved epitopes of PA allowing better antigen presentation to helper T cells. Alternatively, the mutations may have enhanced the immunological processing of PA by altering endosome trafficking of the toxin in antigen-presenting cells.

Humans produce five distinct classes of immunoglobulins: IgM, IgG, IgA, IgD, and IgF. IgG is the major immunoglobulin in normal human serum accounting for 70-75% of the total immunoglobulins present. Human IgG can be further subdivided into four subclasses: IgG1, 2, 3, and 4 which in normal serum occur in the following percentages: 66%, 23%, 7%, and 4% respectively. Baillie *et al.*, (1999) compared the IgG subclass antibody response to PA of individuals infected with anthrax to that of healthy individuals immunized with the UK licensed anthrax vaccine which consists of a protein precipitate from the supernatant fluid of cultures of the Sterne strain of *B. anthracis*. The predominant subclass in both groups was IgG1. In addition, IgG3 was observed in convalescent serum while vaccines produced IgG2, IgG3, and IgG4 subclasses.

The immune system has evolved potent effector cells termed cytotoxic T lymphocytes (CTL). CTL become activated upon recognition of pathogen-encoded peptides, usually 8 to 10 amino acids in length, presented on the surface of the host cell in the context of class I major histocompatibility molecules. Activated CTL respond by lysing the host cell and secreting cytokines so as to disrupt intracellular replication of the pathogen. After clearance of the organism, a subset of these activated CTL differentiate into long-lived memory CTL that can respond more effectively on reexposure to the same organism. Zarokinski, Collier, and Starnbach (2000) fused the amino-terminal 255 residues of LF (LFn) to an 8 amino acid restricted CTL epitope from chicken albumin (OVA257-264) and created a plasmid containing the LFn-epitope fusion as an insertion. The expression of the fusion protein was then induced in *E. coli* with isopropyl β-D thiogalactopyranoside (IPTG) and the cells sonicated to release the fusion protein which was purified on a Ni^{++}-charged column. Mice were then immunized *i.p.* simultaneously with the fusion protein and PA which resulted in the *in vivo* induction of antigen specific CTL in the absence of adjuvant in contrast to other bacterial toxin vaccines that either require adjuvants or have not been demonstrated to function *in vivo* thus limiting their use for humans.

Present human vaccines are delivered parenterally and therefore are not expected to elicit significant mucosal antibody responses. In addition, the only approved human vaccine in the U.S. (Anthrax Vaccine Adsorbed, AVA) is licensed for subcutaneous (*s.c.*) rather than intramuscular (*i.m.*) administration and frequently gives rise to mild to severe cutaneous reactions with accompanying fever and malaise due to the *s.c.* method of delivery (Pittman *et al.*, 2002). Present human vaccines are liquids which require refrigerated storage prior to administration, which may not always be available during transit and prior to field administration. Matyas *et al.*, (2004) reported the induction of long term neutralizing titers of transcutaneous immunization (TCI) of mice with recombinant (rPA) using three immunizations that resulted in superior titers compared to those obtain with aluminum hydroxide adsorbed rPA. In addition, rPA alone exhibited adjuvant activity for TCI. Two days prior to immunization, hair was shaved from the backs of female A/J mice. The cutaneous site was moistened with saline-soaked gauze and mildly abraded with emery paper. Patches were then applied for delivery overnight of 20 µg of rPA mixed with 10.4 – 20 µg of heat-labile enterotoxin from *E. coli* at 0, 2, and 4 wks. After 50 weeks, the mice (15 per group) were challenged *s.c.* with a $1000LD_{50}$ dose of *B. anthracis* spores of the Sterne strain and 100% were found to be completely protected.

Rabbits were vaccinated by Wimer-Mackin *et al.*, (2006) intranasally with PA-based vaccines formulated as dry powders with the glutamate salt of chitosan. Chitosan is a bioadhesive polysaccharide consisting of copolymers of gucosamine and N-acetylglucosamine, derived from partial deacetylation of chitin obtained from the shells of crustacea. A 10-mer peptide (Ac-CGGG-[g-D-glu]$_9$-g-D-glu) where "C" is cysteine, "G" is glycine, and "glu" is glucosamine representing the capsule of vegetative cells of *B. anthracis* was conjugated to PA. Rabbits were immunized intranasally on days O and 28 and aerosol challenged with an average $250LD_{50}$ dose of Ames spores on day 85. Significant anti-PA serum IgG levels were obtained, particularly with the use of chytosan based formulations, containing the PA-peptide conjugate. In contrast, a chytosan formulation containing PA plus the free unconjugated capsule peptide did not yield a significant level of capsule IgG. All immunized rabbits (10 per group) survived challenge, but only rabbits immunized with PA plus the PA-peptide conjugate appeared normal throughout the post-challenge period (14 days). All rabbits that received PA plus the free unconjugated capsule peptide appeared ill at times. Intranasal immunization with a chitosan powder vaccine combining PA and the PA-capsule peptide provided superior protection against infection compared to only PA as a single antigen vaccine with chitosan. Such a vaccine would appear to have considerable advantage over the present U.S. human vaccine.

DelVecchio *et al.,* (2006) studied the differentially expressed and immunogenic spore proteins of the *B. cereus* group consisting of *B. anthracis, B. cereus,* and *B. thuringiensis.* Comparative proteomic profiling of their spore proteins distinguished the three species from each other as well as the virulent from the avirulent strains. A total of 458 proteins encoded by 232 open reading frames were identified by mass spectrometry. A number of highly expressed proteins, including elongation factor Tu (Ef-tu), elongation factor G, a 60-kda chaperonin, enolase, pyruvate dehydrogenase complex, and others were found to exist as charge variants on two-dimensional gels. The majority of identified proteins had cellular functions associated with energy production, such as carbohydrate transport and metabolism, amino acid transport and metabolism, posttranslational modifications, and translation. Novel vaccine candidate proteins were identified using *B. anthracis* polyclonal antisera from humans postinfected with cutaneous anthrax. Fifteen immunoreactive proteins were identified in *B. anthracis* spores, whereas 7, 14, and 7 immunoreactive proteins were identified for *B. cereus* and in the virulent and avirulent strains of *B. thuringiensis* spores, respectively. Some of the immunodominant antigens included charge variants of EF-Tu, glyceradehde-3-posphate dehydrogenase dihydrolipoamide acetyl-transferase, Δ-1-pyrroline-5-carboxylate dehydrogenase, and a dihydrolipoamide dehydrogenase. Alanine racemase and a neural protease were uniquely immunogenic to *B. anthracis.* Comparative analysis of the spore immunome was looked upon by the authors to be of significance for further nucleic acid- and immuno-based detection systems in addition to future vaccine development.

NEW PROTEIN BASED VACCINES

Recombinant Vaccines

The AVA vaccine and a similar British anthrax vaccine contain predominantly PA, but also small amounts of LF and trace amounts of EF (Whiting *et al.,* 2004). These traces of LF and EF are thought to contribute to the vaccine side effects, such as local pain and edema, local and systemic reactions, including inflammation, flu-like symptoms, malaise, rash, arthrolgia, and headache.

Cao *et al.,* (2008) described the production of recombinant LF (rLF) in *E. coli*, its purification, and ability to induce a strong immune response in BALB/c mice. In addition, a novel inactive mutant form of LF was obtained which required a 3,700-fold higher level than that of the wild-type LF to achieve a similar level of cytotoxicity in combination with PA.

Immediately external to the cell wall of *B. anthracis* vegetative cells is an S-layer (surface layer), composed of two proteins designated Sap (surface array protein) encoded on the chromosome by the *sap* gene and EA1 (extractable antigen 1 also known as extracellular antigen 1) encoded on the chromosome by the *eag* gene. *B. anthracis* simultaneously synthesizes these two S-layer proteins. In the amino-terminal region, each protein has three motifs of about 50 residues following the signal peptide referred to as S-layer homology (SLH) motifs. Such motifs have been found to interact *in vitro* with cell walls. Mesnage *et al.,* (1999) found that the SLH motifs bind to the cell envelope and are sufficient to anchor a heterologous antigen on the cell surface. The authors envisioned that such a mechanism could be used to engineer the current animal vaccine strain of *B. anthracis* to deliver antigens which were protective against other animal pathogens. The primers SLH/5SLH5 were used to amplify the sequence encoding the three SLH motifs of EA1. The plasmids Pst SLH/7LH were used to amplify the three SLH motifs of sap. The primers 1SacB/2SacB were used to amplify the normally secreted mature levan sucrase (Lvs) from *B. subtilis*. The three motifs of EA1 and Sap from *B. anthracis* were produced in large amounts from *E. coli* and purified. Each polypeptide was found to bind lightly to purified cell walls of *B. anthracis* resulting in aggregation, reflecting the fact that the three SLH motifs of EA1 and Sap are suitable for *in vitro* binding to purified cell walls. Chimeric genes encoding fusion proteins combining the SLH domains of EA1 or Sap and the Lvs from *B. subtilis* were constructed and integrated into the chromosome of *B. anthracis*. The resulting hybrid proteins with three SLH motifs from EA1 or Sap were stable and exposed at the cell free EA1 or Sap surface of *B. anthracis* strains in which the genes normally encoding EA1 and Sap were inactivated. The cell surface anchored Lvs was found to be enzymatically active. Injection of recombinant strains of *B. anthracis* into Swiss mice resulted in a significant immune response towards purified Lvs. This study clearly indicated that a heterologous protein from another organism could be fused to the EA1 or Aap gene sequences of *B. anthracis* and that the chimeric protein would be expressed and bound to the cell surface of *B. anthracis* for presentation as an immunogen.

Efficacy of Various Adjuvants

Ivins *et al.,* (1992) determined the protective efficacy of immunization against anthrax with PA combined with different adjuvants for Hartly guinea pigs, CBA/J mice, and A/J mice. Adjuvant components derived from microbial products that were tested included threonyl-muramyl dipeptide (threonyl-MDP); monophosphoryl lipid A (MPL); trehalose dimycolate (TDM); and the delipidated, deproteinized, cell wall skeleton (CWS) from either *Mycobacterium phlei* or the BCG vaccine strain of *Mycobacterium bovis*. Non microbially derived adjuvants tested included aluminum hydroxide and the lipid amine CP-20,961. In guinea pigs, all adjuvants and adjuvant mixtures enhanced antibody titers to PA as well as survival after an *i.m.* challenge with spores of the Ames strain of *B. anthracis*. In contrast, PA alone or combined with either aluminum hydroxide or CP-20,961 failed to protect CBA/J mice following a *s.c.* challenge. Consistent with previous observations (Turnbull *et al.,* 1990; Welkos and Friedlander, 1988a) the human vaccine MDPH-PA failed to protect mice against anthrax challenge despite the stimulation of high anti-PA ELISA titers. Vaccines containing PA combined with threonyl-MDP or MPL-TDM-CWS protected a majority of the female CBA/J mice, while lower levels of survival occurred with male mice. PA-MPL-CWS or PA-MPL-TDM-CWS were more protective than the licensed human vaccine MDPH-PA. The authors concluded that the diminished protective efficacy of MDPH-PA is most likely due to its inability to stimulate sufficiently the full complement of immune mechanisms responsible for protection against anthrax. The MDPH-A human vaccine is an aluminum hydroxide PA adjuvant vaccine.

McBride *et al.,* (1998) undertook comparative studies involving aerosol challenge to guinea pigs with the Ames strain of *B. anthracis* following vaccination with two different adjuvants used in combination with recombinant PA (rPA) from *B. subtilis* in addition to vaccination of a third group of guinea pigs with the human UK vaccine. One adjuvant consisted of Aldrogel and the other was the Ribi Trimix (monophosphoryl lipid A, trehalose dimycolate from *Mycobacterium phlei*, and the delipidated deproteinized cell wall skeleton of *M. phle*i). The various vaccines were administered intramuscularly at 0, 2, and 4 weeks and the animals challenged by aerosol 4 weeks after the last vaccination. The human UK vaccine protected only 3/21 (14%) of test animals. The rPA/aldrogel vaccine protected only 5/21 (23%). The rPA/ribi vaccine protected 22/22 (100%) of the test animals. Analysis of immunological parameters in the individual animals revealed significant differences between the rPA/Ribi vaccine group, the rPA/Aldrogel vaccine group, and the human vaccine group for antigen specific lymphocyte proliferation, PA neutralization, and antigen specific IgG2 levels, but indicated no significant difference in PA-specific IgG1 levels. The IgG2 levels for the rPa/ribi group were 30 time higher than the human vaccine group. The rPA/Aldrogel vaccine induced mainly an IgG1 response while the rPA/ribi vaccine produced a predominantly IgG2 response associated with greater protection.

Bielinska *et al.*, (2007) evaluated a novel mucosal adjuvant consisting of a nontoxic, water-in-oil nanoemulsion (NE) which consisted of cetyl pyridinium chloride (1.0%), Tween 20 (5.0%), and ethanol (8.0%) in water with hot-pressed soybean oil (64%) which was emulsified using a high speed emulsifier. The material does not contain a proinflammatory component but penetrates mucosal surfaces to load antigen into dendritic cells. Mice and guinea pigs were intranasally immunized with recombinant *B. anthracis* PA antigen (rPA) in NE as an adjuvant. rPA-NE immunization was effective in inducing both serum anti-PA IgG and bronchial anti-PA IgA and IgG antibodies after either one or two mucosal administrations. Serum anti-PA IgG2a and IgG2b antibodies and PA-specific cytokine induction after immunization indicated a Th1-polarized immune response. rPA-NE immunization also produced high titers of lethal toxin-neutralizing serum antibodies in both mice and guinea pigs. Guinea pigs nasally immunized with the rPA-NE vaccine were completely protected against an intradermal challenge of ~1000LD$_{50}$ of *B. anthracis* Ames spores (1.38 x 10^3 spores) which killed control animals within 96 hrs. Intranasal challenge of guinea pigs (10 per group) with 10LD$_{50}$ and 100LD$_{50}$ spore doses after immunization resulted in survival levels of 70% and 40% respectively, while none of the nonimmunized control animals survived.

PASSIVE IMMUNIZATION

Advantages and Efficacy of Passive Immunization

Passive immunization against anthrax could have advantages over active immunization and antibiotic treatments *via* low toxicity, high specificity, large scale storage of freeze-dried preparations, and immediate protection.

The efficacy of passive immunization as a postexposure prophylactic measure for treatment of guinea pigs intranasally infected with *B. anthracis* spores was evaluated by Kobiler *et al.,* (2002). Rabbit anti-PA serum given 24 hrs. postinfection was found to be most effective when injected *i.p.* into guinea pigs and protected 90% of the infected animals, whereas Sterne spore anti-sera (44% survivors) and anti-LF (25% survivors) were less effective.

The A/J mouse strain is highly susceptible to nonencapsulated strains of *B. anthracis.* Beedham, Turnbull, and Williamson (2001) passively immunized A/5 mice against *i.p.* challenge by the STI veterinary vaccine strain (pXO1$^+$ pxO2$^-$). Lymphocytes were harvested from the spleen of mice immunized with recombinant PA. Pooled immunized murine serum (500 μL) was administered *i.p.* to naive mice 24 hrs. before *B. anthracis i.p.* challenge. Suspensions of activated T and B lymphocytes were also used to immunize naive A/J mice. Animals receiving immune serum survived significantly better (90% survival) than animals receiving control serum (0% survival). The passive transfer of lymphocytes from mice immunized with recombinant PA to naive mice failed to yield protection. The authors concluded that antibodies to PA clearly play a key role in protection against *B. anthracis.*

PA possesses one site for EF/LF binding and one site for cell receptor binding. In live attenuated strains, the cellular receptor binding domain of PA, in contrast to the EF/LF binding site, is more important in generating an effective antibody response to PA (Brossier *et al.*, 2000). Vaccination with PA has been found to induce antibodies to both the EF/LF and cell receptor binding domains that neutralize the cytotoxicity of LT *in vitro* (Little *et al.*, 1997). Zhou, Carney, and Janda (2008) reported on the selection and characterization of several human monoclonal neutralizing antibodies against the PA of *B. anthracis* from a phage displayed human scFv library. A set of neutralizing antibodies were identified, of which one clone recognized the LF/EF binding domain, and three clones recognized the cellular receptor binding domain found within the PA. One cell receptor clone was found to possess the highest affinity to PA, and provided superior protection from LT in a macrophage cytotoxicity assay.

Karginov *et al.*, (2004) undertook studies with mice based on the premise that simultaneous inhibition of LT action by antibodies and blocking of bacterial growth by antibiotics would be beneficial for the treatment of anthrax. The effects of a combination treatment using purified rabbit or sheep anti-PA antibodies and the antibiotic ciprofloxacin were assessed in DBA/2 mice. Among animals infected with 2×10^7 Sterne spores ($10LD_{50}$) by *i.p.* injection and treated with ciprofloxacin (50 mg/kg) alone only 50% survived. Among animals treated with just IgG alone, only 30% survived. Administration of anti-PA IgG in combination with ciprofloxacin resulted in 90 – 100% survival. The results indicated that a combination of antibiotic and IgG therapy was notably more effective than antibiotic or IgG treatment alone over a 9-day course of treatment.

The mortality of cheetahs from anthrax in the Etosha National Park, Namibia prompted vaccination studies by Turnbull *et al.*, (2004). In addition the black rhinoceros has been vaccinated for over three decades in the Park but efficacy had not been previously evaluated. Passive protection tests in A/J mice using sera from 10k cheetahs together with ELISA assays indicated that cheetahs are able to mount seemingly normal primary and secondary humoral responses to the Sterne 34F2 live spore livestock vaccine. Protection conferred on mice by sera from three of four vaccinated rhinoceros was almost complete, however, none of the mice receiving serum from the fourth rhinoceros were protected. Interestingly, sera from three park lions with naturally acquired high antibody titers, included as controls, also conferred high levels of protection with mice.

Mohamed *et al.*, (2005) developed an affinity-enhanced monoclonal antibody (mAn) to PA which when administered *i.v.* to rabbits resulted in significant protection against a lethal aerosolized challenge with anthrax spores from the Ames strain of *B. anthracis.* Lower levels of protection were observed when the mAb was administered 36 and 48 hrs. after challenge compared to 0 and 24 hrs.

post challenge. The time after exposure is therefore a critical factor in passive immunization.

Fully human monoclonal antibodies (mAbs) specific for PA were produced by Vitale *et al.,* (2006) by the development of a murine hybridoma using mice engineered to express human immunoglobulin in response to vaccination with PA. The mAb, designated 1303 or Valortin, did not block attachment of PA to the human monocyte cell surface. Full passive *in vitro* protection was dependant upon interaction with the host cell immunoglobulin (Fc) receptors. This Fc receptor dependence was found to be a dominant feature of the effective polyclonal immune response elicited in humans and animals by PA-based anthrax vaccines. The murine mAb to PA was found to be notably effective in protecting rabbits and monkeys against inhalation anthrax after a 200 x LD_{50} aerosol dose of *B. anthracis* Ames spores was administered. A larger percent of challenged animals survived when *i.v.* injected with 10 mg of mAb within 24 hr. of aerosol challenge compared to 48 hr. after challenge. The mAb against PA recognized both PA 83 and PA63 by ELISA assay. mAb 1303 was found to be unable to neutralize *in vitro* preformed LT.

Prevention of inhalation anthrax is well recognize to require early and extended antibiotic treatment. Peterson *et al.,* (2006) investigated the ability of human monoclonal anti-PA antibody (hmAb) to protect mice, guinea pigs, and rabbits against anthrax. Control animals challenged with *B. anthracis* Ames spores intranasally succumbed within 3 to 7 days. Mice and guinea pigs were challenged with 5×10^4 and 6×10^5 CFU of *B. anthracis* Ames spores respectively, comprising a $5LD_{50}$ dose for each species. Dutch Belted rabbits were challenged with a $100LD_{50}$ spore dose (1×10^7 CFU). Mice were passively immunized with a single *s.c.* dose of hmAb (16.7 mg/kg). Guinea pigs were given a single *i.p.* injection of hmAb (10, 20, or 50 mg/kg). Rabbits were given a single *s.c.* injection of hmAb (0.2, 1.0, 5, 10, or 20 mg/kg). Ciprofloxacin was also administered alone and in combination with hmAb. hmAb alone provided minimal protection of mice (40% survivors). Its efficacy was notably higher with guinea pigs. When mice were given a single injection of hmAb in conjunction with the administration of ciprofloxacin (30 mg/kg/day) for 6 days, starting 24 hrs. after challenge, 100% of the mice were protected for more than 30 days, while ciprofloxacin alone resulted in minimal protection. Interestingly, when hmAb (10 – 50 mg/kg) was administered to guinea pigs with a low, nonproductive dose of ciprofloxacin (3.7 mg/kg/day) for 6 days, synergistic protection against anthrax resulted. In contrast, a single dose of hmAb (1, 5, 10 or 20 mg/kg) protected rabbits completely against a $100LD_{50}$ spore dose, whereas a dose of only 0.2mg/kg alone resulted in only 20% survivors. In addition, administration to rabbits of hmAb 0, 6, and 12 hrs. after challenge resulted in 100% survival. In contrast, the delay of hmAb administration up to 24 and 48 hrs. reduced survival to 80 and 60%.

Revera *et al.,* (2006) generated two monoclonal antibodies (mAbs) to PA, mAbs 7.5G an 10F4. These MAbs did not compete for binding to PA, reflecting specificities for different PA epitopes. The MAbs were tested for their ability to protect a monolayer of cultured murine macrophages against LT-mediated cytotoxicity. MAb 7.5G, the most-neutralizing MAb, bound to domain 1 of PA and reduced LT toxicity in BALB/c mice. Remarkably, MAb 7.5G provided protection without blocking the binding of PA or LF or the formation of the PA heptamer complex. However, MAb 7.5G slowed the proteolytic digestion of PA_{83} by furin *in vitro*, suggesting a potential mechanism for Ab-mediated protection. These observations indicated that some Abs to domain 1 can contribute to host protection.

Peterson *et al.,* (2007) passively immunized Dutch-Belted and New Zealand white rabbits with a human monoclonal antibody to PA (10 mg/kg) at the same time of *B. anthracis* Ames spore challenge using either nasal instillation (100 LD_{50}) or aerosol infection (87 and 100 LD_{50}). Both rabbit strains (12 rabbits per group) were completely protected regardless of the inoculation method. In addition, all but one of the passively immunized animals (23/24) were completely resistant to rechallenge with spores five weeks after primary rechallenge. Analysis of the sera five weeks after primary challenge indicated that residual anti-PA levels had deceased by 85 to 95%. Low levels of rabbit-anti-PA were also found. The authors concluded that both sources of anti-PA could have contributed to protection to rechallenge.

On the basis that passive transfer of antibodies may be useful for preexposure prophylaxis against *B. anthracis*, Staats *et al.,* (2007) undertook studies to evaluate the ability of anti-PA and anti-LF neutralizing monoclonal antibodies (mABs) to protect against a challenge with LT in a mouse model and to identify correlates of immunity to LT challenge. Passive transfer of up to 1.5 mg of anti-PA mAb did not provide significant protection when transferred to mice 24 hrs. before LT challenge. In contrast, transfer of as little as 0.375 mg of anti-LF mAb did provide significant protection. Serum collected 24 hrs. after passive transfer had LT-neutralizing titers using a standard *in vivo* mouse neutralizing assay which did not correlate with protection against LT challenge. However, measurement of LT-neutralizing serum responses using an *in vitro* neutralizing assay employing the addition of LT to J774A.1 macrophages 15 min. before addition of serum did result in neutralization titers that correlated with protection against LT challenge.

Albrecht *et al.,* (2007) isolated and characterized two protective fully human monoclonal antibodies with specificity for PA and LF. The antibodies, designated IQNPA and IQNLF were developed as hybidomas from individuals immunized with the licensed anthrax vaccine. When combined, the antibodies had additive neutralization efficacy. IQNPA bound to domain IV of PA containing the host receptor binding site, while IQNLF recognized domain I containing the PA binding region in LF. A single 180 µg dose of either mAb given to A/J mice 2.5 hrs. before challenge conferred 100% protection against a lethal 24 LD_{50} *i.p.* spore challenge of the Sterne strain and against rechallenge on day 20 with a 41LD_{50} spore challenge dose. Mice treated with either antibody and infected with the sterne strain developed detectable murine antiPA and anti-LF IgG responses by day 17 that were dependant on which human antibody the mice had received. Mice given human IQNPA geneted a significantl higher IgG response to PA than to LF, while mice inoculated with IQNLF generated a significantly higher IgG response to LF than to PA with mice receiving a subsequent spore challenge. These findings indicated that the toxin-neutralizing capacities of IQNPA and IQNLF not only allowed the development of protective endogenous antibody responses against PA and LF but presumably stimulated respective humoral responses to PA and LF resulting from a toxin producing strain. One possible explanation offered by the authors for this humoral immunostimulating effect might be the formation of human antigen-antibody immune complexes and their subsequent binding to Fc recepors on macrophages and natural killer cells so as to specifically enhance PA and LF antigen uptake and presentation of major histocompatibility complex molecules to stimulate T- and B-cell expression. This in turn could result in subsequently enhanced production of endogenous murine polyclonal antibodies to PA and LP as was observed. The currently recommended postexposure therapeutic treatment of anthrax is a combination of antibiotics (ciprofloxacin or doxycycline) and the licensed human AVA vaccine. In severe cases, *i.v.* administration of preformed polyclonal anthrax immunogobulin (AIGIV) derived from immunized donors is recommended (Albrecht *et al.*, 2007). The advantage of using AIGIV is that it provides instant protection, is likely to be effective during mid-to advanced-stage disease, is equally effective against antibiotic resistant strains, results in minimal adverse reactions, has a prolonged serum half-life, and targets multiple epitopes.

Osorio *et al.*, (2009) linked the PA gene to an inducible promoter for maximum expression in the host and fused it to the secretion signal of the *Escherichia coli* alpha-hemolysin protein (HlyA) on a low-copy-number plasmid. This plasmid was then introduced into the licensed typhoid vaccine strain, *Salmonella enterica* serovar Typhi strain Ty21a, and was found to be genetically stable. Immunization of mice with three vaccine doses elicited a strong IgG response to PA. Vaccinated mice demonstrated 100% protection against a lethal intranasal challenge with aerosolized spores of a toxigenic non encapsulated strain of *B. anthracis*. In contrast, all control mice succumbed within five days.

INNATE IMMUNE RESPONSE TO *B. anthracis*

Response of Toll-Like Receptors

The innate immune response is the first line of defense against invading pathogens. One of the most important components of the innate response is comprised of the phagocytic cells such as the macrophages, dendritic cells, and neutrophils. These cells engulf and kill invading pathogens. The ability of these phagocytic cells to recognize microbial products is mediated by the Toll-like receptors (TLRs). TLRs are a family of receptors that recognize pathogen-associated molecular patterns shared by large groups of microorganisms, thereby assisting in microbial recognition by the host immune system. Peptidoglycan (PGN) and lipopotein of Gram-positive bacteria are TLR2 ligands. A central feature of this system is that binding of TLRs by pathogen-associated molecular patterns leads to activation of signaling pathways that are critical for induction of the early host defense against invading pathogens. The TLRs can activate a common signaling pathway leading to the activation of a cascade of defense factors including mitogen-activated protein-kinase (Barton and Medzhito, 2003). This pathway involves an adapter protein, myeloid differentiation factor 88 (MyD88), which contains a Toll/interleukin-1 receptor) domain and a death domain (Yamamoto, Takeda, and Akira, 2004). MyD88 is essential for responses against a broad range of microorganisms or their components which are recognized by TLR2 and additional Tolls.

Hughs *et al.,* (2005) hypothesized that the *B. anthracis* Sterne 34F2 strain, is recognized by TLR2. Heat-killed cells of *B. anthracis* were found to activate TLR2 signaling in human embryonic kidney cells and also stimulated tumor necrosis factor alpha (TNF-α) production in C3H/HeN mice, C3H/HeJ mice, and C57BL/65 bone marrow derived macrophages. The ability of heat-killed vegetative cells of *B. anthracis* 34F2 to induce a TNF-α response was preserved in TLR2$^{-/-}$ 172but not in MyD88$^{-/-}$172 macrophages. *In vivo* studies revealed that TLR-1- mice and TLR4-deficient mice were resistant to challenge with aerosolized Stern strain spores, but MyD88$^{-/-}$ mice were as susceptible as A/J mice. These results indicate that although recognition of *B. anthracis* occurs *via* TLR2, additional MyD88-dependant pathways contribute to the host innate immune response to anthrax infection.

Hahn, Lyons, and Lipsomb (2008) examined the ability of *B. anthracis* spores and vegetative cells to stimulate human monocyte-derived dendritic cells (MDDC), primary myeloid dendritic cells (mDC), and plasmacytoid dendritic cells (pDC) cytokine secretion. Exposure of MDDCs and mDCs to spores or vegetative cells of the genetically complete *B. anthracis* strain UT500 induced significantly increased cytokine secretion. Spores lacking genes required for capsule synthesis stimulated significantly higher cytokine secretion than UT500 spores in mDCs, but not MDDCs. In contrast, vegetative cells lacking capsule synthesis stimulated significantly higher cytokine secretion than UT500 vegetative cells. With both MDDCs and mDCs. Spores or vegetative cells lacking both LT and ET stimulated significantly higher cytokine secretion stimulated significantly higher cytokine section than UT500 spores or vegetative cells, respectively, in both mDCs and MDDCs. pDZCs exposed to spores or vegetative cells did not produce significant amounts of cytokines, even when virulence factors were absent. The authors concluded that *B. anthracis* produced toxins as well capsule to inhibit MDDC and mDC cytokine secretion whereas human pDCs respond poorly even when capsule or both toxins are absent.

Chemokines and Cytokines

Chemokines are a group of structurally related, low-molecular-mass (8- to 10 kDsa) proteins originally defined by their ability to induce directed cell migration among leukocytes (Luster, 1998). Although some chemokines are constitutionally expressed and function in homeostatic roles, the majority of presently recognized chemokines are considered to be potent inflammatory mediators induced in response to Toll-like receptor agonists and/or proinflammatory cytokines (Luster, 2002; Rossi and Zlotnik, 2000). Three members of the interferon (IFN)-inducible tripeptide motif Gu-Leu-Arg-negative (ELR⁻) CXC chemokines, monokine-induced by IFN-α (CXCL9), IFN-α-inducible protein of 10kDa (CXCL10), and IFN-inducible T-cell-activated chemokine (CXCL11), have been shown to be strongly induced by IFN-α and together comprise a family of IFN-inducible ELR-CXC chemokines (Cole *et al.*, 1998; Luster *et al.*, 1998; Wright and Farber, 1991). cXCL9, CXCL10, and CXCL11 are produced and secreted primarily by monocytes, macrophages, fibroblasts, and epithelial cells upon stimulation with proinflammatory cytokines, especially IFN-α (Farber, 1997; Luster and Ravetch, 1987). Each of these chemokines acts primarily in the recruitment of activated CD4[+] and CD8[+] T cells, NK cells, and plasmalytoid dendritic cells to sites of inflammation (Colonna, Trinchieri, and Liu, 2004; Mohan *et al.*, 2005). In addition to their roles in leukocyte recruitment, CXCL9, CXCL10, and CXCL11 have been found to exert dire antimicrobial effects, comparable to the effects mediated by cationic antimicrobial peptides, including defensins (Cole *et al.*, 2001).

Crawford *et al.,* (2009) examined the ability of the interferon-inducible Glu-Leu-Arg-negative CXC chemokines CXCL9, CXCL10, and CXCL11 and found that they exhibited antimicrobial activities against *B. anthraci*s. *In vitro* analysis demonstrated that all three CXC chemokines exerted direct antimicrobial effects against *B. anthracis* spores and bacilli resulting in marked reductions in spore and vegetative cell viability. Electron microscopy revealed that CXCL10-treated spores failed to undergo germination which was reflected by an absence of cytological changes in spore structure that normally occurs during germination. Immunogold labeled CXCL10-treated spores indicated that the chemokine was located internal to the exosporium in association primarily with the spore coat and its interface with the cortex. Upon inhalation challenge with *B. anthracis* spores the lungs of C57BL/6 mice (resistant to inhalation infection by *B. anthracis*) had significantly higher levels of all three chemokines than did he lungs of A/J mice (highly susceptible to infection by *B. anthracis* spores). Increased CXC chemokine levels were associated with significantly reduced levels of spore germination within the lungs as determined by *in vivo* imaging. In addition to observations on spores, CXCL10, the most potent of the CXC chemokines examined was found to result in extensive cell wall and cell membrane disruption of *B. anthracis* vegetative cells.

Drysdale *et al.,* (2007) used a genetically complete (pXO1$^+$ pXO2$^+$) fully virulent *B. anthracis* strain and four isogenic toxin null mutants to determine the effects of ET and LT on the hosts innate immune response during systemic infection of mice. Using the spleen as an indicator for host response, *i.v.* inoculation of LT-deficient mutants into C57BL/6 mice significantly increased production of several cytokines over that observed after infection with the fully virulent parent strain or an EF-deficient mutant. Bacteria producing one or both of the toxins were capable of inducing significant apoptosis of cells present in the spleen, whereas apoptosis was greatly reduced in mice infected with nontoxigenic mutants. Mice infected with toxin-producing strains also showed increased splenic neutrophil recruitment compared to mice infected with nontoxigenic strains and neutrophil deletion prior to infection with toxin producing strains, leading to decreased levels of apoptosis. These observations indicate that LT suppresses cytokine secretion during infection, and that both EF and LF play major roles in inducing neutrophil recruitment and enhanced apoptosis. However, in the absence of LF the effect of EF-induced cell recruitment and apoptosis was enhanced, presumably because LF effectively suppresses the secretion of chemokines.

It was shown by Chakrabarty *et al.,* (2006) that human alveolar macrophages rapidly internalize spores of *B. anthracis*, triggering induction of cytokine and chemokine mRNA and protein through activation of the signaling pathways P38, ERK, and SAB/JNK. Lung epithelial cells have also been suspected of playing a similar role in activating the innate immune response. Chakrabarty *et al.,* (2007) developed a human lung slice model (tissue derived from lung cancer resection patients, tumor free) to study the local lung response to human pathogens. Exposure of human lung tissue to spores of *B. anthracis* Sterne rapidly activated the mitogen-activated protein kinase signaling pathways ERK, p38, and JNK. In addition, an RNase protection assay showed induction of mRNA of several cytokines and chemokines. This finding was reflected at the translational level by protein peak increases of 3-,25-,9-,34, and 5-fold for interleukin-6 (IL-6), tumor necrosis factor alpha, IL-8, macrophage inflammatory protein 1α/β, and monocyte chemoattractant protein 1, respectively. An immunohistological analysis of IL-6 andIL-8 revealed that alveolar cells and macrophages and a few interstitial cells are the source of cytokines and chemokines. These results indicated that lung-epithelial cells actively participate in the innate immune response to *B. anthracis* infection through signal-mediated elaboration of cytokines and chemokines.

Pickering *et al.,* (2004) examined the cytokine response to spore infection in BALB/c mouse primary peritoneal macrophages, primary human dendritic cells, and during a spore aerosol (*B. anthracis* Sterne strain 7702) infection model utilizing the susceptible A/J mouse strain. Both mouse peritoneal macrophages and human dendritic cells exhibited significant intracellular bactericidal activity during the first hours following uptake, providing the necessary time to mount a cytokine response prior to cell lysis. Strong tumor necrosis factor alpha (TNF-α) and interleukin-6 (IL-6) responses were observed in mouse peritoneal macrophages. In addition to TNF-α and IL-6, human dendritic cells produced the cytokines IL-1B, IL-8, and IL-12. A mixture of Th1 – Th2 cytokines were detected in sera obtained from infected animals. These observations provide further evidence of an acute cytokine-response when cells in culture and mice are infected with *B. anthracis* spores. This study also revealed that lung spore doses of the Sterne strain of 2 x 10^3, 3 x 10^5 to 8 x 10^5, and 1 x 10^6 to 4 x 10^6 resulted in 0, 50, and 80 to 100% mortality with A/J mice, further indicating that the host mounts a vigorous innate response to a *B. anthracis* spore challenge. This conclusion is further supported by the relatively high doses of fully virulent aerosolized spores that are required to induce lethality in rabbits (LD_{50} = 10^5, Pitt *et al.*, 2001) and non human primates (rhesus macaques, LD_{50} – 5.5 x 10^4, Ivins, *et al.*, 1998).

The Oxidative Burst

The ability of spores of *B. anthracis* to survive in activated macrophages is critical for germination and the organisms survival. Upon infection, germination of endospores occurs during their internalization within the phagocyte, and the ability to survive exposure to intracellular microbial killing mechanisms such as superoxide ($O_2^{\cdot -}$), nitric oxide ($NO^{\cdot}$), and H_2O_2 is an important initial event in the infection process. Macrophages generate $NO^{\cdot}$ from oxidation of L-arginine to L-citrulline using the isoform of nitric oxide synthase (NOS 2). Rainer *et al.,* (2006) found that exposure of murine macrophages to *B. anthracis* endospores increased the expression of NOS 2, 12 hours after exposure and production of $NO^{\cdot}$ was comparable to that achieved following other bacterial infections. Spore-killing assays demonstrated an $NO^{\cdot}$ -dependant bactericidal response that was significantly decreased in the presence of the NOS 2 inhibitor L-N^6-(1-imino-ethyl) lysine and in L-arginine-depleted media. It was also found that *B. anthracis* vegetative cells and endospores exhibited arginase activity, possibly competing with host NOS 2 for its substrate, L-arginine, resulting in reduced levels of NO**.**

Weaver *et al.,* (2007) demonstrated that macrophage-generated ·NO, but not ONOO⁻, is a critical intermediate in the killing of *B. anthracis*. Further, in addition to SODs, the exosporium was found to contain an arginase, which allows this organism, by attenuating ·NO generation, to subvert the consequences of host production of this free radical. Therefore, the ability of this bacterium to limit ·NO generation may have important implications for the virulence of *B. anthracis* infections.

INCOMPATIBIITY OF LIVE VACCINES WITH ANTIBIOTIC TREATMENT

The incompatibility of live vaccines with simultaneous administration of antibiotics led Stepanov *et al.,* (1996) to obtain a variant of the Russian human vaccine strain STI-1 designated STI-AR resistant to six antibiotics including doxycycline by introducing a recombinant plasmid into STI-1. When hamsters were infected with a lethal strain of *B. anthracis* and simultaneously administered STI-AR for vaccination plus doxicycline, 100% of these animals survived When the same test animals were subjected to a second infection, 80% of the animals survived. In contrast, when hamsters were infected with the lethal strain and simultaneously administered the parental vaccine strain STI-1, although 100% of the animals survived the initial infection, none survived the second infection. These results clearly indicated that the use of an antibiotic resistant live vaccine strain allowed immunization to be initiated at the time antibiotic therapy was administered. The authors also noted that the nutrient composition of the culture medium was critical for appropriate cultivation of both STI-1 and STI-AR.

Although antibiotics are effective against the vegetative cells of *B. anthracis*, they do not exhibit activity against the spore form of the organism. There are currently no therapeutic agents that target the spores of *B. anthracis* so as to prevent infection. This is particularly critical with dormant spores in the lungs that may germinate and cause infection several weeks after termination of antibiotic therapy.

NONSPECIFIC IMMUNOLOGICAL PROTECTION

Leffel, Twenhafel, and Whitehouse (2008) described the potential efficacy of *Serratia marcescens* infection resulting in protection against aerosolized *B. anthracis* spore infection. Five African green monkeys (AGMs) had intravenous catheters implanted. All were multiply exposed to lethal aerosol doses of *B. anthracis* spores. One animal's catheter was accidentally pulled out The AGM with no catheter succumbed from anthrax on day 10 the other four AGMs had *S. marcescens* contamination in the catheters; with pure colonies grown from blood samples. None of these four AGMs showed signs of illness, had *B. anthracis* or a detectable level of PA in the blood stream. Of interest were the observations that with the AGM that succumbed to anthrax, concentrations of both interferon-gamma (IFN-γ) and interleukin-2 (IC-2) were below detectable levels. In contrast, among the four AGMs with sub-clinical infections with S. *marcescens,* concentrations of serum IFNγ ranged from 119 to 1170 pg/ml and IL-2 from 2346 to 6951 pg/ml. These observations suggested that the presence of a subclinical infection by *S. marcescens* may have resulted in an innate immune response that protected the animals from otherwise fatal anthrax inhalation doses.

REQUIREMENT FOR COMPLEMENT

Susceptibility to infection by a Sterne strain(pXO1+ pXO2⁻) of *B. anthracis* varies considerably among various strains of mice. The C57BL6 mouse strain is completely resistant to Sterne. In contrast, the A/5 and DBA/2J mouse strains are notably susceptible to infection by Sterne. Harville *et al.,* (2005) hypothesized that the major role of the capsule is to protect *B. anthracis* vegetative cells from the effects of complement, and that an acapsular Sterne strain would be virulent in normally resistant mice if they lacked complement. Cobra venom factor (CVF) has been shown to result in nearly complete depletion of complement by enzymatically degrading this immunological co-factor (Vogel *et al.*, 1991). When Harville *et al.,* (2005) injected C57Bl/6 mice *i.p.* with CVF so as to result in complete depletion of complement, they became as nearly susceptible (93% death) to lethal aerosol infection by spores of the Stern 7702 strain as the A/5 strain with 100% of A/J mice succumbing with or without CVF treatment. These results indicated that depletion of complement alone is sufficient to render the resistant C57BL6 mouse strain susceptible to the Sterne strain.

MACROPHAGE STUDIES

Animal Models

Gold *et al.,* (2004) established an *in vitro* infectious model of *B. anthracis* sterne (34F2). Strain 34F2 was found lethal to murine and human macrophages. Treatment with exogenous interferons (IFNs) significantly improved cell viability and reduced the number of germinated intracellular spores. Infection with 34F2 failed to induce the latent transcription factors signal transducer and activators of transcription 1 (STAT1) and ISGF-3, which are central to the IFN response. In addition, strain 34F2 reduced STST1 activation in response to exogenous alpha/beta IFN, suggesting direct inhibition of IFN signaling. There are two major categories of IFNs, alpha/beta IFN (IFN-α/β) and gamma IFN (IFN-γ). IFN responses induce a large set of genes including a number of genes with antibacterial activity, including the inducible nitric oxide synthetase gene. The authors concluded that IFNs may have an important role in the control of *B. anthracis* and in the potential benefit of using exogenous IFN as an immunoadjuvant for therapy.

Kang *et al.,* (2005) compared the killing of germination-proficient (gP) and germination-deficient (*ΔgerH*) Sterne 34F2 strain spores by murine peritoneal macrophages. Although macrophages similarly ingested both spore types, only the gP spores were killed. Pretreatment of macrophages with gamma interferon or opsonization with immunoglobulin g (IgG) isolated from a human subject immunized with anthrax vaccine enhanced the macrophage killing of the gP spores but had no effect on the ΔgerH spores. These results indicated that macrophages are unable to kill the spore form of *B. anthracis* and that the exosporium may play a role in the protection of spores from macrophages.

Chakrabarty *et al.,* (2006) studied the initial events after exposure of macrophages to spores beginning with the rapid internalization of spores by macrophages. Spore exposure rapidly activated the mitogen-activate protein kinase signaling pathways extracellular signal-regulated kinase, cJun-NH$_2$-terminal kinase, and p38. This was followed by the transcriptional activation of cytokine and primary monocyte chemokine genes. Transcriptional induction at the translational level, as interleukin-1α (IL-α), IL-1β, IL-6, and tumor necrosis factor alpha (TNF-α-cytokine protein levels were markedly elevated. Induction of IL-6 and TNF-a protein kinases, while the complete inhibition of cytokine induction was achieved when multiple signaling pathway inhibitors were used. These observations collectively indicated activation of the innate immune system in human alveolar macrophages by *B. anthracis* spores and that multiple signaling pathways are involved in this cytokine response.

Cote, Rooijen, and Welkos (2006) investigated the roles that macrophages and neutrophils play in the progression of infection by *B. anthracis* in Balb/c mice. Mice were treated with a macrophage depletion agent (liposome-encapsulated clodronate) or rat anti-mouse granulocyte monoclonal antibody RB6-8C5), and the animals were then infected *i.p.* or by aerosol challenge with fully virulent Ames spores. The macrophage-depleted mice were significantly more susceptible to the ensuing infection than the saline-pretreated mice, whereas the difference observed between the neutropenic mice and the saline pretreated controls were generally not significant. In addition, augmented peritoneal neutrophil populations before spore challenge did not increase resistance of the mice to infection. These results indicate that neutrophils play a relatively minor role in the early host response to spores, whereas macrophages play a more dominant role in early host defense against infection by anthrax spores.

Parent *et al.,* (2006) found that both active and passive immunity to pulmonary infections is dependant on IFN-gamma and TNF-alpha cytokines which are classically associated with type 1 cellular immunity. With both protocols, abrogating IFN-gamma or TNF-alpha activity significantly decreased survival of mice and increased the bacterial burden in pulmonary, spleenic, and hepatic tissues. Neutralization of either cytokine also counteracted challenging-induced, vaccination-dependant upregulation of nitric oxide synthase 2 (NOS2). In addition, genetic depletion of NOS2 suppressed protection conferred by serotherapy. IFN-gamma, TNF-alpha and NOS2 are therefore key elements of cellular immunity and perform critical protective functions during humoral defense against *Y. pestis* challenge. These observations strongly suggest that plague vaccines should strive to maximally stimulate both cellular and humoral immunity.

RAPID DETECTION OF *B. anthracis* BY IMMUNOLOGICAL METHODS

Immunological Detection of *B. anthracis* Spores

Detection of *B. anthracis* in environmental samples is invariably based on detection of spores. Spores of *B. anthracis* are surrounded by a layer called the exosporium which is composed of a paracrystalline basal layer and an external hairlike nape (Daubenspeck *et al.,* (2004)). The key component of the latter is a glycoprotein called BclA (*Bacillus* collagen-like protein of *anthracis*) which has been shown to react with most antibodies raised against anthrax spores. It has been shown that multiple copies of a tetrasaccharide are linked to the collagen-like region of BclA (Daubenspeck *et al.,* (2004)). This linear tetrasaccharide contains a unique nonreducing terminal sugar known as anthrose.

Starting with galactose, Dhénin *et al.,* (2008) synthesized anthrose and conjugated it to keyhole limpet hemocyanin (LLH) through a spacer arm using gluteraldehyde as the conjugation reagent. Rabbits immunized with the anthrose-KLH conjugate yielded high antibody titers directed against anthrose in an ELISA assay. Presumably the antibody to anthrose can be used to immunologically detect spores of *B. anthracis.*

Uithoven, Schmidt, and Ballman (2000) developed a light addressable potentiometric sensor and a flow-through immunofiltration-enzyme assay system for the rapid and specific identification of spores of *B. subtilis* equally applicable to spores of *B. anthracis.* The limit of detection was 3 x 10^3 CFU/ml for *B. subtilis* with an assay time of 15 min. for a pure suspension of spores.

Spores of *B. anthracis* are enclosed by a prominent loose-fitting layer called the exosporium. A highly immunogenic glycoprotein BclA is the major component of the nap-like hairs that protrude from the exosporium. BclA contains two-O-linked oligosaccharides, a 715 Da tetrasaccharide (anthrose) present only in *B. anthracis,* and a 324 Da disaccharide. Parthasarathy *et al.,* (2008) developed a microarray platform for detection of several pathogenic bacteria including *B. anthracis* spores by immobilizing bacterial "signature" carbohydrates onto epoxide modified glass slides. The carbohydrate microarray platform was probed with sera from rabbits vaccinated with *B. anthracis* spores. This microarray was able to detect antibodies in infected or vaccinated animals simultaneously for several different microorganisms.

Hao *et al.,* (2009) reported on the development of a quartz crystal microbalance biosensor for *B. anthracis* spores and vegetative cells utilizing a single monoclonal antibody derived from the subcutaneous injection of a formaldehyde inactivated spores into mice. A quartz crystal was coated on both sides with gold and a mixed self assembled monolayer (SAM) formed on one gold surface. Protein A was then bound to the activated SAM and the mAB applied. Suspensions of spores or vegetative cells were then applied, the sensor rinsed and dried, and the resonant frequency shift of the sensor determined in the gas phase before and after application of the spore or vegetative cell suspensions. The limit of detection was 10^3 CFU or spores/ml in less than 30 min. It is of interest to note that the mAB was able to detect both spores and vegetative cells. The sensor would not be expected to detect encapsulated vegetative cells from clinical tissue samples.

Law enforcement and Public Health laboratories in Australia presently have available a commercial hand held analytical devise (RMPO) for detection of spores of *B. anthracis* (Hoile *et al.*, 2007) The RAMP Anthrax Assay test cartridge contains an analyte-specific immunochromatographic strip, which employs latex particles tagged with fluorescene-labeled antibodies, specific for the surface antigen of *B. anthracis* spores. The RsAMP instrument detects the presence of fluorescent beads that attach to the test line when *B. anthracis* spores are present. Results are obtained in 15 min. The assay system was found to be highly specific for spores *of B. anthracis* with no cross reactivity with spores of other *Bacillus* species. The limit of detection was ~6.2 x 10^5 spores/ml (~6200 spores/10 µl).

Love, Redmond, and Mayers (2008) described a targeted approach for the production of biological recognition elements capable of rapid and specific detection of anthrax spores on a biosensor surface. Single chain antibodies (scFvs) were developed to EA1, a *B. anthracis* S-layer protein for detection of spores of *B. anthracis* which would not detect the spores of other *Bacillus* species. EA1 is a surface layer protein of vegetative cells, also present in spores. BALB/c mice were immunized with irradiated *B. anthracis* spores. Splenic mRNA was isolated and used to produce an scFv library. PCR amplification of antibody sequences, overlap extension PCR, cloning of the assembled scFv sequences into plasmid AK100 and production of phage displayed scFvs was as described by Krebber *et al.*, (1977). Competitive panning (pe-adsorption or subtractive antibody screening) was used to eliminate cross-reactive antibodies. Coating of the surface of a resonance mirror biosensor resulted in real-dime detection of *B. anthracis* spores. The minimum level of detection of sonicated spores of *B. anthracis* with an scFv antibody was 100-fold lower than with a mAb. The sensitivity of detection with intact spores was notably lower than with sonicated spores, undoubtedly due to the release or unmasking of EA1 of spores.

Bruno and Yu (1996) developed an immunomagnetic chemiluminescent assay for detection of *B. anthracis* spores in soil suspensions. Streptavidin-coated paramagnetic microbeads were incubated with biotinylated capture antiserum. Strain-dependant detection limits were in the range of 10^2 to 10^5 spores in buffer. Detection was up to three orders of magnitude less sensitive in soil suspensions. With spores in soil suspensions a minimum of 1 x 10^5 spores were required for detection. The rank order of sensitivity for the assay was Stern > Ames > Vollum 1B.

Edelstein *et al.,* (2000) applied a Bead ARray Counter (BARC) to the detection of *B. anthracis* DNA encoding LF. BARC is a multi-analyte biosensor that uses DNA hybridization, magnetic microbeads and giant magnetoresistance (GMR) sensors. The unit is a table-top instrument consisting of a microfabricated chip with an array of GMR sensors; a chip carrier board, a fluidics cell and cartridge, and an electromagnet. DNA probes are patterned onto the solid substrate chip directly above the GMR sensor, a sample analyte containing complementary DNA hybridizes with the biotinylated DNA probes on the surface. Labeled, micron-sized streptavidin magnetic beads are then injected that specifically bind to the sample DNA. A magnetic field is applied, removing any beads that are not specifically bound to the surface. The beads remaining on the surface are detected by the GMR sensors, and the intensity and location of the signal indicate the concentration and identity of the pathogen present in the sample. The BARC chip contained a 64-element sensory array. Each GMR sensor is capable of detecting a single magnetic bead and is therefore theoretically capable of detecting a single analyte molecule.

Wan *et al.,* (2007) reported on the detection of *B. anthracis* spores in liquids using free standing microfabricated magnetoelastic resonator particles MERP as small as 500 µm x 100 µm x 4 µm. These MERPs featured wireless signal transmission and were suitable for both air and liquids. Affinity-selected filamentous bacteriophage were employed as the bio-molecular recognition agent and were immobilized onto all surfaces of the resonators. A detection limit of 10^3 CU/ml was obtained. The phage-coated biosensors were found to have a better preference for spores of *B. anthracis* than for spores of other bacilli, and did exhibit some cross reactivity with spores of *B. cereus* and *B. megaterium* when present at 10^8 CFU/ml which was not deemed significant.

Williams *et al.,* (2003) screened a phage display peptide library and identified a family of peptides, with the consensus sequence TYPXPXP, that bind selectively to *B. anthracis* spores. This work was extended to identify a peptide variant, ATYPLPIR, with enhanced ability to bind to *B. anthracis* spores and an additional peptide, SLLPGLP, that preferentially binds to spores of species phylogenetically similar to, but distinct from *B. anthracis.* The peptides were conjugated to R-phycoerythrin and mixed with spore suspensions and unbound conjugate washed away and the fluorescence intensity of spores measured.

Tims and Limn (2004) developed an evanescent wave biosensor for detection of *B. anthracis* spores based on antibodies to the spores, a sandwich immunoassay system, and a fiber optic waveguide to transmit and receive light information. A biotinylated capture antibody was attached to the fiber *via* streptavidin, and the target analyte was detected with a Cy-5-labeled detection antibody. A commercial evanescent wave fiber-optic biosensor was used to detect spores suspended in PBS. The limit of detection was 10^4 spores/ml with an assay time of less than 1 hr.

Campbell and Mutharasan (2006) fabricated piezolelectric-excited 2.48 mm^2 cantilever (PEMC) sensors for detection of *B. anthracis* spores. The sensors consisted of two layers, a lead zirconate titanate (PZT) and borosilicate glass cover slip layer cemented together with a non-conductive adhesive so that a 2 mm length of the glass cover slip protruded to provide surface for antibody immobilization. Antibodies specific to *B. anthracis* Sterne spores were immobilized onto the PEMC sensors and exposed to spores suspended in PBS for attachment. The resonant frequency was found to decrease at a rate proportional to the spore concentration. The limit of sensitivity was 300 spores/ml and assay time was 3 three to 5 min. depending on the concentration of spores.

Davila *et al.,* (2007) reported on the application of microcantilevers as resonant mass sensors for detection of *B. anthracis* Sterne spores in air and liquid. The detection scheme was based on measuring the resonance frequency decrease driven by thermally induce oscillations, as a result of the added mass of the spores with the use of a laser Doppler vibrometer. For detection of spores in water, the cantilevers were functionalized with antibodies to fix the spores onto the surface. As few as few 50 spore on the cantilever could be detected in water. Measurement sensitivity of 9.23 Hz/fg for air and 0.1 Hz/fg for water were obtained. The authors speculated that by driving the cantilevers and using higher order modes, detection of a single spore in liquids could be achievable.

Campbell, deLesdernier, and Mutharasan (2007) described the development of an integrated air sampler and cantilever immunosensor for detection of *B. anthracis* spores. The methodology employed a commercial air sampler, a 10 min, air sample at 267 L/min. to capture airborne particles containing *B. anthracis* spores and concentration in 5 ml of PBS. The sample was then injected into a flow cell containing an antibody-functionalized cantilever sensor. The resonant frequency of the sensor at 925.1 kHz decreased exponentially as spores attached to the sensor surface producing a positive response beyond the noise level in 2 min and reached a steady state value in 20 min. The minimum level of detection was ~5 spores/L of air.

Aucoin *et al.,* (2009) developed a quantitative latex agglutination assay for rapid detection of the poly-γ-D-glutamic acid (γDPGA) capsule of *B. anthracis* vegetative cells. A carbodiamide kit was used to conjugate 0.982μ diam. carboxylated latex microparticles to a mAb. A linear relationship was obtained between the intensity of agglutination and the concentration of polyglutamic acid. A quantitative ELISA assay was also developed for detection of gamma-DPGA. The gamma-DPGA concentrations determined by latex agglutination closely correlated with the quantitative ELISA an could be used for quantification of gamma-DPGA in the serum of mice infected with pulmonary anthrax. The latex agglutination assay lost significant sensitivity to the presence of degraded gamma-DPGA (10 and 25 mer) occurring in the urine of mice.

Zahavy *et al.,* (2003) developed a flow cytometry FRET immunoassay for spores of *B. anthracis*. IgG was conjugated to the fluorophore alexa488 yielding IgG-alexa488 as the FRET donor. IgG was also conjugated to the fluorophore alexa594 yielding IgG-alexa594 as the FRET acceptor and fluorescence emitter. Spores were labeled with both IgG-alexa488 and IgG-alexa594 in the development of a highly selective gate for immunodiagnosis of *B. anthracis* spores by flow cytometry with enhanced discrimination between *B. anthracis* and the related species *B. thuringiensis* and *B. subtilis*.

Schumacher *et al.,* (2008) developed a two-color flow cytometric assay capable of simultaneously identifying *B. anthracis* spores and the presence of spore-associated PA. The peptide ATYP when conjugated to R-phycoerythrin (RPE) was found to bind only to *B. anthracis* spores (Williams, Benedek, and Turnbough, 2003). In addition, PA has been identified as a spore surface antigen and is thought to be noncovalently entrapped in the spore coat and exosporium (Cote *et al.*, 2005). A fluorescein isothiocyanate (FITC)-conjugated antibody-based PA assay was coupled to the *B. anthracis* spore specific ATYP-RF assay. The assay was able to differentiate PA-positive *B. anthracis* spores from PA-negative *B. cereus* and *B. thuringiensis* spores.

Shen *et al.*, (2009) developed a freestanding magnetoelectric biosensor coated with phage for the real-time *in vitro* detection of *B. anthracis spores*. The sensor consisted of a magnetoelastic alloy with a thin layer of gold on the surface to prevent corrosion. After fabrication of the platform, phage were immobilize onto the gold surface by immersing the gold-coated sensor in a phage suspension of 10^{12} phage/ml for 1 hr. The sensor was then washed thoroughly with distilled water and air dried. The sensor exhibited a characteristic resonance frequency upon the application of an alternating external magnetic field. Spore detection was realized by measuring the resonance frequency change due to the change in mass as spores are captured onto the sensor surface (1000μ x 200μ x 15μ). The frequency *versus* mass sensitivity increased significantly with a decrease in sensor length. Spore detection was achieve by measuring the change in resonance frequency due to the change in mass as spores are captured onto the sensor surface. A flowing liquid sample in contact with the sensor was used to achieve a more rapid response than with a static liquid sample. The detection limit of the sensor in a flowing fluid at a rate of 40 ul/min. was 50 spores/ml.

Single-Chain Fv Antibodies for Detection of *B. anthracis* Spores

Recombinant DNA technology has enabled the production of engineered antibody fragments such as Fabs on single chain Fv (scFv) antibodies. An scFv antibody is a small engineered antibody in which the variable heavy chain and light chain of the antibody molecule are connected by a short, flexible polypeptide linker. Phage display technology enables the presentation of scFv antibody on the phage surface and has been used successfully for the isolation of specific scFv antibodies from animal repertoire libraries. Using scFv antibodies for antigen detection has several advantages. scFv antibodies can be produced in large quantities in bacterial expression systems and can be manipulated genetically for improved specificity. The recombinant antibody can also be fused to marker molecules to facilitate detection. Mechaly, Zahavy, and Fisher (2008) developed a single-chain fv antibody for specific detection of *B. anthracis* spores. The antibody was isolated from a phage display library prepared from the spleens of mice immunized with a water-soluble extract of the outer membrane of *B. anthracis* spores (exosporium). The library contained 7×10^6 independent scFv recombinant PFU and was biopanned against a suspension of live, native *B. anthracis* spores, resulting in a unique soluble scFv antibody. Preliminary characterization of scFv specificity indicated that the scFv antibody did not cross-react with spores of *B. subtilis*.

ANIMAL MODELS OF ANTHRAX

The Mouse and Rat

Mice are considered among the laboratory animals most susceptible to infection by *B. anthracis*. However, inbred mouse strains differ in their susceptibilities to lethal infection with different strains of *B. anthracis*.

Macrophages from certain inbred strains of mice are uniquely sensitive to rapid lysis by LT, while those from other strains of mice are resistant (Friedlander *et al.*, 1993). Moayeri *et al.,* (2004) found that LT-mediated lethality in mice is influenced by genetic factors in addition to those controlling macrophage lysis. *B. anthracis* lethal toxin (LT) causes vascular collapse and high lethality in BALB/cJ mice, intermediate lethality in C57BL/6J mice, and no lethality in DBA/2J mice (Moayeri *et al.*, 2005).

Lyons *et al.,* (2004) found that the LD_{50} dose (number of spores) of the Ames strain of *B. anthracis* for six inbred strains of mice varied with respect to the route and method of infection. The LD_{50} values also differed among the mouse strains, but within a given mouse strain, neither the age nor the sex of the mice influenced the LD_{50} values. The subcutaneous route yielded the lowest LD_{50} values with all six mouse strains, with no significant differences observed among the six strains. In contrast significant LD_{50} differences were observed among different strains of mice after pulmonary challenge. BALB/c mice were the most resistant, whereas DBA/2, C3H, and A/J were the most sensitive.

The lethality of nontoxigenic encapsulated (pXO1⁻ pXO2⁺) strains of anthrax in mice and the failure of PA-based vaccines to protect mice after challenge with virulent strains of *B. anthracis* spores have led to the view that the mouse model does not accurately reflect the disease observed in humans (Welkos and Friedlander, 1988a). The murine lethality of PXO2⁺ strains is due to capsule synthesis and does not occur with the Sterne strain (pXO1⁺ pXO2⁻) (Welkos and Friedlander, 1988b). Certain strains of mice such as C57BL/6 and BALB/c, are resistant to challenge with nonencapsulated strains of *B. anthracis*. In contrast, complement deficient A/J mice or C57BL/6 mice treated with Cobra venom factor that destroys complement, are sensitive to aerosol challenge with nonencapsulated Stern spores (Harville *et al.*, 2005; Welkos and Friedlander, 1988b). Loving *et al.*, (2007) in a highly detailed study of anthrax infectious pathology documented that A/J mice (complement deficient) when subjected to aerosol challenge with the pXO1⁺ pXO2⁻ Sterne spores of *B. anthracis* resulted in pathology similar to that observed when rabbits and non-human primates are challenged with a fully virulent strain. An LD_{50} for aerosol challenge of 2 x 10^5 spores for A/J mice was obtained. When a spore dose of $20LD_{50}$ was delivered to three different groups of A/J mice, the mean time to death was 4.7 ± 1.6 days. Death before day 2 and after day 10 was never observed. Interestingly, the infected A/J mice remained asymptomatic until about 12 to 24 hrs. prior to death at which time piloerection (raised hair), edematous swelling of the chest, neck, or head; hind limb paralysis, and lethargy occurred. As early as day 1 postchallenge CFXU were detected in the draining lymph nodes of 70 % of mice and in the livers of nearly 60% of mice. Only 21% of the spleens from mice day 1 postchallenge were positive for CFU. These results clearly indicated that following aerosol challenge of A/J mice with Sterne spores, dissemination to distal organs can occur as early as 1 day after challenge. Mice surviving aerosol challenge were found to retain dormant spores in their lungs for at least 25 days post challenge. The heart tissue of 4/5 moribund mice contained large clusters of bacilli in the myocardium with mild lymphocytic to no inflammatory response. Splenic lesions ranged from lymphocytic depletion (apoptosis) to areas of hemorrhage. In liver tissue, 3/5 mice had numerous bacteria in liver vessels and sinusoids, while two mice had mild hemorrhage with no bacteria observed. Lung lesions ranged from little inflammatory response to bacterial colonies with little inflammatory response to an animal with extensive hemorrhage, necrosis, and numerous bacterial colonies. Homogenates of lungs, lymph nodes, and livers were collected on days 1 and 3 postchallenge and the lung found to posses 1.5 times higher TNF-α levels than uninfected controls. Livers contained a 3-fold higher level of TNF-α than controls, while lymph nodes contained a 7-fold increase on 1 day postchallenge which declined to a 3-fold increase 3 days postchallenge. Similar findings were noted in an inhalation anthrax model using cynomolgus macaques where 100% of the lung sections showed hemorrhages (Vasconcelos *et al.,* (2003)). Vegetative *B. anthracis* bacilli were present in 79% of lungs, while congestion (64%), edema (50%), and inflammation (29%) were noted. In rhesus monkey model of inhalation anthrax, 15% of the infected animals showed anthrax-related pneumonia and bacilli in the alveoli with alveolar hemorrhages in 31% of the infected monkeys (Fritz *et al.,* (1995)). Interestingly, although the total number of CFU in the lungs of moribund mice was high, the number of heat-resistant CFU (spores) was not reduced relative to day 1. This observation suggests that the increase in the lung CFU in moribund mice was due to the presence of vegetative bacilli in the pulmonary vasculature. The liver histopathology of inhalation infected mice was similar to that of macaques with 60% of mice and 70% of the macaques exhibiting bacteria in hepatic sinusoids. Hepatic necrosis was exhibited by 20% of the infected mice and 30% the macaques. A notable difference was the presence of acute inflammation (leukocytosis) in nearly all livers of infected rhesus monkeys, which was not observed with mice. Loving *et al.,* (2007) suggested that the difference between the hepatic findings observed in mice in their study and other animal models may be due to the sensitivity of anthrax infection, as these mice do not survive long enough to mount a strong, inflammatory immune response. These authors concluded that the course of anthrax disease in complement-deficient A/J mice challenged with aerosolized Sterne spores (pXO1$^+$ pXO2$^-$) is similar to that observed in other animal species (rabbits, guinea pigs, and non-human primates) challenged with fully virulent *B. anthracis*.

In contrast to humans, with the mouse model of pulmonary infection, *B. anthracis* virulence is considered to be capsule dependent and toxin independent so that vaccines such as AVA based on PA are of reduced efficacy with mice (Welkos, 1991; Welkos, Vietri, and Gibbs 1993). Rats however are extremely sensitive to the effects of the toxins but are relatively resistant to infection (Young, Zelle, and Lincoln *et al.,* 1946). Toxigenic acapsular derivatives of the Ames and Vollum 1B strains have been found to have significantly higher LD_{50} doses for mice than nontoxigenic capsular derivatives (Welkos, Vieteri, and Gibbs, 1993). Chand *et al.,* (2009) examined the role of toxins in *s.c.* murine infections using two genetically complete (pXO1$^+$ pXO2$^+$) strains of *B. anthracis* (Ames and UT500) and toxin negative derivatives of these strains. Toxin was confirmed as not required for *s.c.* infection by the Ames strain and was concluded from the observation that an LD_{50} dose of a PA deficient (PA$^-$) Ames mutant was identical to that of the parent Ames strain with BALB/c mice. However, PA was required for efficient *s.c.* infection by the UT5009 strain, because the *s.c.* LD_{50} of a UT500 PA- mutant was 10,000-fold higher than the LD_{50} of the parent UT500 strain. These observations were in contrast to findings with BALB/c mice for toxin-deficient mutants of the Ames strain or the UT500 strain (Heninger *et al.,* 2006) infected *via* the respiratory route, where the toxin-deficient mutants of both strains were as virulent as the parent strain. These findings indicate that the dominant virulence factors used to establish infection by *B. anthracis* in mice depend on the route of infection and the bacterial strain.

To further understand the role of anthrax toxins in pathogenesis *in vivo* and to investigate the contribution of antibodies to toxin proteins in protection, Loving *et al.,* (2009) examined a murine aerosol challenge model involving A/J mice (the most sensitive murine strain) and a collection of in-frame deletion mutants lacking toxin components and the Sterne 7702 strain. After aerosol exposure to *B. anthracis* spores, anthrax LT was required for outgrowth of bacilli in the draining lymph nodes and subsequent. After pulmonary exposure to anthrax spores, toxin expression was required for the development of protective immunity to a subsequent lethal challenge. However, IgG titers to toxin proteins, prior to secondary challenge did not correlate with the protection observed upon secondary challenge with wild-type spores. A correlation was observed between survival after secondary challenge and rapid anamnestic responses directed against toxin proteins. These studies indicated that anthrax toxins are required for dissemination of bacteria beyond the draining lymphoid tissue, leading to full virulence in the mouse aerosol challenge model, and that the primary and anamnestic immune responses to toxin proteins provide protection against subsequent lethal challenge.

The IOM report (2002) concluded that neither mice nor rats are good models of human anthrax and that AVA affords incomplete protection in mice.

Guinea Pigs

Guinea pigs have been used extensively in anthrax vaccine development and serve as the standard test system for evaluating the potency of anthrax vaccines. In the standard potency test, guinea pigs are immunized parentally and are challenged with 1000 spores of the Vollum strain. Reuveny *et al.,* (2001) found that with guinea pigs administered PA, TNA titers were a better correlate of protection compared to anti-PA igG antibody titers determined by ELISA.

The pathological changes in guinea pigs afflicted with inhalation anthrax are characterized by widespread edema and hemorrhage, particularly in the spleen, lungs and lymph nodes. In addition fulminating septicemia rather than a primary pulmonary infection results, similar to that in humans (Ross, 1957).

The IOM committee (2002) concluded that AVA affords incomplete protection in guinea pigs from inhalation exposure to anthrax.

Rabbits

In rabbits, the effects observed for inhalation anthrax are similar to those seen in humans and rhesus monkeys (Zauchy *et al.*, 1998). Hemorrhage and edema are sometimes found in the meninges of the brain, but with less inflammation that that seen in humans and rhesus monkeys (IOM, 2002). Other differences between rabbits and humans or rhesus monkeys include milder mediastinal lesions and a lower incidence of lung lesions.

Survival of rabbits administered AVA and then challenged with ~1 x 107 spores (100LD$_{50}$) of the Ames strain was correlated with the levels of anti-PA IgG and with toxin neutralizing antibody (TNA) (IOM, 2002; Pitt *et al.*, 1999, 2001).

Long-term protection of rabbits that had been vaccinated with two doses (four weeks apart) of a recombinant protective antigen (rPA) vaccine was examined by little *et al.,*

(2006) against an aerosol spore challenge with the Ames strain of *B. anthracis* at 6 and 12 months. At 6 months after primary vaccination, survival was 74.1% (20/27). At 12 months after primary vaccination only 37.5% (9/24) of the rabbits were protected. The authors concluded that booster vaccinations with rPA may be required for long term protection of rabbits against anthrax. These results are consistent with the recommended annual booster vaccination of humans with PA-based vaccines.

The IOM report (2002) concluded that AVA is effective in protecting rabbits from inhalation exposure to anthrax.

Chimpanzees

The pathological changes involving chimpanzees with inhalation anthrax have been found to resemble those observed in guinea pigs, mice, and monkeys, with widespread edema and hemorrhage, particularly in the spleen, lungs, and lymph nodes, with death resulting from fulminating sepsis rather than a primary lung infection (Albrink and Goodlow, 1959).

Rhesus Monkeys (Macaques)

The pathology of inhalation anthrax in rhesus monkeys is similar to that in humans. Berdjis *et al.,* (1962) observed edema, hemorrhage, necrosis, and inflammatory infiltrates in the lungs, lymph nodes, spleen, and liver. Gleiser *et al.,* (1963) reported hemorrhagic meningitis in a third of the animals and that the affected lymph nodes were predominantly intrathoracic (hilar, mediastinal, tracheobronchial), whereas Fritz *et al.,* (1995) reported hemorrhagic meningitis in half of the animals and that the mesenteric lymph nodes were more commonly involved. Studies have indicated that AVA is highly protective for rhesus monkeys (Ivins *et al.,* 1996; Pitt *et al.,* 1996; Ivins *et al.,* 1998; Fellows *et al.,* 2001.

The IOM report (2002) concluded that AVA is effective in protecting rhesus monkeys from inhalation exposure to anthrax.

REFERENCES

Alani-Grinstein, Gat, O., Altboum, Z., Velan, B., Cohen, S., Shafferman, A. (2005). Oral spore vaccine based on live attenuated nontoxigenic *Bacillus anthracis* expressing recombinant mutant protective antigen. Infect. Immun. 73:4043-4053.

Albrecht, M., Li, H., Williamson, E., LeButt, C., Flick-Smith, H., Quinn, C., Westra, H., Galloway, D., Mateczun, A., Goldman, S., Groen, H., Baille, L. (2007). Human monoclonal antibodies against anthrax lethal factor and protective antigen act independantly to protect against *Bacillus anthracis* infection and enhance endogenous immunity to anthrax. Infect. Immun. 75:5425-5433.

Albrink, W., Goodlow, R. (1959). Experimental inhalation anthrax in the chimpanzee. Amer. J. Pathol. 35:1055-1065.

Ariel, N., Grosfeld, H., Gat, O., Inbar, Y., Velan, B., Cohen, S., Shafferman, A. (2002). Search for potential vaccine candidate open reading frames in the *Bacillus anthracis* virulence plasmid pXO1: *In silico* and *in vitro* screening. Infect. Immun. 70:6817-6827.

Aucoin, D., Sutherland, M., Percival, A., Lyons, C., Lovcik, J., Kozel, T. (2009). Rapid detection of the poly-γ-D-glutamic acid capsular antigen of *Bacillus anthracis* by latex agglutination. Diag. Microbiol. Infect. Dis. 64:229-232.

Auerbach, S., Wright, G. (1955). Studies on immunity in anthrax. VI. Immunizing activity of protective antigen against various strains of *Bacillus anthracis*. J. Immunol. 75:129-133.

Aulinger, B., Roehrl, M., Mekalanos, J., Collier, R., Wang, J. (2005). Combining anthrax vaccine and therapy: a dominant-negative inhibition of anthrax toxin is also a potent and safe immunogen for vaccines. Infect. Immun. 73:3408-3414.

Baillie, L., Moore, P., McBride, B. (1998). A heat-inducible *B*acillus subtilis φ105 bacteriophage system. FEMS Microbiol. Lett. 163:43-47.

Baillie, L, Fowler, K., Turnbull, P. (1999). Human immune responses to the UK human anthrax vaccine. J. Appl. Microbiol. 87:306-308.

Ballas, Z., Rasmussen, W., Krieg, A. (1996). Induction of NK activity in murine and human cells by CpG motifs in oligonucleotides and bacterial DNA. J. Immunol. 157:1840-1847.

Barnard, J., Friedlander, A. (1999). Vaccination against anthrax with attenuated recombinant strains of *Bacillus anthracis* that produce protective antigen. Infect. Immun. 67:562-567.

Barton, G., Medzhito, R. (2003). Toll-like receptor signaling. Nat. Rev. Immunol. 4:499-1525.

Beedham, R., Turnbull, P., Williamson, E. (2001). Passive transfer of protection against *Bacillus anthracis* infection in a murine model. Vaccine. 19:4409-4416.

Bielinska, A., Janczak, K., Landers, J., Makidon, P., Sower, L., Peterson, J., Baker, J., Jr. (2007). Mucosal immunization with a novel nanoemulsion-based recombinant anthrax protective antigen vaccine protects against *Bacillus anthracis* spore challenge. Infect. Immun. 75:40209-4029.

Berdjis, C., Gleiser, C., Hartman, H., Kuchne, R., Gochenour, W. (1962). Pathogenesis of respiratory anthrax in Macaca mulatta. Brit. J. Exp. Pathol. 43:515-524.

Brachman, P., Gold, H., Plotkin, S., Fekery, F., Werrin, M., Ingraham, N. (1962). Field evaluation of a human anthrax vaccine. Amer. J. P. H. 52:632-645.

Brahmbhatt, T., Janes, B., Stibitz, Darnell, S., E., Sanz, P., R., Rasmussen, S., O'Brien, A. (2007a). *B*. anthracis exosporium protein BclA affects spore germination, interaction with exracellular matrix proteins, and hydrophobicity. Infect. Immun. 75:5233-5239.

Brahmbhatt, T., Darnell, S., Carvalho, H., Sanz, P., Kang,T., Bull, R., Rasmussen, S., Cross, A., O'Brien, A. (2007b). Recombinant exosporium protein BclA of *Bacillus anthracis* is effective as a booster for mice primed with suboptimal amounts of protective antigen. Infect. Immun. 75:5240-5347.

Brockstedt, G., Bahjat, K., Giedlin, M., Liu, W., Leong, M., Luckett, W., Gao, Y., Schnupf, P., Kapadia, D., Castro, G., Lim, J., Sampson-Johannes, A., Herskovits, A., Stassinopoulos,

A., Bower, H., Hearst, J., Portnoy, D., Cook, N., Dubensky, Jr., T. (2005). Killed but metabolically active microbes: a new vaccine paradigm for eliciting effector T-cell responses and protective immunity. Nat. Med. 11:853-860.

Brossier, F., Levy, M., Landlier, A., Lafaye, P., Mock, M. (2004). Functional analysis of *Bacillus anthracis* protective antigen by using neutralizing monoclonal antibodies. Infect. Immun. 72:6313-6317.

Brossier, F., Levy, M., Mock, M. (2002). Anthrax spores make an essential contribution to vaccine efficacy. Infect. Immun. 70:661-664.

Brossier, F., Weber-Levy, M., Mock, M., Sirard, J., (2000). Role of functional domains in anthrax pathogenesis. Infect. Immun. 68:1781-1786.

Brossier, F., Sirard, J., Guidi-Rontani, C., Duflot, E., Mock, M. (1999a). Functional analysis of the carboxy-terminal domain of *Bacillus anthracis* protective antigen. Infect. Immun. 67:964-967.

Brossier, F., Mock, M., Sirard, J. (1999b). Antigen delivery by attenuated *Bacillus anthracis*: new prospects in veterinary vaccines. J. Appl. Microbiol. 87:298-302.

Brossier, F., Weber-Levy, M., Mock, M., Sirard, J. (2000). Role of toxin functional domains in anthrax pathogenesis. Infect. Immun. 68:1781-1786.

Broster, M., Hibbs, S. (1990). Protective efficacy of anthrax vaccines against aerosol challenge. Salisbury Med. Bull. 68 (Suppl.):91-92.

Bruno, J., Yu, H. (1996). Immunomagnetic-electrochemiluminescent detection for *Bacillus anthracis* spores in soil matrices. Appl. Environ. Microbiol. 62:3474-3476.

Campbell, G., deLesdernier, D., Mutharasan, R. (2007). Detection of airborne *Bacillus anthracis* spores by an integrated system of an air sampler and a cantilever immuno- sensor. Sensors and Actuators B. 127:376-382.

Campbell, G., Mutharasan, R. (2006). Piezoelectric-excited millimeter-size canti- lever (PEMC) sensors detect *Bacillus anthracis* at 300 spores/mL. Biosensors & Bioelecronics. 21:1684-1692.

Cote, C., Rossi, C., Kang, A., Morrow, P. (2005). The detection of protective antigen (PA) associated with spores of *Bacillus anthracis* and the effects of anti-PA antibodies on spore germination and macrophage interactions. Microb. Pathogen. 38:209-225.

Cao, A., Liu, Z., Guo, A., Li, Y., Zhang, C., Gaobing, W., Chunfang, F., Tan, Y., Chen, H. (2008). Efficient production and characterization of *Bacillus anthracis* lethal factor and a novel inactive mutant rLFm-Y236F. Protein Expr. Purif. 59:25-30.

Chakrabarty, K., Wu, W., Booth, J., Duggan, E., Coggeshall, K., Metcalf, J. (2006). *Bacillus anthracis* spores stimulate cytokine and chemokine innate immune responses inhuman alveolar macrophages through multiple mitogen-activate protein kinase pathways. Infect. Immun. 74:4430-4438.

Chakrabarty, K., Wu, W., Booth, J., Duggan, E., Nale, N., Coggeshall, K., Metcalf, J. (2007). Human lung innate response to *Bacillus anthracis* spore infection. Infect. Immun. 75:3729-3738.

Chand, H., Drysdale, M., Lovchik, J., Koehler, T., Lipscomb, M., Lysons, C. (2009). Discriminating virulence mechanisms among *Bacillus anthracis* strains by using a murine subcutaneous infection model. Infect. Immun. 77:429-435.

Chauhan, V., Singh, A., Wahed, S., Singh, S., Bhatnagar, R. 2001. Consitutive expression of protective antigen gene of *Bacillus anthracis* in *Escherichia coli*. Biochem. Biophys. Res. Comm. 283:308-315.

Chichester, J., Musiychuk, K., Rosa, P., Horsey, A., Stevenson, N., Ugulava, N., Rabindran, S., Palmer, G., Mett, V., Yusoibov, V. (2007). Immunogenicity of a subunit vaccine against *Bacillus anthracis*. Vaccine. 25:3111-3114.

Chiocco, D., Sobrero, L. (1985). Profilassi vaccinale del carbonchio ematico Inicazioni CEE e vaccino ad uso veterinario preparato dal'Instituto Zooprofilattico Sperimetal della Pugglia edela Basilicata. Atti SIS Vet. 39:657-659.

Cohen, S., Mendelson, I., Altboum, Z., Kobiler, D., Elhanany, E., Bino, T., Leitner, M., Inbar, Rosenberg, H., Gozes, Y., Barak, R., Fisher, M., Kronman, C., Velan, B., Shaferman, A. (2000). Attenuated nontoxigenic and nonencapsulated recombinant *Bacillus anthracis* spore vaccines protect against anthrax. Infect. Immun. 68:4549-4558.

Cole, A., Ganz, T., Liese, A., Burdick, M., Liu, L., Strieter. (2001). IFN-inducible LR-CXC chemokines display defensin-like antimicrobial activity. J. Immunol. 167:623-627.

Cole, K., Strick, C., Paradis, T., Ogborne, K., Loetscher, M., Gladue, R., Lin, W., Boyd, J., Moser, B., Wood, D., Sahagan, Neote, K. (1998). Interferon-inducible T cell alpha chemoattractant (I-TAC): a novel non-ELR CXC chemokine with potent activity on activated T cells through selective high affinity binding to CXCR3. J. Exp. Med. 187:2009-2021.

Colonna, M., Trinchieri, Liu, Y. (2004). Plasmacytoidal dendritic cells in immunity. Nat. Immunol. 5:1219-1226.

Cote, C., Jooijen, N., Welkos, S. (2006). Roles of macrophages and neutrophils in the early host response to *Bacillus anthracis* in a mouse model of infection. Infect. Immun. 74:469-480.

Cowan, G., Atkins, H., Johnson, L., Titball, R., Mitchell, T. (2007). Immunization with anthrolysin O or a genetic toxoid protects against challenge with the toxin but not against *Bacillus anthracis*. Vaccine. 25:7197-7205.

Crawford, M., Zhu, Y., Green, C., Burdick, M., Sanze, P., Alem, F., O'Brien, A., Mehra, B., Strieter, R., Hughes, M. (2009). Antimicrobial effects of interferon-inducible CXC chemokines against *Bacillus anthracis* and bacilli. Infect. Immun. 77:1664-1678.

Cybulski, Jr., R., Sanz, P., McDaniel, D., Darnell, S., Bull, R., O'Brien, A. (2008). Recombinant *Bacillus anthracis* spore proteins enhance protection of mice primed with suboptimal amounts of protecctive antigen. Vaccine. 26:4927-4939.

Dal Molin, F., Fasanella, A., Simonato, M., Garofolo, G., Montecucco, C., Tonello, F. (2008). Ratio of lethal and edema factors in rabbit systemic anthrax. Toxicon. 52:82828.

Davila, A., Jang, J., Gupta, A., Walter, T., Aronson, A., Bashir, R. (2007). Microresonator mass sensors for detection of *Bacillus anthracis* Sterne spores in air and water. Biosens. and Bioelectron. 22:3028-3035.

DelVecchio, V., Connolly, J., Alefantis, T., Walz, A., Quan, M., Para, G., Ashton, J., Wittington, J., Chafin, R., Liang, X., Grewal, P., Khan, A., Mumer, C. (2006). Proteomic profiling and identification of immunodominant spore antigens of *Bacillus anthracis*, Bacillus cereus, and Bacillus thuringiensis. Appl. Environ. Microbiol. 72:6355- 6363.

Daubenspeck, J., Zeng, H., Chen, P., Dong, S., Steichen, C., Krishna, N., Pritchard, D., Turnbough, Jr., C. (2004). Novel oligosaccharide side chains of the collagen-like region of BclA, the major glycoprotein of the *Bacillus anthracis* exosporium. J. Biol. Chem. 279:30945-30953.

Dhénin, S., Moreau, V., Morel, N., Nevers, M., Voland, H., Creminon, C., Djedaini- Pillard, F. (2008). Synthesis of an anthrose derivative and production of polyclonal antibodies for the detection of anthrax spores. Carbohydrate Res. 343:2101-2110.

Drysdale, M., Olson, G., Koehler, T., Lipscomb, M., Lyons, C. (2007). Murine innate response to virulent toxigenic and nontoxigenic Bacilus anthracis strains. Infect. Immun. 75:1757-1764.

Edelstein, R., Tamanaha, C., Sheehan, P., Miller, M., Baselt, D., Whitman, L., Colton, R. (2000). The BARC biosensor applied to the detection of biological warfare agents. Biosens. Bioelectron. 14:805-813.

Ezzell, Jr., J., Abshire, T. (1988). Immunological analysis of cell-associated antigens of *Bacillus anthracis*. Infect. Immun. 56:349-356.

Ezzell Jr., J., Abshire, T. (1992). Serum protease cleavage of *Bacillus anthracis* protective antigen. J. Gen Microbiol. 138:543-549.

Farber, J. (1997). Mig and IP-10: CXC chemokines that target lymphocytes. J. Leukoc. Biol. 61:246-257.

Fasanella, A., Losito, S., Trott, A, T., Adone, R., Massas, S., Ciucini, F., Chiocco, D. (2001). Detection of anthrax vaccine virulence factors by polymerase chain reaction. Vaccine. 19:4214-4218.

Fasanella, A., Scasiamacchia, S., Garafolo, G. (2009). The behavior of virulent *Bacillus anthracis* strain A0843. Vet. Microbiol. 133:208-209.

Fellows, P., Linscott, M., Ivins, B., Pitt, M., Rossi, C., Gibbs, P., Friedlander, A. (2001). Efficacy of a human anthrax vaccine in guinea pigs, rabbits, and rhesus macaques against challenge by *Bacillus anthracis* isolates of diverse geographic origin. Vacccine. 19:3241-3247.

Fellows, P., Linscott, M., Little, Gibbs, P., S., Ivins, B. (2002). Anthrax vaccine efficiency in golden Syrian hamsters. Vaccine. 20:1421-1424.

Fowler, K., McBride, B., Turnbull, P., Baillie, L. (1999). Immune correlates of protection against anthrax. J. Appl. Microbiol. 87:305.

Friedlander, A., Bhatnagar, R., Leppla, S., Johnson, L., Singh, Y. (1993). Characterization of macrophage sensitivity and resistance to anthrax lethal toxin. Infect. Immun. 61:245-252.

Friedlander, A., Welkos, S., and Ivins, B. (2002). Anthrax vaccines. Curr. topics Microbiol Immun. 271:33-60.

Fritz, D., Jaax, N., Lawrence, W., Davis, K., Pitt, M., Ezell, J., Friedlander, A. (1995). Pathology of experimental inhalation anthrax in the rhesus monkey. Lab. Invest. 73:691-702.

Galen, J., Zhao, L., Chinchilla, M., Wang, J., Pasetti, M., Green, J., Levine, M. (2004). Adaptation of the endogenous *Salmonella enterica* serovar Typhi clyA-encoded hemolysin for antigen export enhances the immunogenicity of anthrax protective antigen domain 4 expressed by the attenuated live-vector vaccine strain CVD 908- *h*trA. Infect. Immun. 72:7096-7106.

Gat, O, Gosfeld, H., Ariel, N., Inbar, I., Zaide, G., Broder, Y., Zvi, A., Chitlaru, T., Altbum, Z., Stein, D., Cohen, S., Shafferman, A. (2006). Search for *Bacillus anthracis* potential vaccine candidates by a functional genomic-serological screen. Infect Immun. 74:3987-4001.

Garmory, H., Titball, R., Griffin, K., Hahn, U., Bohm, R., Beyer, W. (2003). *Salmonella enterica* serovar Typhimurium expressing a chromosomally integrated copy of the *Bacillus anthracis* protective antigen gene protects mice against an anthrax spore challenge. Infect. Immun. 71:3831-3836.

Gaur, R., Gupta, P., Banerjea, C., Singh, Y. (2002). Effect of nasal immunization with protective antigen of *Bacillus anthracis* on protective immune response against anthrax toxin. Vaccine. 20:2836-2839.

Gerhardt, P., Ribi, E. (1964). Ultrastructure of the exosporium enveloping spores of *B*acilus cereus. J. Bacteriol. 88:1774-1789.

Gleiser, C., Berdjis, C., Hartman, H., Gochenour, W. (1963). Pathology of experimental respiratory anthrax in Macaca mulatta. Brit. J. Exp. Pathol. 4:416-426.

Gold, J., Hoshino, Y., Hoshino, S., Jones, M., Nolan, A., Weiden, M. (2004). Exogenous gamma and alpha/beta interferon rescues human macrophages from cell death induced by *Bacillus anthracis*. Infect. Immun. 72:1291-1297.

Goldman, D., Zeng, W., Rivera, J., Nakouzzi, A., Casadevall, A. (2008). Human serum contains a protease that protects against cytotoxic activity of *Bacillus anthracis* lethal toxin *in vitro*. Clin. Vaccine Immunol. 15:970-973.

Gorse, G., Keitel, W., Keyserling, H., Taylor, D., Lockl, M., Alves, K., Kenner, J., Deans, L., Gurwith, M. (2006). Immunogenicity and tolerance of ascending doses of a recombinant protective antigen (rPA102) anthrax vaccine: a randomized, double- blinded, controlled, multicenter trial. Vaccine. 24:5950-5959.

Gu, M., Leppla, S., Klinman, D. (1999). Protection against toxin by vaccination with a DNA plasmid encoding anthrax protective antigen. Vaccine. 17:340-344.

Gupta, P., Waheed, S., Bhatnagar, R. (1999). Expression and purification of the recomb- inant protective antigen of *Bacillus anthracis*. Protein Expr. Purif. 16:369-376.

Gwinn, W., Zhang, M., Mon, S., Sampey, D., Zukauskas, Kassebaum, C., Zmuda, J., Tsai, A., Laird, M. (2006). Scalable purification of Bacilus anthracis protective antigen from *Escherichia coli*. Protein Expr. Purif. 45:30-36.

Hahn, U., Alex, M., Czerny, C., Böhm, R., Beyer, W. (2004). Protection of mice against challenge with *Bacillus anthracis* STI spores after DNA vaccination. Int. J. Med. Microbiol. 294:35-44.

Hahn, A., Lyons, C., Lipsomb, M. (2008). Effect of *Bacillus anthracis* virulence factors on human dendritic cell activation. Human Immunol. 69:552-561.

Halpern, M., Kulander, R., Pisetsky, D. (1996). Bacterial DNA induces murine interferon- gamma production by simulation of IL-12 and tumor necrosis factor-alpha. Cell Immunol. 167:72-78.

Hao, R., Wang, D., Zhang, X., Zuo, G., Wei, H, Yang, R., Zhanag, Z., Cheng, Z., Guo, Y., Cui, Z., Zhou, Y. (2009). Rapid detection of *Bacillus anthracis* using monoclonal antibody functionalized QCM sensor. Biosensors and Bioelectronics. 24:1330-1335.

Harville, E., Lee, G., Grippe, V., Merkel, T. (2005). Complement depletion renders C57BL/6 mice sensitive to the *Bacillus anthracis* Sterne strain. Infect. Immun. 73:4420- 4422.

Henderson, D., Peacock, S., and Belton, F. (1956). Observations on prophylaxis of experimental pulmonary anthrax in the monkey. J. Hyg. 54:28-36.

Heninger, S, Drysdale, M., Lovchik, J., Hutt, J., Lipscomb, M., Koehler, T., Lyons, C. (2006). Toxin-deficient mutants of *Bacillus anthracis* are lethal in a murine model for pulmonary anthrax. Infect. Immun. 74: 6067-6074.

Hepler, R., Kelly, R., McNeesly, T., Fan, H., Losada, M., George, H., Woods, A., Cope, L., Bansal, A., Cook, J., Zang, G., Cohen, S., Wei, X., Keller, P., Leffel, E., Joyce, J., Pitt, L., Schultz, L., Jansen, K., Kurtz, M. (2006). A recombinant 63-kDa form of *Bacillus anthracis* protective antigen produced in the yeast *Saccharomyces cerevisiae* provides protection in rabbit and primate inhalation challenge models of anthrax infection. Vaccine. 24:1501-1514.

Herrmann, J., Wang, S., Zhang, C., Panchal, R., Bavari, S., Lyons, C., Lovchik, J., Golding, B., Shiloach, J., Lu, S. (2006). Passive immunotherapy of *Bacillus anthracis* pulmonary infection in mice with antisera produced by DNA immunization. Vaccine. 24:5872-5880.

Hodgson, A. (1941). Cutaneous anthrax. Lancet. 237:811-813.

Hoile, R., Yuen, M., James, G., Gilbert, G. (2007). Evaluation of the rapid analyte measurement platform (RAMP) for the detection of *Bacillus anthracis* at a crime scene. Forensic Sci. Intl. 171:1-4.

Hughs, M., Green, C., Llowchyj, L., Lee, G., Grippe, V., Smith, M., Jr., Huang, L., Harville, E., Merkel, T. (2005). MyD88-dependant signaling contributes to protection following *Bacillus anthracis* spore challenge of mice: implications for Toll- like receptor signaling. Infect. Immun. 73:7535-7540.

Henderson, D., Peacock, S., and Belton, F. (1956). Observations on prophylaxis of experimental pulmonary anthrax in the monkey. J. Hyg. 54:28-36.

Heninger, S, Drysdale, M., Lovchik, J., Hutt, J., Lipscomb, M., Koehler, T., Lyons, C. (2006). Toxin-deficient mutants of *Bacillus anthracis* are lethal in a murine model for pulmonary anthrax. Infect. Immun. 74: 6067-6074.

Hepler, R., Kelly, R., McNeesly, T., Fan, H., Losada, M., George, H., Woods, A., Cope, L., Bansal, A., Cook, J., Zang, G., Cohen, S., Wei, X., Keller, P., Leffel, E., Joyce, J., Pitt, L., Schultz, L., Jansen, K., Kurtz, M. (2006). A recombinant 63-kDa form of *Bacillus anthracis* protective antigen produced in the yeast *Saccharomyces cerevisiae* provides protection in rabbit and primate inhalation challenge models of anthrax infection. Vaccine. 24:1501-1514.

Herrmann, J., Wang, S., Zhang, C., Panchal, R., Bavari, S., Lyons, C., Lovchik, J., Golding, B., Shiloach, J., Lu, S. (2006). Passive immunotherapy of *Bacillus anthracis* pulmonary infection in mice with antisera produced by DNA immunization. Vaccine:5872-5880.

Hodgson, A. (1941). Cutaneous anthrax. Lancet. 237:811-813.

Hoile, R., Yuen, M., James, G., Gilbert, G. (2007). Evaluation of the rapid analyte measurement platform (RAMP) for the detection of *Bacillus anthracis* at a crime scene. Forensic Sci. Intl. 171:1-4.

Iacono-Connors, L., Welkos, S., Ivins, B., Dalrymple, J. (1991). Protection against anthrax with recombinant virus-expressed protective antigen in experimental animals. Infect. Immun. 59:1961-1965.

IOM. (2002). Institute of Medicine Report. The Anthrax Vaccine: Is it Safe? Does It Work? Lois M. Joellenbeck, Lee L. Zwanziger, Jane S. Durch, and Brian L. Strom, eds. National Academy Press, Washington, DC. 265 p. Ipatenko, N., Sedov, V., Gushchin, V. (1990). Pathogenesis of anthrax in farming animals. Veterinaria. 7:35-38.

Ivins, B., Pitt, M., Fellows, P., Farchaus, J., Benner, G., Waag, D., Little, Z., Anderson, G., Jr., Gibbs, P., Friedlander, A. (1998). Comparative efficacy of experimental anthrax vaccine candidates against inhalation anthrax in rhesus macaques. Vaccine. 16:1141-1148.

Ivins, B., Ezzell, Jr., Jemski, J., Hedlung, K., Ristroph, J., Leppla, S. 1986a. Immunization studies with attenuated strains of *Bacillus anthracis*. Infect. Immun. 52:454-458.

Ivins, B., Fellows, P., Pitt, M., Estep, J., Welkos, S., Worsham, P., Friedlander, A. (1996). Effects of a standard human anthrax vaccine against *Bacillus anthracis* aeosol spore challenge in rhesus monkeys. Salisbury Med. Bull. 87 (Suppl.):125-126.

Ivins, B., Welkos, S. (1986b). Cloning and expression of the *Bacillus anthracis* protective antigen gene in Bacillus subtilis. Infect. Immun. 54:537-542.

Ivins, B., Pitt, M., Fellows, P., Farchaus, J., Benner, G., Waag, D., Little, S., Anderson, Jr., G., Gibbs, P., Friedlander, A. (1998). Comparative efficacy of experimental anthrax vaccine candidates against inhalation anthrax in rhesus monkeys. Vaccine. 16:1141-1148.

Ivins, B., Welkos, S., Knudson, G., Little, S. (1990a). Immunization against anthrax with aromatic compound-dependant (Aro-) mutants of *Bacillus anthracis* and with recombinant strains of Bacillus subtilis that produce anthrax protective antigen. Infect. Immun. 58:303-308.

Ivins, B., Welkos, S., Little, S., Knudson, G. (1990b). Cloned protective activity and progress in development of improved anthrax vaccines. Proc. Intl. Workshop on Anthrax, 11-13, April 1989, Winchester, UK, Salisbury Med. Bull. Special Suppl. No. 68, pp. 86-88.

Ivins, B., Welkos, S., Little, S., Crumrine, M., Nelson, G. (1992). Immunization against anthrax with *Bacillus anthracis* protective antigen combined with adjuvants. Infect. Immun. 60:662-668.

Ivins, B., Fellows, P., Pitt,L., Estep, J., Farchaus, J., Friedlander, A., Gibbs, P. (1995). Experimental anthrax vaccines: efficacy of adjuvants combined with protective antigen against aerosol *Bacillus anthracis* spore challenge in guinea pigs. Vaccine. 13:1779-1784.

Jackson, F., Wright, G., Armstrong, J. (1957). Immunization of cattle against experimental anthrax with alum-precipitated protective antigen or spore vaccine. Am. J. Vet. Res. 18:771-777.

Jernigan, T., Stephens, D., Ashford, D., Omenaca, C., Topiel, M., Galbraith, M., Tapper, M., Fisk, T., Zaki, S., Popovic, T., Meyer, R., Quinn, C., Harper, S., Fridkin, S. Sejvar, J., Shepard, C., McConnell, M., Guarner, J., Shieh, W., Malecki, J., Gerberding, J., Hughes, Jones, T., Obadia, N., Gramsinski, R., Charoenvit, Y., Kolodny, N., Kitov, S., Davis, H., Kieg, A., Hoffman, S. (1999). Synthetic oligodeoxynucleoides containing CpG motifs enhance immunogenic vaccine in Aous monkeys. Vaccine: 17:3065-3071.

Jernigan, T., Stephens, D., Ashford, D., Omenaca, C., Topiel, M., Galbraith, M., Tapper, M., Fisk, T., Zaki, S., Popovic, T., Meyer, R., Quinn, C., Harper, S., Fridkin, S. Sejvar, J., Shepard, C., McConnell, M., Guarner, J., Shieh, W., Malecki, J., Gerberding, J., Hughes, J., Perkins, B. (2001). Bioterrorism-related inhalation anthrax: the first 10 cases reported in the United States. Emerg. Infect. Dis. 7:933-944.

Kang, T., Fenton, M., Weiner, M., Hibbs, S., Basu, S., Baille, L., Cross, A. (2005). Murine macrophages kill vegetative form of *Bacillus anthracis*. Infect. Immun. 73:7495-7501.

Karginov, V., Robinson, T., Riemenscheider, J., Golding, B., Kennedy, M., Siloach, J., Alibek, K. (2004). Treatment of anthrax infection with combination of ciprofloxacin and antibodies to protective antigen of *Bacillus anthracis*. FEMS Immun. Med. Microbiol. 40:71-74.

Klas, S., Petrie, C., Warwood, S., Williams, M., Olds, C., Stenz, J., Cheff, A., Hinchcliffe, M., Richardson, C., Wimer, S. (2008). A single immunization with a dry powder anthrax vaccine protects rabbits against lethal aerosol challenge. Vaccine. 26:5494- 5502.

Klichko, V., Miller, J., Wu, A., Popov, G., Alibek, K. (2003). Anaerobic induction of *Bacillus anthracis* hemolytic activity. Biochem. Biophys. Res. Commun. 303:855-862.

Klinman, D., Tross, D. (2009). A single dose-dose combination therapy that both prevents and treats anthrax infection. Vaccine. 27:1811-1815.

Klinman, D., Xie, H., Litle, S., Curie, D., Ivins, B. (2004). CpG oligonucleotides improve the protective immune response induced by the anthrax vaccination of rhesus macaques. Vaccine. 22:281-2886.

Klinman, D., Yi, A., Beaucage, S., Conover, J., Krieg, A. (1996). CpG motifs expressed by bacterial DNA rapidly induce lymphocytes to secrete IL-6, IL-12, and IFNg. Proc. Natl. Acad. Sci. USA. 93:2879-2883.

Kobiler, D., Gozes, Y. Rosenberg, H., Marcus, D. Reuveny, S., Altboum, Z. (2002). Efficiency of protection of guinea pigs against infection with *Bacillus anthracis* by passive immunization. Infect. Immun. 70:544-550.

Kolesov, S., Mikhailov, N., Presnov, I. (1962). Dissemination and elimination of B. *a*nthracis in vaccinated animals. Proc. GNCIVP. Moscow. T11, pp124-131.

Koya, V., Moayeri, M., Leppla, S., Daniell, H. (2005). Plant-based vaccine: Mice immunized with chloroplast-derived anthrax protective antigen survive anthrax letha challenge. Infect. Immun. 73:8266-8274.

Krieg, A., Yi, A., Matson, S., Waldschmidt, T., Bishop, G., Teasdale, R., Koretsky, G., Klinman, D. (1995). Cp motifs in bacterial DNA trigger direct B-cell activation. Nature. 374:546-548.

Kozel, T., Thorkildson, P., Brandt, S., Welch, W., Lovchik, J., AuCoin, D., Vilai, J., Lyons, C. (2007). Protective and immunochemical activities of monoclonal antibodies reactive with the *Bacillus anthracis* polypeptide capsule. Infect. Immun. 75:1562-163.

Kozel, T., Murphy, W., Brandt, S., Blazar, B., Lovchik, J., Thorkildson, P., Percivalk, A., Lyons, C. (2004). mAbs to *Bacillus anthracis* capsular antigen for immunoprotection in anthrax and detection of antigenemia. Proc. Natl. Acad. Sci. USA. 101:5042-5047.

Krebber, A., Bomhauser, S., Bumester, J., Honeggar, A., Wilunda, J, Bosshard, H., Plückthun, A. (1977). Reliable cloning of functional antibody variable domains from hybridomas and spleen cell repertoires employing a reengineered phage display system. J. Immunol. Methods. 201:35-55.

Kubler-Kielb, J., Liu, T., Mocca, C., Jamadl, F., Robbins, J., Schneerson, R. (2006). Additional conjugation methods and immunogenicity of *Bacillus anthracis* poly-g-D- glutamic acid-protein conjugates. Infect. Immun. 754:4744-4749.

Kudva, I., Griffin, R., Garren, J., Calderwood, S., John, M. (2005). Identification of a protein subset of the anthrax spore immune in humans immunized with the anthrax vaccine adsorbed preparation. Infect. Immun. 73:5685-5696.

Laird, M., Zukauski, D., Johnson, K., Sampy, G., Olsen, H., Garcia, A., Karwoski, J., Cooksey, B., Choi, G., Askins, J., Tsai, A., Pierre, J., Gwinn, W. (2004). Production and purification of *Bacillus anthracis* protective antigen from *Escherichia coli*. Protein Expr. Purif. 38:145-152.

Lee, J., Hadjipanayis, A., Welkos, S. (2003). Venezuelan equine encephalitis virus- vectored-vaccines protect mice against spore challenge. Infect. Immun. 71:1491-1496.

Leffel, L, Twenhafel, N., and Whitehouse, C. (2008). Nosocomial infection of *Serratia marcescens* may induce a protective effect in monkeys exposed to *Bacillus anthracis*. J. Infect. 57:162-164.

Little, S., Ivins, B., Fellows, P., Friedlander, A., (1997). Passive protection by polyclonal antibodies against *Bacillus anthracis* infection in guinea pigs. Infect. Immun. 65:5171-5175.

Little, S., Knudsen, G. (1986). Comparative efficacy of *Bacillus anthracis* live spore vaccine and protective antigen vaccine against anthrax in the guinea pig. Infect. Immun. 52:509-512.

Little, S., Ivins, B., Webster, W., Fellows, P., Pitt, M., Norris, S., Andrews, G. (2006). Duration of protection of rabbits after vaccination with *Bacillus anthracis* recombinant protective antigen vaccine. Vaccine. 24:2530-2536.

Longchamp, P., Leighton, T. (1999). Molecular recognition specificity of *Bacillus anthracis* spore antibodies. J. Appl. Microbiol. 87:246-249.

Luxembourge, A., Hannaman, D., Nolan, E., Ellefsen, B., Nakamura, G., Chau, L., Tellez, O., Little, S., Bernard, R. (2008). Potentiation of an anthrax DNA vaccine with electroporation. Vaccine. 26:5216-5222.

Love, T., Redmond, C., Mayers, C. (2008). Real time detection of anthrax spores using highly specific anti EA1 recombinant antibodies produced by competitive panning. J. Immunol. Methods. 334:1-10.

Loving, K, Kennett, M., Lee, G., Grippe, X., Merkel, T. (2007). Murine aerosol challenge model of anthrax. Infect. Immun. 75:2689-2698.

Loving, L, Khurana, T., Osorio, M., Lee, G., Kelly, V., Stibitz, S., Merkel, T. (2009). Role of anthrax toxins in dissemination, disease progression, and induction of protective adaptive immunity in the mouse aerosol challenge model. Infect. Immun. 77:255-265.

Lui, Y., Lin, S., Huang, C., Huang C. (2008). A novel immunogenic spore coat-associated protein in *Bacillus anthracis*: characterization *via* proteomics approaches and vector- based vaccine system. Prot. Expr. Purif. 57:72-80.

Luster, A., Ravetch, J. (1987). Biochemical characterization of a gamma interferon- inducible cytokine (IP-10). J. Exp. Med. 166:1084-1097.

Luster, A. (1998). Chemokines—chemotactic cytokines that mediate inflammation. N. Engl. J. Med. 338:436-445.

Luster, A. (2002). The role of chemokines in linking innate and adaptive immunity. Cur. Opin. Immunol. 14:129-135.

Lyons, C., Lovchik, J., Hutt, J., Lipscomb, M.,Wang, E., Heninger, S., Berliba, L., Garrison K. (2004). Murine model of pulmonary anthrax: kinetics of dissemination, histo- pathology, and mouse strain susceptibility. Infect. Immun. 72:4801-4809.

Mabry, R., Rani, M., Geiger, R., Hubbard, G., Carrion, Jr., R., Brasky, K., Paterson, J., Georgiou, G., Iverson, B. (2005). Passive protection against anthrax by using a high- affinity antitoxin antibody fragment lacking an Fc region. Infect. Immun. 73:8362-8368.

Marcus, H., Danieli, R., Epstein, E., Velan, B., Shafferman, A., Reuveny, S. (2004). Contribution of immunological memory to protective immunity conferred by a *B. a*nthracis protective antigen-based vaccine. Infect. Immun. 72:3471-3477.

Matyas, G., Friedander, A., Glenn, G., Little, S., Yu, J., Alving, C. (2004). Needle-free skin patch vaccination method for anthrax. Infect. Immun. 72:1181-1183.

McBride, B., Mogg, A., Telfer, J., Lever, M., Miller, J., Turnbull, P., Baillie, L. (1998). Protective efficacy of a recombinant protective antigen against *Bacillus anthracis* challenge and assessment of immunological markers. Vaccine. 16:810-817.

McConnell, M., Hanna, P., Imperiale, M. (2006). Cytokine response and survival of mice immunized with an adenovirus expressing *Bacillus anthracis* protective antigen domain 4. Infect. Immun. 74:1009-1015.

Mechaly, A., Zahavy, E., Fisher, M. (2008). Development and implementation of a single- chain Fv antibody for specific detection of *Bacillus anthracis* spores. Appl. Environ. Microbiol. 74:818-822.

Mesnage, S., Tosi-Couture,E., Mock, M., Fouet, A. (1999). The S-layer homology domain as a means for anchoring heterologous proteins on the cell surface of *Bacillus anthracis*. J. Appl. Microbiol. 87:256-260.

Mikesell, P., Ivins, B., Ristroph, J., Dreier, T. (1983). Evidence for plasmid-mediated toxin production in *Bacillus anthracis*. Infect. Immun. 38:371-376.

Moayeri, M., Haines, D., Young, H., Leppla, S. (2003). *Bacillus anthracis* lethal toxin induces TNF-a-independant hypoxia-mediated toxicity in mice. J. Clin. Invest. 112:670-682.

Moayeri, M., Martinez, N., Wiggins, J., Young, H., Leppla, S. (2004). Mouse susceptibility to anthrax lethal toxin is influenced by genetic factors in addition to those controlling macrophage sensitivity. Infect. Immun. 72:4439-4447.

Mohamed, N., Clagett, M., Li, J., Jones, S., Pincus, S., D'Alia, G., Nardone, L., Babin, M., Spitalny, G., Casey, L. (2005). A high-affinity monoclonal antibody to anthrax protective antigen passively protects rabbits before and after aerosolized *Bacillus anthracis* spore challenge. Infect. Immun. 73:795-802.

Mohamed, N., Claudia, J., Ferreira, C., Little, S., Friedlander, A., Spitalny, G., Casey, L. (2004). Enhancement of anthrax lethal toxin cytoxicity: a subset of monoclonal antibodies against protective antigen increases lethal toxin-mediated killing of murine macrophages. Infect. Immun. 72:3276-3283.

Mohan, K., Cordeiro, E., Vaci, M., McMaster, C., Issekutz, T. (2005). CXCR3 is required for migration to dermal inflammation by normal and *in vivo* activated T cells: differential requirements by CD4 and CD8 memory subsets. Eur. J. Immunol. 35:1702–1711.

Nicolson, G., Nass, M., Nicolson, N. (2000). Anthrax vaccine: controversy over safety and efficacy. Antimicrobics and Infect. Dis. Newslett. 18:1-6.

Niu, M., Ball, R., Woo, E., Burwen, D., Knippen, M., Braun, M., the VAERS working group. (2009). Adverse events after vaccination reported to the Vaccine Adverse Event Reporting System (VAERS), 1990-2007. Vaccine. 27:290-297.

Osorio, M., Wu, Y., Singh, S., Merkel, T., Bhattqaharyya, S., Blake, M., Kopcko, D. (2009). Anthrax protective antigen delivered by *Salmonella enterica* serovar Typhi Ty21a protects mice from a lethal anthrax spore challenge. Infect. Immun. 77:1475-1482.

Parthasarathy, N., Saksena, R., Kovac, P., DeShazer, D., Peacock, S., Wuthiekanum, V., Heine, Parthasarathy, N., Saksena, R., Kovac, P., DeShazer, D., Peacock, S., Wuthiekanum, V., Heine, H., Friedlander, A., Cote, C, Welkos, S, Adamovicz, J., Bavari, S., Waag, D. (2008). Application of microarray technology for the detection of *Burkholderia pseudomalei, Bacilus anthracis, and Francisella tularensis* antibodies. Carbohydr. Res. 343:2783-2788.

Peterson, J., Comer, J., Nofsinger, D., Wenglikowski, A, Walberg, K., Chatuev, B., Chopra, A., Stanberry, L., Kang, A., Scholz, W., Walberg, K., Sircar, J. (2006). Human monoclonal anti-protective antigen antibody completely protects rabbits and is synergistic with ciprofloxacin in protecting mice and guinea pigs against inhalation anthrax. Infect. Immun. 74:1016-1024.

Peterson, J., Comer, W., Baze, J., Nofsinger, D., Wenglikowski, A, Walberg, K., Hardcastle, J., Pawlik, J., Bush, K., Taormina, T., Moen, S., Thomas, J., Chatuev, B., Sower, L., Chopra, A., Stanberry, L., Sawada, R., Scholz, W., Sircar, J. (2007). Human monoclonal antibody AVP-21D9 to protective antigen reduces dissemination of the *Bacillus anthracis* Ames strain from the lungs in a rabbit model. Infect. Immun. 2007. 75:3412-3424.

Pezard, C., Berche, P., Mock, M. (1991). Contribution of individual toxin components to virulence of *Bacillus anthracis*. Infect. Immun. 59:3472–3477.

Pezard, C., Duflot, E., Mock, M. (1993). Construction of *Bacillus anthracis* mutant strains producing a single toxin component. J. Gen. Microbiol. 139:2459–2463.

Pezard, C., Weber, M., Sirard, J., Berche, P., Mock, M. (1995). Protective immunity induced by *Bacillus anthracis* toxin-deficient strains. Infect. Immun. 63:1369-1372.

Phelps, J., Dang, N., Rasochova, L. (2007). Inactivation and purification of cowpea mosaic virus-like particles displaying peptide antigens from *Bacillus anthracis*. J. Virol. Meth. 141:146-153.

Pickering, A., Osorio, M., Lee, G., Grippe, V., Bay, M., Merkel, T. (2004). Cytokine response to infection with *Bacillus anthracis* spores. Infect. Immun. 72:6382-6389.

Pitt, M., Little, S., Ivins, B., Fellows, P., Barth, J., Hewetson, J., Gibbs M., Dertzbaugh, M., Friedlander, A. (2001). *In vitro* correlate of immunity in a rabbit model of inhalation anthrax. Vaccine. 19:4768-4773.

Pitt, M., Ivins, B., Estep, J., Farchaus, J., Friedlander, A. (1996). Comparison of the efficacy of purified protective antigen and MDPH to protect non-human primates from inhalation anthrax. Salisbury Med. Bull. 87 (Suppl.):130.

Pitt, M., Little, S., Ivins, B., Fellows, P., Boles, J., Barth, J., Hewetson, J., Friedlander, A. (1999). *In vitro* correlate of immunity in an animal model of inhalation anthrax. J. Appl. Microbiol. 87:304.

Pittman, P., Gibbs, P., Cannon, T., Friedlander, A. (2001). Anthrax vaccine: short-term safety experience in humans. Vaccine. 290:972-978.

Pittman, P., Kim-Ahn, G., Pifat, D., Coonan, K., Gibbs, P., Little, S., Pace-Templeton, J., Myers, R., Parker, G., Friedlander, A. (2002). Anthrax vaccine: immunogenicity and safety of a dose-reduction route-change comparison study in humans. Vaccine. 20:1412-1420.

Pombo, M., Berthold, I., Gingrich, E., Jaramillo, M., Leef, M., Sirota, L., Hsu, H., Arciniega, J. (2004). Validation of an anti-PA-ELISA for the potency testing of anthrax vaccine in mice. Biologicals. 32:157-163.

Price, B., Liner, A., Park, S., Leppla, S., Mateczun, A., Galloway, D. (2001). Protection against anthrax lethal toxin challenge by genetic immunization with a plasmid encoding the lethal factor protein. Infect. Immun. 69:4509-4515.

Raines, K., Kang, T., Hibbs, S., Cao, G., Weaver, J., Tsai, P., Baillie, L., Cross, A., Rosen, G. (2006). Importance of nitric oxide synthase in the control of infection by *Bacillus anthracis*. Infect. Immun. 74:2268-2276.

Reason, D., Liberato, J., Sun, J., Keitel, W., Zhou, J. (2009). Frequency and domain specificity of toxin-neutralizing paratopes in the human antibody response to anthrax vaccine adsorbed. Infect. Immun. 77:2030-2035.

Reimenschneider, J., Garrison, A., Geisbert, J., Jahring, P., Hevey, M., Negley, D., Schmaljohn, A., Lee, J., Hart, M., Vanderzanden, L., Custer, D., Bray, M., Ruff, A., Ivins, B., Bassett, A., Rossi, C., Schmaljohn, C. (2003). Vaccine. 21:4071-4080.

Reuveny, S., White, Moshe, Adar, Y., Kafri, Y., Altboum, Z., Gozes, Y., Kobiler, D., Shaferman, A., Velan, B. (2001). Search for correlates of protective immunity conferred by anthrax vaccine. Infect. Immun. 69:2888-2893.

Revera, J., Nakouzi, A., Abboud, N., Revskaya, E., Goldman, D., Collier, R., Dadachova, E. Casadavall, A. (2006). A monoclonal antibody to *Bacillus anthracis* protective antigen defines a neutralizing epitope in domain 1. Infect. Immun. 74:4149- 4156.

Rhie, G., Roehrl, M., Mourez, M., Collier, R., McKalanos, J., Wang, J. (2003). A dually active anthrax vaccine that confers protection against both bacilli and toxins. Proc. Natl. Acad. Sci. USA. 100:10925-10930.

Rodriquez, V., Centeno, M., Ulrich, M. (1996). The IgG isotypes of specific antibodies in patients with American cutaneus leishmaniasis; relationship to the cell-mediated immune response. Parasite Immunol. 18:341-345.

Ross, J. (1957). The pathogenesis of anthrax following the administration of spores by the respiratory route. J. Pathol. Bacteriol. 73:485-494.

Rossi, D., Zlotnik, A. (2000). The biology of chemokines and their receptors. Ann. Rev. Immunol. 18:217–242.

Sancar, A., Sancar, G. (1988). DNA repair enzymes. Ann. Rev. Biochem. 57:29-67.

Schumacher, W., Storozuk, C., Dutta, P., Phipps, A. (2008). Identification and characterization of *Bacillus anthracis* spores by multiparameter flow cytometry. Appl. Environ. Microbiol. ss74:5220-5223.

Sellman, B., Mourez, M., Collier, R. (2001). Dominant-negative mutants of a toxin subunit: an approach to therapy of anthrax. Science. 292:695-697.

Shen, W., Lakshmanan, R., Mathison, L., Petrenko, V., Chin, B. (2009). Phage coated magnetoeleastic micro-biosensors for real-time detection of *Bacillus anthracis* spores. Sensors and Actuators B. 137:501-506.

Shlyakhov, E., Rubinstein, E., (1994). Human live anthrax vaccine in the former USSR. Vaccine. 12:727-730.

Shlyakhov, E., Rubinstein, E., and Novikov, I. (1997). Anthrax post-vaccinal cell- mediated immunity in humans: kinetics pattern. Vaccine. 15:631-636.

Singer, D., Schneerson, R., Bautista, C., Ruberton, M., Robbins, J., Taylor, D. (2008). Serum IgG antibody response to the protective anigen (PA) of *Bacillus anthracis* induced by anthrax vaccine adsorbed (AVA) among U.S. military personnel. Vaccine. 26:869-873.

Singh, Y., Ivins, B., Leppla, S. (1998). Study of immunization against anthrax with the purified recombinant protective antigen of *Bacillus anthracis*. Infect. Immun. 66:3447- 3448.

Skoble, J., Beaber, J., Gao, Y., Lovchik, J., Sower, L., Liu, W., Luckett, W., Peterson, J., Calendar, R., Portnoy, D., Lyons, C., Dubensky, Jr., W. (2009). Killed but metabolically active *Bacillus anthracis* vaccines induce broad and protective immunity against anthrax. Infect. Immun. 77:1649-1663.

Staats, H., Alam, S., Scearce, R., Kirwan, S., Zhang, J., Gwinn, W., Hanes, B. (2007). *In vitro* and *in vivo* characterization of anthrax anti-protective antigen and anti-lethal factor monoclonal antibodies after passive transfer in a mouse lethal challenge model to define correlates of immunity. Infect. Immun. 75:5443-5452.

Steichen, C., Chen, P., Kearney, J., Turnbough, Jr., C. (2003). Identification of the immuno- dominant protein and other proteins of the *Bacillus anthracis* exosporium. J. Bacteriol. 185:1903-1910.

Stepanov, A., Marinin, L., Pomerantsev, A., Staritsin, N. (1996). Development of novel vaccines against anthrax in man. J. Biotechnol. 44:155-160.

Sterne, M. (1939a). The use of anthrax vaccines from avirulent (unencapsulated) variants of *Bacillus anthracis*. Onderstepoort J. of Vet. Sci. Ann. Index. 13:307-312.

Sterne, M (1939b). Immunization of laboratory animals against anthrax. J. South Afric. Vet. Med. Assoc. 13:53-57.

Sterne, M. (1937). Variation in *Bacillus anthracis*. Onderstepoort J. Vet. Sci. Anim. Ind. 8:271.

Sterne, M, Nichol, J., Lambrechts, N. (1942). The effect of large-scale active immunization against anthrax. J. S. Afric. Vet. Med. Assoc. 13:53-63.

Stokes, M., Titdball, R., Neeson, B., Galen, J., Walker, N., Stagg, A., Jenner, D., Tshwaite, J., Nataaro, J., Bailie, L., Atkins, H. (2007). Oral administration of a *Salmonella enterica*- based vaccine expressing *Bacillus anthracis* protective antigen confers protection against aerosolized *B*. anthracis. Infect. Immun. 75:1827-1834.

Swiderski, R. (2004). Anthrax-A History. McFarland Press, Jefferson, NC, U.S.A. 266 Pg. Sylvestre, P., Couture-Tosi, E., Mock, M. (2002). A collagen-like surface glycoprotein is a structural component of the *Bacillus anthracis* exosporium. Mol. Microbiol. 45:169-178.

Tims, T., Lim, D. (2004). Rapid detection of *Bacillus anthracis* spores directly from powders with an evanescent wave fiber-optic biosensor. J. Microbiol. Meth. 59:127-130.

Turnbull, P., Broster, M., Carman, J., Manchee, R., Melling, J. (1986). Development of antibodies to protective antigen and lethal factor components of anthrax toxin in humans and guinea pigs and their relevance to protective immunity. Infect. Immun. 52:356-363.

Turnbull, P. (1991). Anthrax vaccines: past, present and future. Vaccine. 9:533-539.

Turnbull, P., Quinn, C., Hewson, R., Stockbridge, C., Melling, J. (1990). Protection conferred by microbially-supplemented UK and purified PA vaccines. P. 89-91.

In J. R. H. Inkerton (ed.), Proceedings of the International Workshop on Anthrax, Winchester, England, Salisbury Medical Bulletin no. 68, Salisbury Medical Society, Salisbury, England. Turnbull, P., Tindall, B., Coetzee, J., Conradie, C., Bull, R., Lindeque, P., Huebschle, O. (2004). Vaccine-induced protection against anthrax in cheetah (Acinonyx jubatus) and black rhinoceros (Diceros bicornis). Vaccine. 22:3340-3347.

Uchida, I., Hashimoto, K., Terakado, N. (1986). Virulence and immunogenicity in experimental animals of *Bacillus anthracis* strains harboring or lacking 110 MDa and 60 MDa plasmids. J. Gen. Microbiol. 132:557-559.

Uithoven, K., Schmidt, J, and Ballman, M. (2000). Rapid identification of biological warfare agents using an instrument employing a light addressable potentiometric sensor and a flow-through immunofiltration-enzyme assay system. Biosens. Bioelectron. 14:761-770.

Vasconcelos, D., Barnewall, R., Babin, M., Hunt, R., Estep, J, Nielse, R., Carnes, R., Carney, J. (2003). Pathology of inhalation anthrax in cynomolgus monkeys (macaca *fascicularis*). Lab Invest. 83:1201-1209.

Verthelyi, D., Kenney, R., Seder, R., Gam, A., Friedag, B., Klinman, D. (2002). CpG oligo-deoxynucleotides as vaccine adjuvants in primates. J. Immunol. 168:1659-1663.

Vitale, L., Blanset, D., Lowry, I., O'Neill, T., Goldstein, J, Little, S. Adrews, G., Dorough, G, Taylor, R., Keler, T. (2006) Prophylaxis and therapy of inhalation anthrax by a novel monoclonal antibody to protective antigen that mimics vaccine-induced immunity. Infect. Immun. 74:5840-5847.

Vogel, C. (1991). Cobra venom factor: The complement-activating protein of cobra venom. In A. T. Tu, (ed.), Handbook of natural toxins: reptile venoms and toxins, vol 5. Dekker, New York, N.Y. pp. 147-188.

Wan, J., Johnson, M., Guntupalli, R., Petrenko, V., Chin, B. (2007). Detection of *Bacillus anthracis* spores in liquid using phage-based magnetoelastic micro- resonators. Sensors and Actuators B. 127:559-566.

Williams, D., Benedek, O., Turnbough, Jr., C. (2003). Species-specific peptide ligands for the detection of *Bacillus anthracis* spores. Appl. Environ. Microbiol. 69:6288-6293.

Wagner, S., Neuberger, M. (1996). Somatic hypermutation of immunoglobulin genes. Ann. Rev. Immunol. 14:441-457.

Wang, T., Fellows, P., Leighton, T., Lucas, A. 2004. (Induction of opsonic antibodies to the g-D-glutamic acid capsule of *Bacillus anthracis* by immunization with a synthetic peptide-carrier protein conjugate. FEMS Immun. Med. Microbiol. 40:231-237.

Ward, M., McGann, V., Hogge, Jr. A., Huff, M., Kanode, Jr., R., Roberts, E. (1965). Studies on anthrax infections in immunized guinea pigs. J. Infect. Dis. 115:59-67.

Watson, J. Koya, V., Leppla, S, Danniell, H. (2004). Expression of *Bacillus anthracis* protective antigen in transgenic chloroplasts of tobacco, a non-food/feed crop. Vaccine. 22:4374-4384.

Weaver, J., Kang, T., Raines, K., Cao, G., Tsai, S., Baillie, L., Rosen, G., Cross, A. (2007). Protective Role of *Bacillus anthracis* exosporium in macrophage- mediated killing by nitric oxide. Infect. Immun. 2007. 75:3894-3901.

Weaver, J., Kang, T, Raines, K., Cao, G., Hibbs, S., Tsai, P., Baillie, L., Rosen, G., Cross. (2007). Protective role of *Bacillus anthracis* exosporium in macrophage-mediated killing by nitric oxide. Infect. Immun. 75:3894-3901 Weiss, S., Kobiler, D., Levy, H., Mrcus, H., Pass, A., Rothchild, N., Altboum, Z. (2006). Immunological correlates for protection against intranasal challenge of *Bacillus anthracis* spores conferred by a protective antigen-based vaccine in rabbits. Infect. Immun. 74:394-398.

Welkos, S. (1991). Plasmid-associated virulence factors of non-toxigenic (pXO1-) *Bacillus anthracis*. Microb. Pathogen. 10:183-198.

Welkos, S., Friedlander, A. (1988a.) Comparative safety and efficacy against *Bacillus anthracis* of protective antigen and live vaccines in mice. Microb. Pathog. 5:127-139.

Welkos, S., Friedlander, M. (1988b). Comparative safety and efficacy against *Bacillus anthracis* of protective antigen and live vaccines in mice. Microb. Pathog. 5:127–139.

Welkos, S., Friedlander, A. (1988b). Pathogenesis and genetic control of resistance to the Sterne strain of *Bacillus anthracis*. Microb. Path. 4:53-69.

Welkos, S. Vietri, N., Gibbs, P. (1993). Non-toxigenic derivatives of the Ames strain of *Bacillus anthracis* are fully virulent for mice: role of plasmid pXO2 and chromosome in strain-dependent virulence. Microb. Path. 14:14:381-388.

Whiting, G., Rijpkema, T., Adams, T., Corbel, M. (2004). Characterization of adsorbed anthrax vaccine by two-dimensional gel electrophoresis. Vaccine. 22:4245-4251.

WHO. (2008). Anthrax in Humans and Animals. World Health Organization. 4th ed. WHO Press, Geneva, Switzland. 209 Pg. Williamson, E., Beedham, R., Bennet, A., Perkins, S., Baillie, L. (1999). Presentation of protective antigen to mouse immune system: immune sequelae. J. Appl. Microbiol. 87:315-317.

Williamson, E., Hodgson, I., Walker, N., Topping, A., Duchars, M., Mott, J., Estep, J., LeButt, C., Flick-Smith, H., Jones, H., Li, H., Quinn, C. (2005). Immunogenicity of recombinant protective antigen and efficacy against aerosol challenge with anthrax. Infect. Immun. 73:5978-5987.

Wimer-Mackin, S., Hinchcliffe, M., Petrie, C., Warwood, S., Tino, W., Williams, M., Stenz, J., Cheff, A., Richardson, C. (2006). An intranasal vaccine targeting both *Bacillus anthracis* toxin and bacterium provides protection against aerosol spore challenge in rabbits. Vaccine. 24:3953-3963.

Wright, T., Farber, J. (1991). 5' regulatory region of a novel cytokine gene mediates selective activation by interferon gamma. J. Exp. Med. 173:417-422.

Xie, H., Gursel, I., Ivins, B., Singh, M., O'Hagan, D., Ulmer, J., Klinman, D. (2005). CpG oligonucleotides adsorbed onto polyactide-co-gycolide microparticles improve the immunogenicity and protective activity of the licensed anthrax vaccine. Infect. Immun. 73:828-833.

Yamamoto, M., Takeda, K., Akira, S. (2004). TIR domain-containing adaptors define the specificity of TLR signaling. Mol. Immunol. 40:861.

Yin Y., Zhang, J., Dong, D., Liu, S., Guo, Q. (2008). Chimeric hepatitis B virus core particles carrying an epitope of anthrax protective antigen induce protective immunity against *B. anthracis*. Vaccine. 26:5814-5821.

Young, G., Zelle, M., Lincoln, R. (1946). Respiratory pathogenicity of *Bacillus anthracis* spores. Methods of study and observations on pathogenesis. J. Infect. Dis. 79:233-246.

Zahavy, E., Fisher, M., Bromberg, A., Olshevsky, U. (2003). Detection of frequency resonance energy transfer pair on double-labeled microsphere and *Bacillus anthracis* spores by flow cytometry. Appl. Environ. Microbiol. 69:2330-2339.

Zarokinski, C., Collier, R., Starnbach, M. (2000). Use of anthrax toxin fusions to stimulate immune responses. Meth. Enzymol. 326:542-551.

Zegers, N., Kluter, E., van der Stap, H., van Dura, E., van Dalen, P., Shaw, M., Baillie. (1999). Expression of the protective antigen of *Bacillus anthracis* by Lactobacillus casei towards the development of an oral vaccine against anthrax. J. Appl. Microbiol. 87:309-314.

Zhou, B., Carney, C., Janda, K. (2008). Selection and characterization of human antibodies neutralizing *Bacillus anthracis* toxin. Bioorg. Med. Chem. 16:1903-1913.

Zhou, J., Ullal, A., Liberato, J., Sun, J., Keitel, W., Reason, D. (2008). Paratope diversity in the human antibody response to *Bacillus anthracis* protective antigen. Molec. Immunol. 45:338-347.

Zaucha, G., Pitt, L., Estep, J, Ivins, B., Friedlander, A. (1998). The pathology of experimental anthrax in rabbits exposed by inhalation and subcutaneous inoculation. Arch. Pathol. & Lab. Med. 122:982-992.

CHAPTER 6

Genetics and Molecular Techniques for Detection and Identification of *Bacillus anthracis*

Abstract: All isolates of *Bacillus anthracis* have been found to be genetically homogeneous and are considered to have arisen from a common clone derived from *B. cereus*. Major virulence factors are the protective antigen (PA), the edema toxin (ET), the lethal toxin (LT) and the poly-D-glutamic capsule. The plasmid pXO1 carries the noncontiguous toxin genes *pagA*, *cya*, and *lef* that encode the PA, ET, and LT proteins respectively. The capsule is produced at 37 °C but not at temperatures below 28 °C and only under condition of elevated CO_2. The plasmid pXO2 encodes the contiguous genes *capBCA* for capsule synthesis in addition to *acpA* which is a capsule gene activator. The gene *acpA* carried by the pXO2 plasmid encodes a positive *trans*-acting protein involved in the bicarbonate-mediated regulation of capsule synthesis. Molecular techniques used to distinguish isolates of *B. anthracis* include variable number tandem repeat (VNTR) loci, single nucleotide repeat (SNR) analysis, amplified fragment length polymorphism analysis (AFLP), random amplified polymorphic DNA (RAPD) analysis, multilocus sequence typing (MLST), and long-range repetitive element-PCR (LR REP-PCR), in addition to whole genome typing. Molecular techniques used for detection of *B. anthracis* include conventional PCR and real-time PCR and those used for detection and characterization include DNA microarrays and pulsed-field gel electrophoresis (PFGE).

Keywords: Genetic diversity; molecular typing; phenotypic characteristics; plasmid pXO1; Chinese isolates; plasmid pXO2; capsule genes; poly-D-glutamic acid; PCR primers; protective antigen; lethal toxin; edema toxin; CO_2 induced capsule; VNTR loci; SNRs; AFLP; RAPD, MLDT; LR REP-PCR, conventional & real-time PCR; forensic microbiology.

INTRODUCTION

B. anthracis is a member of the *B. cereus* group, does not exhibit great genetic diversity, and appears to constitute one of the most homogeneous bacterial species. DNA:DNA hybridization studies do not distinguish strains of *B. anthracis* from strains of *B. cereus*. The argument has been made that *B. anthracis* is not a distinct species but is merely a clonal lineage of *B. cereus*. A wide variety of molecular techniques have been used to distinguish *B. anthracis* strains. Differences among the variable number tandem repeat (VNTR) loci has been used to identify phylogenetic relationships among global isolates of *B. anthracis*. The

Robert E. Levin

B. anthracis genome consists of the chromosome and the two plasmids pXO1 and pXO2 plasmids. The presence of both plasmids is required for full virulence. Genetic analysis of *Bacillus anthracis* strains has focused primarily on the genes that encode the anthrax protein toxins, *pagA, lef,* and *cya,* the genes involved in capsule synthesis *capBCA.* The *atxA* gene, which has also been extensively studied is an important regulator of virulence gene expression in *B. anthracis* and positively regulates expression of the three toxin genes and at least one gene required for capsule production. The conventional polymerase chain reaction (PCR) and real-time PCR (Rti-PCR) have proven to be invaluable for rapid detection of *B. anthracis* in environmental samples utilizing numerous primers. This chapter deals extensively with these various aspects of the genetics and molecular techniques that have been applied to *B. anthracis.*

GENETIC HOMOGENEITY OF *B. anthracis* ISOLATES

Genetic Homogeneity and Relationship to *Bacillus cereus*

Although *B. anthracis* is a globally distributed pathogen, it does not exhibit great genetic diversity. The organism appears to constitute one of the most homogeneous species because of (1) its notably narrow taxonomic definition, (2) its slow evolutionary progression, and (3) the fact that it is probably a recent common ancestor to all known isolates (Keim and Smith, 1992). The only highly variable sequences are associated with variable number tandem repeated (VNTR) sequences. The difference among the VNTRs has been used to identify phylogenetic relationships among global isolates (Fig. **1**). The A group is the most common worldwide. The "B" lineage accounts for only about 11% of all known strains and its distribution is restricted primarily to Southern Africa (Keim and Smith, 2002). An association between toxicity of colonies and virulence has been suggested. "B" type isolates are consistently more tenacious than "A" isolates and even appear slimy. In the Kruger National Park, South Africa, the animal mortality rate is 13 times higher in areas where the Kruger "B" strains are found compared with areas where Kruger "A" strains are located (Keim and Smith, 2002).

When large numbers of isolates from diverse sources are examined, strains are encountered that are resistant to the diagnostic gamma phage. In addition, flagellated isolates have been reported (Liang and Yu, 1999) and wild-type isolates have been reported to be resistant to penicillin (Pomerantsev *et al.*, 1993). Helgason *et al.*, (2000) has presented the argument that *B. anthracis* is not a distinct species but is merely a clonal lineage of *B. cereus*. Using both multiple enzyme electrophoresis (MEE) and multiple locus sequence typing (MLST) *B. anthracis* strains were found to be associated with particular *B. cereus* strains.

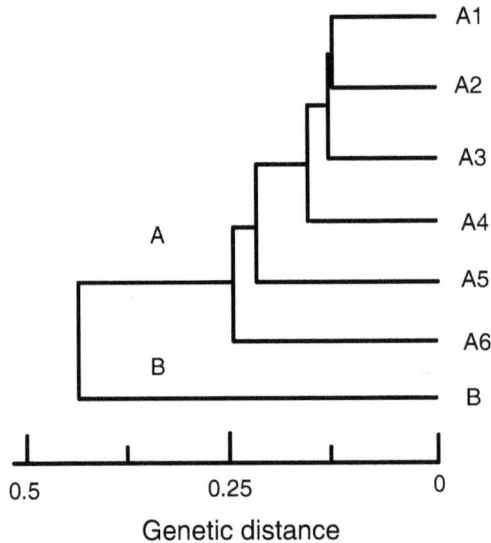

Figure 1: Major *Bacillus anthracis* diversity groups. Molecular typing of *B. anthracis* isolates using amplified fragment length polymorphisms and variable number tandem repeat markers was used to estimate the differences among isolates. Seven distinct clusters were observed when analyzed using unrooted UPGMA cluster analysis. The genetic distance is a simple matching coefficient that is indicative of the percentage of marker loci that differ between groups. Redrawn from Keim *et al.*, (1999) with permission.

La Duc *et al.*, (2004) compared the results of DNA:DNA hybridization and *gyrB* sequence analysis among members of the *B. cereus* group. In a total of 900 DNA:DNA hybridizations performed on 35 strains of the *B. cereus* group, ~90% showed reassociation values of more than 50% to one another including *B. anthracis* isolates reflecting the extremely low level of genetic variability within this group. The DNA:DNA reassociation values yielded the same four groupings as derived from *gyrB* phylotyping with the exception of the *B. thuringiensis* group.

Liang and Yu (1999) performed genetic, phenotypic, and biochemical analyses on 84 strains of *B. anthracis*, 81 of Chinese origin from various sources. These

Chinese isolates differed from those of other geographic origins by: possessing single polar flagella with resulting motility and fimbriae, 64 exhibiting self agglutination, and 60 being unable to ferment maltose. All 84 strains grew well on LB agar. The colony fringe was curled like hair. Virulent strains grew rapidly were rough while attenuated strains grew slowly and were smooth. All 84 strains were non-haemolytic on sheep blood agar. Broth cultures were translucent with no surface film and exhibited pellets on the bottom. Among the 84 strains, only two failed to ferment sucrose. All 84 isolates were lysed by phage AP631, the diagnostic phage isolated in 1963. Motility was exhibited by 77 strains. When growth on LB agar was emulsified with saline, 64 of the 84 strains exhibited self-agglutination which did not occur when isolates were cultured on an $NaHCO_3$ medium with 5-10% CO2 which would have resulted in capsule formation. When PCR products derived from unspecified primers that amplified 16S and 23S rDNA sequences were restricted with *Msp*I and *Hae*II and fragments resolved by polyacrylamide gel electrophoresis, identical restriction patterns for the respective sequences were obtained from the *B. anthracis* strains which were distinct from *B. cereus*. These results indicated that the 16s and 23s rRNA are highly conserved among *B. anthracis* strains and are useful for species identification.

THE *B. anthracis* GENOME

Genomic Composition of *B. anthracis*

Plasmid Associated Virulence of B. anthracis

The *B. anthracis* genome consists of the chromosome and two plasmids. The presence of both plasmids is required for full virulence. Strains have been isolated that contain the plasmids but which are non virulent (Keim and Smith, 2002). The plasmid pXO1 carries the toxin genes and has been sequenced (Okinaka *et al.*, 1999a) and consists of 181,654-bp with 143 open reading frames (ORFs). The pXO1 plasmid carries the noncontiguous toxin genes *pagA*, *cya*, and *lef* that encode the PA, ET, and the LT proteins respectively. These three genes are located in a 44.8-kbp region flanked by inverted IS1627 elements, which has been termed a "pathogenicity island". This pathogenicity island is considered potentially mobile *via* horizontal transfer with 15 putative genes on pXO1 that are implicated in horizontal transfer, including topoisomerases, integrases, and solvases.

Lamonica *et al.,* (2005) used a comparative proteomic approach to identify the differences among the extracellular secretions of three isogenic strains of *B. anthracis* that differed solely in their plasmid contents. The strains utilized were the wild-type virulent *B. anthracis* RA3 (pXO1$^+$ pXO2$^+$), and its two nonpathogenic derivative strains: the toxigenic, nonencapsulated RA3R (pXO1$^+$ pXO2$^-$) and the totally plasmid cured nontoxigenic, nonencapsulated, RA3:00 (pXO1$^-$ pXO2$^-$). A total of twenty different proteins encoded on the chromosome or pXO1 were found to be secreted into an *in vitro* culture medium. S-layer-derived proteins, such as sap and EA1, were most frequently observed. Many sporulation-associated enzymes were found to be overexpressed in strains containing pXO1. Experimental evidence also revealed that pXO2 is necessary for the maximum expression of pXO1-encoded LF, EF, and PA. Several newly identified putative virulence factors were also observed; these included the cell surface protein A, enolase, isocitrate lyase, a high affinity zinc uptake transporter, and the peroxide stress-related alkyl hydroxide reductase.

Organization of the pXO1 Plasmid

Okinaka *et al.,* (1999a) sequenced the 110-MDa pXO1 Sterne plasmid and found it to consist of 181,656-bp. One-hundred forty-three open reading frames (ORFs) were predicted, comprising only 61% (110,817-bp) of the pXO1 DNA sequence. The overall G+C content of the plasmid was found to be 32.5%. The most recognizable feature of this plasmid was a "pathogenicity island" defined by a 44.8-bp region bordered by inverted IS1627 elements at each end. This region was found to contain the three toxin genes *cya, lef, and pagA,* regulatory elements controlling the toxin genes, three germination response genes, a gene encoding a type I topoisomerase, and 19 additional ORFs. Among the ORFs with a high degree of similarity to known sequences is a collection of putative transposases, resolvases, and integrases, suggesting lateral evolutionary movement of DNA among species. A 70-kbp region of pXO1 including the toxin genes (*cya, lef,* and *pag*) is distinct from the remainder of the pXO1 sequence in (1) having a lower gene density (58% *versus* 70%), (2) contains all but one of the co-regulated transcriptional fusions, and (3) it contains a significantly higher proportion of positive BLAST scores (62% *versus* 20%). This observation suggests different origins for the two regions.

Organization of the pXO2 Plasmid

The second virulence plasmid pXO2 carries three contiguous genes *capBCA* which are required for synthesis of the poly-D-glutamic acid capsule which is antiphagocytic and greatly enhances virulence. The pXO2 plasmid has been sequenced Okinaka *et al.,* 1999b) and consists of 96,231-bp with 85 predicted ORFs. In addition to the toxin phenotype, a number of other characteristics are also associated with the pXO1 plasmid. pXO1 cured strains exhibit enhanced sporulation, exhibit altered phage sensitivity compared to pXO1$^+$ strains, and exhibit different nutritional requirements on certain mineral media (Thorne, 1993). The Stern strain is pXO1$^+$ pXO2$^-$ and the Pasteur strain is pXO1$^-$ pXO2$^+$. pXO2 also harbors a *trans*-acting gene regulator of capsule gene expression.

The gene *acpA* was found by Vietri *et al.,* (1995) to encode a positive *trans*-acting protein (57 kDa) on pXO2 which is involved in the bicarbonate-mediated regulation of capsule synthesis. Cell extracts of a strain with *acpA* deleted, grown either with or without bicarbonate did not contain detectable RNA specific for *capB* or *acpA*. In contrast RNA specific for both the *acpA* and *capB* structural genes was detectable only in the presence of bicarbonate with an *acpA*$^-$ strain returned to the *acpA*$^+$ state by complementation. CO_2/bicarbonate-enhanced toxin and capsule gene expression is therefore at the level of transcription.

The genes *atxA* on pXO1 and *acpA*, located on pXO2, encode positive *trans*-acting proteins that are involved in bicarbonate-mediated regulation of toxin and capsule production respectively. Uchida *et al.,* (1997) found that a strain of *B. anthracis* cured of pXO1 produced less capsular substance that the parent strain harboring both pXO1 and pXO2. Complementation of the pXO1$^-$ pXO2$^+$ strain with a plasmid containing the cloned *atxA* gene resulted in an increased level of capsule production. An *acpA*-null mutant was complemented by not only *acpA* but also by the *atxA* gene. The *atxA* gene was also found to stimulate capsule synthesis and to activate expression of *cap in trans* at the transcriptional level. These results indicate that the *atxa* gene located on pXO1 is able to regulate the *cap* region genes on pXO2.

Luna *et al.,* (2006) reported the presence of 10 genes (*acpA, capA, capB, capC, capR, capD,* IS*1627*, ORF 48, ORF 61, and *repA*) and the sequence for the capsular promoter normally found on pXO2 in a *B. circulans* strain designated 118 and a *Bacillus* isolate closely related to *B. luciferensis* designated 119. These sequences were found on a large plasmid in each isolate. The 11 sequences matched to the *B. anthracis* plasmid pXO2. The percent nucleotide identities for *capD* and the capsule promoter were 99.9 and 99.7%, respectively, and for the remaining nine genes, the nucleotide identity was 100% for both isolates. The large plasmid in *B. circulans* was found to be different from the pXO2 plasmid of *B. anthracis* by restriction enzyme patterns using a total of five restriction nucleases. Interestingly, the plasmid in the *Bacillus* 119 isolate was found to yield a restriction enzyme pattern similar to that of pXO2. This is the first study that identified the presence of a plasmid carrying the pXO2 genes in a *Bacillus* isolate, unrelated to the *B. cereus* group. These results lend credence to the concept that horizontal transfer of plasmids occurs in *Bacillus* spp. most probably by conjugation and transduction. Capsule expression by the two bacilli was however not observed when grown with and without 0.8% sodium bicarbonate in 5% CO_2 at 35 °C and in a 20% CO_2 atmosphere.

To obtain a better understanding of pXO2 replication by *B. anthracis* Tinsley and Khan (2007) developed a cell-free system for replication of plasmids using a cell extract of *B. anthracis*. A pXO2 minireplicon was isolated containing the *repS* gene and the origin of replication (*ori*) of pXO2 and was used in conjunction with a plasmid negative strain of *B. anthracis*. pXO2 replication was found to require RNA synthesis through its *ori* and that increased transcription through the *ori* increased the efficiency of replication.

Exclusionary Aspects of pX01 and pXO2 Plasmids

The segregational stability of a small non-mobilizable shuttle plasmid pAX-5E was determined by Bowen and Quinn (1999) in fully virulent (pXO1$^+$/pXO2$^+$), partially cured (pXO1$^+$/pXO2$^-$), and (pXO1$^-$/pXO2$^+$) and fully cured (pXO1$^-$/pXO2$^-$) derivatives of *B. anthracis* var. New Hampshire. pAEx-5E remained segregationally stable in pXO1$^-$/pXO2$^+$ and pXO1$^-$/pXO2$^-$ far in excess of 100 generations, but was expelled from pXO1$^+$/pXO2$^+$ and pXO1$^+$/pXO2$^-$ derivatives. In the presence of erythromycin to maintain selectivity pressure for pAEx-5E no comparable loss of pXO1 or pXO2 was observed for over 100 generations among any of the derivatives. These results may reflect an ability of *B. anthracis* to specifically exclude extrachromosomal DNA other than pXO1 and pXO2 since these are the only plasmids found harbored in *B. anthracis* strains.

Genetic Factors Controlling Capsule Synthesis

The capsule of *B. anthracis* has been shown to be critically required for virulence (Vietri *et al.*, 1995). The biosynthetic genes for capsule synthesis are located contiguously on pXO2 and are designated *capBCA* in addition to *acpA* which is a capsule gene activator (Vietri *et al.*, 1995). The *cap* genes are transcribed as a single operon and are predicted to encode proteins for biosynthesis, transport, and attachment of the D-polyglutamic aid residues on the vegetative cell surface. A fourth gene of the *cap* operon *capD,* is unique to *B. anthracis* and encodes an enzyme that depolymerizes high molecular weight polymers of D-glutamic acid so as to release low molecular weight D-glutamic acid polymers into the environment (Uchida *et al.*, 1993b). The protein product of *capB* is thought to catalyze the polymerization of D-glutamic acid. Makino *et al.,* (2002) demonstrated that a *capD*-null mutant of *B. anthracis* was avirulent in mice. Virulence was restored with the addition of lower-molecular weight capsule polymers at the time of vegetative cell injection.

The *atxA* gene is an important regulator of virulence gene expression in *B. anthracis* and positively regulates expression of the three toxin genes and at least one gene required for capsule production. Hoffmaster and Koehler (1999b) found that an *atxA*-null mutant exhibited phenotypes unrelated to toxin and capsule synthesis, in that it grew poorly on minimal media and sporulated more efficiently than the parent strain when grown in a rich media.

atxA is a *trans*-acting regulator gene located on pXO1 that activates expression of *capB* (Guignot, Mock, and Fouet, 1997; Uchida *et al.*, 1997) and the toxin genes *pagA, lef,* and *cya* (Uchida *et al.*, 1993b; Koehler, Dai, and Kaufman-Yarbray, 1994; Dai *et al.*, 1995. *atxA* is required for CO_2/bicaronate-enhanced transcription of the toxin genes in cells growing in laboratory media and also regulates toxin gene expression during infection. An *atx*-null mutant has been shown to be avirulent in mice (Dai *et al.*, 1995).

The *atxA*-activated *pagA* gene was found by Hoffmaster and Koehler (1999b) to be co-transcribed with a 300-bp gene, designated *pagR* (for *pag* regulator), located downstream of *pagA*. *pagR* was found to repress expression of *atxA* and *pagA* and to thereby function as a negative regulator of these genes (Fig. **2**).

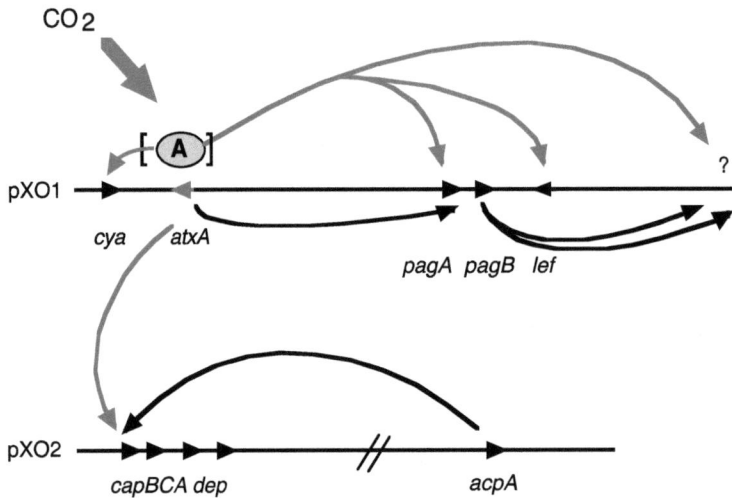

Figure 2: Model depicting *atxA, acpA*, and *pagR*-regulated gene expression in *Bacillus anthracis.* Adapted from Hoffmaster and Koehler, 1999b.

Control of LF, EF, PA and capsule synthesis has been found to be associated with the two regulatory genes *atxA* and *acpA*, located on the virulence plasmids pXO1 and pXO2, respectively. The *atxA* gene on pXO1 has been found to play a significant role in *cap* gene regulation and capsule synthesis while the *acpA* gene on pXO2 plays a minor role (Bourgogne *et al.*, 2003). *acpA* is not essential for capsule gene expression in a pXO1$^+$ pXO2$^+$ strain in which an *acpA*-null mutant (*acpA*$^-$) produced encapsulated cells similar to the parent strain. However, an *atxA* null mutant failed to produce the capsules. Bourgogne *et al.,* (2003) found that expression of the capsule biosynthesis operon *capBCAD*, on pXO2 and *amiA*, a pXO2 gene encoding an amidase that hydrolyzes peptidoglycan, were the only genes whose transcription differed in an *acpA* null mutant. Expression of the *amiA* gene was 10- to 25-fold higher in the wild type than in the *acpA* null mutant, while expression of the *cap* operon genes was only 2-fold higher in the wild type strain. However, in a pXO1⁻ pXO2⁺ strain *acpA* has been found essential for capsule gene expression (Vietri *et al.*, 1995).

In addition to the toxin and capsule genes Bourgogne *et al.*, (2003) found that *atxA* controls expression of various other genes on the chromosome and both plasmids. These observations were derived from a total of 3290 chromosomal and 138 plasmid-borne genes represented as PCR products on a genomic microarray. Generally, plasmid-encoded genes were more highly regulated genes than chromosomal genes, and both positive and negative effects we observed. ORFs showing similarity to genes encoding biosynthetic enzymes for aromatic amino acids *aroA, aroF, hisC* and branched chain amino acids, *ilvC* and *ilvB. tyrA* and *aroF* were negatively regulated by *atxA*.

With the use of reverse transcription-real-time PCR (RT-Rti-PCR) Drysdale *et al.*, (2004) found that *atxA* positively regulates transcription of *acpA* and an *acpA* homolog designated *acpB* and that both *acpA* and *acpB* play a role in capsule gene expression and capsule synthesis. Deletion of *acpA* or *acpB* resulted in cells that were encapsulated throughout growth. However, the average diameter of the capsules with the *acpA-* strain was smaller than that of the parent and the *acpB-* strain throughout growth, indicating that *acpA* has a greater effect on capsule synthesis than does *acpB*. Unlike the individual *acpA⁻* and *acpB⁻* mutants, the double mutant *acpA⁻ acpB⁻* resulted in less than 0.1% encapsulated cells. These observations suggested that the effect of *atxA* on capsule synthesis can be attributed to *atxA* control of *acpA* and *acpB*.

In addition, Drysdale *et al.*, (2004) found that all three regulators *atxA, atxA,* and *acpB*, affected capsule synthesis at the level of transcription. *capB* transcript levels were 32-fold lower in the *atxA⁻* strain than in the wild type, indicating that *atxA* has a significant effect on *capB* gene expression. *capB* transcript levels in the *acpB⁻* strain throughout growth did not differ significantly from those of the wild type, while the *acpA⁻* strain exhibited a two-fold reduction in the *capB* transcript compared to the wild type. The double mutant non-encapsulated strains *atxA⁻ acpA⁻* and *acpA⁻ acpB⁻* exhibited a dramatic decrease in *capB* *t*ranscript levels of more than 100-fold. The *atxA⁻* strain resulted in a 40-fold decrease in the expression of *acpA* and a 15-fold decrease on *acpB*. These observation suggested that *atxA* acts upstream of both *acpA* and *acpB*.

Collectively, these observation indicated that the positive effect of *atxA* on capsule gene expression and capsule synthesis can be attributed to *atxA* control of *acpA* and *acpB*, with *acpA* and *acpB* appearing to be functionally similar. The Rti-PCR primer pairs for the *atxA*, *atxA, acp,* and *capB* genes in addition to primer pairs for the control gene *gyrB* and a sequence derived from the 16S rRNA are given in Table **1**. Results are summarized in Table **2**.

Major phenotypic differences between *B. anthracis* and other members of the *B. cereus* group, excluding the direct pXO1 and pXO2 resulting plasmid phenotypes appear to be the result of altered chromosomal gene expression resulting, at least in part, from the presence of the *AtxA* global regulator gene residing on the pXO1 plasmid.

Genetics of *B. anthracis*

Genetic analysis has been primarily on the genes involved in capsule synthesis c*apB, capC,* and *capA,* a gene associated with depolymerization.

Table 1: PCR primers and DNA probes

Primer or Probe	Sequence (5' —> 3')[a]	Size of Amplified (bp's)	Gene	References
CAP-1	TAG-GAG-TTA-CAC-TGA-GCC	347	*cap-C*	Lee *et al.,* (1999)
CAP-2	TAA-TG-TA-CCC-TTG-TCT-TTG			"
RAPD primer M13	GAG-GGT-GGC-GGC-TCT	-	-	Henderson *et al.,* (1994)
rpoB-s	CCA-CCA-ACA-GTA-GAA-AAT-GCC	175	*rpoB*	Ellerbrok *et al.,* (2002)[d]
rpob-R	AAA-TTT-CAC-CAG-TTC-CTG-GAT-CT			"
rpob-TM	ACT-TGT-GTC-TCG-TTT-CTT-CGA-TCC-AAA-GCG			"
PA-S	CGG-ATC-AAG-TAT-ATG-GGA-ATA-TAG-CAA	204	*pag*	"
PA-R	CCG-GTT-TAG-TCG-TTT-CTA-ATG-GAT			"
PA-TM	CTC-GAA-CTG-GAG-TGA-AGT-GTT-ACC-GCA-AAT			"
Cap-s	ACG-TAT-GGT-GTT-TCA-AGA-TTC-ATG	291	*cap-C*	"

Table 1: contd...

Cap-R	ATT-TTC-GTC-TCA-TTC-TAC-CTC-ACC			"
Cap-TM	CCA-CGG-AAT-TCA-AAA-ATC-TCA-AAT-GGC-AT			"
sspE1-F	GAG-AAA-GAT-GAG-TAA-AAA-ACA-ACA-A	188 & 71	*sspE*	Kim *et al.* (2005)
sspE1-R	CAT-TTG-TGC-TTT-GAA-TGC-TAG			"
lef4-R	TGA-ACC-CGT-ACT-TGT-AAT-CCA-ATC	475	*lef*	"
lef4-R	ATC-GCT-CCA-GTG-TTG-ATA-GTG-CT			"
cap29-F	GTT-GTA-CCT-GGT-TAT-TTA-GCA-CTC	318	*capC*	"
cap29-R	ACC-ACT-TAA-CAA-AAT-TGT-AGT-TCC			"
BACA1F1	ACA-ACT-GGT-ACA-TCT-GCG-CG	623	*capB*	Reif *et al.*, (1994)
BACA6RI	GAT-GAG-GGA-TCA-TTC-GCT-GC			"
PACABL1	Bio-CTG-ACG-AGG-AGC-AAC-CGA-TTA-AGC-GCG-GTA			"
PACAFL1	Fl-CTT-GCT-TTA-GCG-GTA-GCA-GAG-GCT-CTT-GGG			"
g-F	ACT-GGG-GAA-GAT-GAT-GGA-GGA-CT	117	g phage	Reiman *et al.* (2007)
g-R	ACG-TGT-ACG-CAC-TGG-TTC-AA			"
rpoBF1	CCA-CCA-ACA-GTA-GAA-AAT-GCC	175	*rpoB*	QI *et al.*, (2001)
rpoBR1	AAA-TTT-CAC-CAG-TTC-CTG-GAT-CT			"
BaP1 probe	TCC-AAA-GCG-CTA-TGA-TTT-AGC-AAA-TGT-Fl			"
BaP2 probe	Cy5-GGT-CGC-TAC-AG-ATC-AAC-AAG-AAG-AAG-TTA-CAC-P			"
R1	TTA-ATT-CAC-TTG-CAA-CTG-ATG-GG	152	*rpoB*	Patra *et al.*, (1996)
R2	AAC-GAT-AGC-TCC-TAC-ATT-TGG-AG			"
C1 capt. probe	GCC-AGG-TTC-TAT-ACC-GTA-TCA-GCA-A			"
D3 det. probe	AAT-TTG-AAG-CAT-TAA-CGA-GTT-ACT-C			"
vrrA-fl	FAM-CAC-AAC-TAC-CAC-TGG-CAC-A	-	*vrrA*	Keim *et al.*, (2000)
vrrA-rl	GCG-CGT-TTC-GTT-TGA-TTC-ATA-C			"
vrrB$_1$-fl	FAM-ATA-GGT-GGT-TTT-CCG-CAA-GTT-ATT-C	-	*vrrB*$_1$	"
vrrB$_1$-rl	GAT-GAG-TTT-GAT-AAA-GAA-TAG-CCT-GTG			"
vrrB$_2$-fl	FAM-CAC-AGG-CTA-TTC-TTT-ATC-AAA-CTC-ATC	-	*vrrB*$_2$	"
vrrB$_1$-rl	GAT-GAG-TTT-GAT-AAA-GAA-TAG-CCT-GTG			"
vrrC$_1$-fl	GAA-GCA-AGA-AAG-TGA-TGT-AGT-GGA-C	-	*vrrC*$_1$	"
vrrC$_1$-rl	FAM-CAT-TTC-CTC-AAG-TGC-TAC-AGG-TTC			"

Table 1: contd…

vrrC₂-fl	HEX-CCA-GAA-GAA-GTG-GAA-CCT-GTA-GCA-C	-	*vrrC₂*	"
vrrc₂-rl	GTC-TTT-CCA-TTA-ATC-GCG-CTC-TAT-C			"
CG3-fl	NED-TGT-CGT-TTT-ACT-TCT-CTC-TCC-AAT-AC	-	CG3	"
CG3-rl	AGT-CAT-TGT-TCT-GTA-TAA-AGG-GCA-T			"
pXO1-aat-f3	FAMCAA-TTT-ATT-AAC-GAT-CAG-ATT-AAG-TTC-A	-	pX01-aat	"
pXO1-aat-rl	TCT-AGA-ATT-AGT-TGC-TTC-ATA-ATG-GC			"
pXO2-at-fl	HEX-TCA-TCC-TCT-TTT-AAG-TCT-TGG-GT	-	pXO2-at	"
pXO2-at-rl	GTG-TGA-TGA-ACT-CCG-ACG-ACA			"
CAPC-F	ATT-CAT-GAT-TTT-ATA-TGG-CCG-TA	96	*capC*	Wattiau *et al.* (2008)
CAPC-R	TGG-CAT-AAC-AGG-ATA-ACA-AAT-AAT-C			"
CAPC-1	HEX-CGG-$\underline{C}$AA-$\underline{C}$GC-$\underline{T}$AA-$\underline{T}$TA-CAG-GTA-BHQ1ᵃ			"
PAGA-F	TCT-GAG-TTA-GAA-AAT-ATT-CCA-TCG	97	*pagA*	"
PAGA-R	GCA-AAT-GTA-TAT-TCA-TCA-CTC-TTC			"
PAGA-1	FAM-TGA-TAA-$\underline{A}$TC-$\underline{C}$TG-ACC-AAA-TAG-CAG-A-BHQ1ᵃ			"
CL7F	Cy5-TCC-CCC-AAT-ACA-CTC-CCA-TA	-	CL7	"
CL7R	AAA-TTG-GTT-CTG-CAG-CTG-GT			"
CL10F	TTC-GAA-AAC-GGT-AGA-ACA-ACA	-	CL10	"
CL10R2	Cy5-TCG-TTT-TCA-TAC-GTT-TCA-TTA-CC			"
CL12F	Cy5-TCT-CAC-TGT-GCC-TCG-CTA-AA	-	CL12	"
CL12R	AAG-CCA-GGT-GCA-AAA-ACA-GT			"
CL33F	TGG-GGT-ATA-TTC-CCA-TCG-AA	-	CL33	"
CL33R	Cy5-TGT-ACC-GCA-GAT-ACC-AAC-CA			"
LL-Rep1LR	GAT-ATC-TAA-TAC-AAA-CAA-AAC-AAA-AAC	-	-	Brumlik *et al.,* (2001)
adk-f	CAG-CTA-TGA-AGG-CTG-AAA-CTG	450	*adk*	Olsen *et al.,* (2007)
adk-r	CTA-AGC-CTC-CGA-TGA-GAA-CA			" f
ccpA-f	GTT-TAG-GAT-ACC-GCC-CAA-ATG	418	*ccpA*	" f
ccpA-r	TGT-AAC-TTC-TTC-GCG-CTT-CC			" f
ftsA-f	TCT-TGA-CAT-CGG-TAC-ATC-CA	401	*ftsA*	" f
ftsA-r	GCC-TGT-AAT-AAG-TGT-ACC-TTC-C			" f
glpT-f	TGC-GGC-TGG-ATG-AGT-GA	330	*glpT*	" f
glpT-r	AAG-TAA-GAG-CAA-GGA-AGA			" f
pyrE-f	TCG-CAT-CGC-ATT-TAT-TAG-AA	404	*pyrE*	" f

Table 1: contd...

pyrE-r	CCT-GCT-TCA-AGC-TCG-TAT-G			" f
recF-f	GCG-ATG-GCG-AAA-TCT-CAT-AG	470	*recF*	" f
recF-r	CAA-ATC-CAT-TGA-TTC-TGA-TAC-ATC			" f
sucC-f	GGC-GGA-ACA-GAA-ATT-GAA-GA	504	sucC	" f
sucC-r	TCA-CAC-TTC-ATA-ATG-CCA-CCA			" f
BA813-f	TTA-ATT-CAC-TTG-CAA-CTG-ATG-GG	152	*rpoB*	"
BA813-r	AAC-GAT-AGC-TCC-TAC-ATT-TGG-AG			"
BAlef-f	GCA-GAT-TCC-TAT-TGA-GCC-AAA	156	*lef*	"
BAlef-r	GAA-TCA-CGA-ATA-TCA-ATT-TGT-AGC			"
BAcap-1	ACT-CGT-TTT-TAA-TCA-GCC-CG	126	cap	"
BAcap-2	GTT-GCC-GCA-AAT-TTT-CTA-CG			"
plcR-f	CCA-ATC-AAT-GTC-ATA-CTA-TTA-ATT-TGA-CAC	103	plcR	"
plcR-r	ATG-CAA-AAG-CAT-TAT-ACT-TGG-ACA-AT			"
BA5510-1	CTG-CAT-TGA-TAG-CAA-TTT-CTT-CA	162	*BA5510*	"
BA5510-2	CAG-GTT-GAT-ACA-TAA-ACT-TTC-CA			"
BA5510-FL	GTA-ATT-CCC-ATC-ATT-AAA-CCT-TTT-AAT-TCG-ATA-T-FL	-	*BA5510*	"
BA5510-640	CAA-TCC-CTG-TTA-ATT-GAC-CAT-TAA-GCC-640			"
HM1/pXO2-F	PET-GAA-AAC-TTT-GCA-ACC-CAC-C	-	-	Kenefic *et al.* (2008)
HM1/pXO2-R	GTC-GAA-CGT-GTT-GCT-ACA-G			"
HM13/pXO2-F	NED-GGA-TTG-CTT-AAG-GTA-TAT-AAT-GGA-TTT	-	-	"
HM13-pXO2-R	GTT-GTG-TTC-CAT-ATG-TAT-CCC-TCC			"
HM2-2/chr.-F	FAM-TAA-AAA-GAC-AGA-ATT-TTC-AAT-TTT-ATC-AAC-AAC	-	-	"
HM2-2/chr.-R	GTG-GAA-ACT-AAT-GTG-AGT-TAT-ATA-TGT-TAG-TTA-AG			"
HM6-F/chr.-F	VIC-GCT-ATT-CTC-ACT-GTG-CCT-CG	-	-	"
HM6-R/chr.-R	GTT-AAA-ACG-AAG-TAA-AGA-AAA-GTG-GG			"
CL30-F	FAM-CGT-GGA-CCT-CAA-GCT-ACA-AA	-	-	Stratilo *et al.*, (2006)
CL30-R	CGG-ATC-TTC-TGA-TTT-ATA-CGG-C			"
CL37-F	HEX-CTC-GGC-AAT-TTT-CAA-ACG-AT	-	-	"
CL37-R	CCG-CCG-GCA-TAA-AGA-TAG-TA			"
CL66-F	FAM-AAA-CAC-TTC-CGC-TGG-CAT-AC	-	-	"
CL66-R	AAC-CTT-GGG-GCT-TTT-ATG-CT			"
CL23-F	FAM-CAC-CGT-AGC-AAT-AAC-CAC-GA	-	-	"

Table 1: contd...

CL23-R	ACT-CCA-CCT-CCT-CCA-CAA-AA			"
CL1-F	FAM-TTC-TCG-GAG-ATG-ATT-TTC-GG	-	-	"
CL1-R	CTC-CCA-TTT-TAC-ATC-CCC-CT			"
CL28-F	FAM-CGC-AGC-AAT-TGC-AAA-TAA-AA	-	-	"
CL28-R	AGT-GGC-AGG-AGA-TGC-AGA-AG			"
CL67-F	HEX-CAT-GTT-CGC-ATT-TCC-TCC-TT	-	-	"
CL67-R	CCT-GCA-CCA-CAA-CAA-ACA-AC			"
CL31-F	FAM-AAT-GTT-CGC-TAC-GGC-AAA-CT	-	-	"
CL31-R	ATG-GGA-GAC-GGG-AAA-GAA-CT			"
CL38-F	HEX-GTA-GCG-ATG-GCA-GAA-CAA-CA	-	-	"
CL38-R	TGG-CTT-CGA-AAC-CTT-TCA-TC			"
CL34-F	HEX-CAT-TTT-TGA-ACA-GAT-TGC-TGA	-	-	"
CL34-R	TAA-AAG-CCT-TCA-AAC-CCG-TG			"
CL42-F	FAM-GCA-TGC-CAA-CTG-TAT-TCC-CT	-	-	"
CL42-R	GAG-CTC-GTC-ATT-GCA-TCC-TA			"
CL61-F	FAM-TCC-ACC-TTC-TCG-AAT-ATG-CC	-	-	"
CL61-R	GAA-GCG-AAA-TTA-GCT-CAC-CG			"
CL58-F	FAM-GGC-GAG-AAA-ACA-GAG-ACC-TG	-	-	"
CL58-R	CAA-TCA-CTT-ACT-CCA-TTA-TTT-TTC-AA			"
CL76-F	HEX-TGA-AGA-AAT-GCC-TTT-CCT-TTT-T	-	-	"
CL76-R	CAC-CAA-TTA-TTT-GGC-ATG-GA			"
CL60-F	HEX-TGA-GAA-CGA-TTC-CTC-ACC-AG	-	-	"
CL60-R	CAT-GGT-TGT-CTG-GCT-CCT-TT			"
CL33-F	FAM-TGG-GGT-ATA-TTC-CCA-TCG-AA	-	-	"
CL33-R	TGT-ACC-GCA-GAT-ACC-AAC-CA			"
CL51-F	HEX-TGG-GAC-GAA-AAG-TGG-AAT-TTA	-	-	"
CL51-R	GCC-ATC-TTC-AGA-ACC-CAA-AA			"
CL12-F	FAM- TCT-CAC-TGT-GCC-TCG-CTA-AA	-	-	"
CL12-R	AAG-CCA-GGT-GCA-AAA-ACA-GT			"
CL7-F	HEX-TCC-CCC-AAT-ACA-CTC-CCA-TA	-	-	"
CL7-R	AAA-CCA-GGT-GCA-AAA-ACA-GT			"
CL68-F	FAM-CCC-CTT-ACA-TAG-ATG-GCG-AA	-	-	"
CL68-R	CAG-CGT-CGA-TTT-CAT-TAG-CA			"
CL10-F	HEX-TTC-GAA-AAC-GGT-AGA-ACA-ACA	-	-	"
CL10-R	TTT-CGA-AAA-GGG-TAG-AAC-AAC-A			"
CL56-F	HEX-CGA-AAC-CCA-ATA-CGG-TAA-ATG	-	-	"
CL56-R	CCA-TCT-CCC-TTA-TTC-CCT-CC			"

Table 1: contd...

BA1-F	CAT-TTA-GCG-AAG-ATC-CAG-T	168	Ba813	Wang *et al.,* (2004)
BA2-R	CTT-GCT-GAT-ACG-GTA-AGA-AC			"
BA11-F	TTTTTTTTTT-CAT-TTA-GCG-AAG-ATC-CAG-T			"
PAG1-F	ACG-GCT-CCA-ATC-TAC-AAC	140	*pag*	"
PAG2-R	TTT-GCG-GTA-ACA-CTT-CAC			"
PAG11-F	TTTTTTTTTT-ACG-GCT-CCA-ATC-TAC-AAC			"
CAP1-F	CTT-TAG-CGG-TAG-CAG-AGG	428	*cap*	"
CAP2-R	TGG-ACG-CAT-ACG-AGA-CAT			"
CAPP11-F	TTTTTTTTTT-CTT-TAG-CGG-TAG-CAG-AGG			"
dhP77.002-F	TGA-TAT-TTA-TGA-CCA-AGA-TTC-AAT-ATA-CG	133	-	"
dhP77.002-R	GCA-ATA-GCT-CAA-GGT-CAA-TAG-G			"
dhP61.183-F	GAA-GGA-CGA-TAC-AGA-CAT-TTA-TTG-G	163	-	Radnedge *et al.* (2003)
dhP61.183-R	ACC-GCA-AGT-TGA-ATA-GCA-AG			"
dhP73.019-F	TGT-AAA-TGA-ACG-CCT-TGA-CC	196	-	"
dhP73.019-R	CCG-ACT-CCT-TCT-ATC-AAT-TCC			"
dhP73.017-F	AAA-GGC-GGT-TTA-GAA-TTT-GG	241	-	"
dhP73.017-R	TGC-TGC-TCT-TTA-CCC-ATC-C			"
23S-F	CTA-CCT-TAG-GAC-CGT-TAT-AGT-TAC	288	-	"
23S-R	AGG-TAG-GCG-AGG-AGA-GAA-TCC			"
1099-F	Cy5-GCA-ACG-AGC-GCA-ACC-C	560	-	Nübel *et al.,* (2004)
ILE-R	CTG-AGC-TAT-AGS-CCC-ATA[b]			"
RC-BWM-111	Cy5-ATA-AAT-TCC-GCA-ATT-TGT-ATG	-	-	"
Ba-112	Biotin-CCA-TAC-AAA-TTT-CAG-GAT-TTA	-	-	"
Ba-113	Biotin-CCA-TAC-AAA-TTT-CAG-GAT-TT	-	-	"
Ba-113-B	Biotin-CAT-ACA-AAT-TTC-AGG-ATT-T	-	-	"
Bactwmp-1242 Biotin-(20 As)-TGC-AGC-TCT-TTG-TAC-CGT	-	-		"
PS-1-F	TGA-TGG-TTT-TGA-TTT-CTT-AGG-CTT-T	-	-	Van Ert *et al.* (2007)

Table 1: contd…

PS-1-R	CAC-TTT-GGT-TGG-ATG-GTT-TAA-TGA	-	-	"
PS-1 probe	FAM-AAG-GGT-C**A**C-AAC-TC-BHQ[c]	-	-	"
PS-1 probe	VIC-AAG-GGT-C**G**C-AAC-TC-BHQ[c]	-	-	"
PS-52-F	GTA-TCC-TGA-AAT-ATA-AAA-GTG-TAA-AAG-GTA—AAA-AAT-GGA	-	-	"
PS-52-R	GAT-TCT-TCA-ACG-CAA-TAT-ACC-CTA-CTA-AAA-TTA-TAC-TAT	-	-	"
PS-52 probe	VIC-ATT-AAG-GAC-TCC-CT**C**-TTG-GTT[c]	-	-	"
Ps-52 probe	FAM-AAG-GAC-TCC-CT**A**-TTG-GTT			
Branch1-7-F	TCA-CCT-CAA-TGA-CAT-CGC-CA	-	-	"
Branch1-7-R	TTG-TTG-TGA-AGA-CGG-ATA-ACT-TTT-ATG	-	-	"
Branch1-7 probe	FAM-CAA-ACC-AAT-A**C**C-CCT-TTT-BHQ[c]	-	-	"
Branch1-7 probe	VIC-CAA-ACC-AAT-A**A**C-CCT-TT-BHQ[c]	-	-	"
Branch1-26-F	GAC-GGG-AGC-CAA-CCA-GAA	-	-	"
Branch1-26-R	CCG-TTG-AAT-AAG-CAG-TAT-GAA-ATT-TC	-	-	"
Branch1-26 probe FAM-ATA-GCT-TTT-TTT-CTA-TTC-C-BHQ[c]	-		-	"
Branch1-26 probe VIC-ATA-GCT-TTT-T**C**T-CTA-TTC-C-BHQ[c]	-		-	"
Branch1-28-F	AAT-ATG-TTT-CAT-ACA-AGG-CGC-ACT-ACT	-	-	"
Branch1-28-R	CCA-TAA-TCG-TGC-TTG-TCC-AAA-TC	-	-	"
Branch1-28 probe FAM-CGT-TGT-AGT-TAT-TTT-AC-BHQ[c]	-		-	"
Branch1-28 probe VIC-CGT-TGT-AG**G**-TAT-TTT-AC-BHQ[c]	-		-	"

Table 1: contd…

Branch1-31-F	GAA-GAA-CAA-GCG-AA-GAC-GTA-CCT	-	-	"
Branch1-31-R	GTA-GTT-CAT-AAC-GTT-TGA-AAA-AGT-AGG-GAT-A	-	-	"
Branch1-31 probe	VIC-CGG-TTC-ACA-T**GG**-CAT-BHQ^c	-	-	"
Branch1-31-probe	FAM-TCG-GTT-CAC-AT**A**-GCA-T-BHQ^c	-	-	"
Dhp61_183_113F	CGT-AAG-GAC-AAT-AAA-AGC-CGT-TGT	96	*dhp61.1 83*	Antwerpen *et al.* (2008)
Dhp61_183_113R	CGA-TAC-AGA-CAT-TTA-TTG-GGA-ACT-ACA-C			"
Dhp61_183_143T	FAM-TGC-AAT-CGA-TGA-GCT-AAT-GAA-CAA-TGA-CCC-T-TAMRA			"
LEF-1F	GGT-GCG-GAT-TTA-GTT-GAT-TC	851	*lef*	Volokhov *et al.,* (2004)
LEF-1P	GGC-TTC-ATT-TGT-TCT-CCC			"
CYA-1F	GCG-ATG-AAA-ACA-ACG-AAG-TA	720	*cyaA*	"
CYA-P	TCG-TCT-TTG-TCG-CCA-ACT-ATC			"
PA-1F	CCA-GAC-CGT-GAC-AAT-GAT-G	508	*pagA*	"
PA-1R	CAA-GTT-CTT-TCC-CCT-GCT-A			"
CAPAF	TGC-GAC-ATG-GGT-ACA-ACG-TAC-AGA-AGC	525	*capA*	"
CAPAR	TGT-TCT-TTC-GTT-GCA-ATA-GCT-CCT-GCT-AC			"
CAPBF	CAT-CGC-TTG-CTT-TAG-CGG-TAG-CAG-AG	432	*capB*	"
CAPBR	GCA-TAC-GAG-ACA-TAA-TTT-CAC-TTG-TTG-ACC			"
CAPCF	TTG-CAT-TAG-TAT-TAG-GAG-TTA-CAC-TGA-GCC	534	*capc*	"
CAPCR	AGA-GGT-AAT-ACG-ATT-GCT-ACA-TAA-CGA-GG			"
BA-F1	CAT-GGA-ATC-AAG-CTG-CAA-ATT-ATA-AAG	1017	BA-5449	"
BA-R1	CTA-AAT-TGT-CAT-TTG-GCA-AAT-TGA-AAT-TC			"
EWA2	ATG-GTT-CCG-CCT-TAT-CG	~100-168	*vvrA*	Anderson *et al.,* (1996)
EWA1	TAT-CCT-TGG-TAT-TGC-TG			"
Lef 3	CTT-TTG-CAT-ATT-ATA-TCG-AGC	385	*lef*	Ramisse *et al.,* (1996
Lef 4	GAA-TCA-CGA-ATA-TCA-ATT-TGT-AGC			"
Lef 59	GGA-TAT-GAA-CCC-GTA-CTT-GTA-A	993	*lef*	"
Lef 60	TAA-ATC-CGC-ACC-TAG-GGT-TGC			"
pag 67	CAG-AAT-CAA-GTT-CCC-AGG-GG	747	*pag*	"
pag 68	TCG-GAT-AAG-CTG-CCA-CAA-GG			"

Table 1: contd...

pag 23	CTA-CAG-GGG-ATT-TAT-CTA-TTC-C	151	*pag*	"
pag 24	ATT-GTT-ACA-TGA-TTA-TCA-GCG-G			"
cya 25	GGT-TTA-GTA-CCA-GAA-CAT-GC	546	*cya*	"
cya 26	CGG-CTT-CAA-GAC-CCC			"
cya 65	CAG-CAT-GCG-TTT-TCT-TTA-GC	929	*cya*	"
cya 66	CCC-TTA-GTT-GAA-TCC-GGT-TT			"
cap (B,C,A) 17	GAA-ATA-GTT-ATT-GCG-ATT-GG	873	*cap (B,C,A)*	"
cap (B,C,A) 20	GGT-GCT-ACT-GCT-TCT-GTA-CG			"
cap (C) 57	ACT-CGT-TTT-TAA-TCA-GCC-CG	264	*capC*	"
cap (C) 58	GGT-AAC-CCT-TGT-CTT-TGA-AT			"
AnlA-RT-F	GCA-AAG-CAG-GCA-AAA-GAA-ACT	~500	*anlA*	Klichko *et al.,* (2003)
AnlA-RT-R	CCG-TAC-GTA-TCA-AAC-CAA-AGC			"
AnlB-RT-F	CGG-AAA-TGA-ATG-GGA-TAA-AAC	~400	*anlB*	"
AnlB-Rt-R	CAC-TGT-GGA-GAT-TTC-GGT-TGT			"
AnlC-RT-F	CCT-GAT-AAT-ATC-CCG-TTA-GCA	~300	*anlC*	"
AnlC-RT-R	TTC-CCA-CGA-GCA-TCT-CC			"
AnlO-RT-F	AGT-AAT-TCG-GCT-CCA-CCT-GTT	~500	*anlO*	"
AnlO-RT-R	CAA-ATG-TAA-ATT-CAT-CCC-AAG			"
Anl III-RT-F	TTA-TGA-CTG-AAA-AAA-TGA-CAC	~400	*anl III*	"
Anl III-RT-R	TTA-TAA-AAG-CGA-CGT-ACA-A			"
Eag-RT-F	AAA-GTT-TCT-GAG-CCA-ACG-A	~400	*eag*	"
Eag-RT-R	TGC-CAA-CAG-CTA-GTT-TAC-C			"
Sap-RT-F	AGC-AGC-AAT-GGT-AGC-AGG-T	~300	*sap*	"
Sap-RT-R	GAT-TTT-TCC-GTT-TGG-CTC-A			"
PA-RT-F	CCT-TGT-GGC-AGC-TTA-TCC	~450	*pag*	"
PA-RT-R	TTC-CTT-AGC-TTT-AAT-TGT-CG			"
gyrB-F	ACT-TGA-AGG-ACT-AGA-AGC-AG	-	*gyrB*	Drysdale *et al.,* (2004)
gyrB-R	TCC-TTT-TCC-ACT-TGT-AGA-TC			"
gyrB probe	FAM-CGA-AAA-GGC-CCT-GGT-ATG-TAT-A-Q			"
atxA-F	ATT-TTT-AGC-CCT-TGC-AC	-	*atxA*	"
atxA-R	AAG-TTA-ATG-TTT-TAT-TGC-TGT-C			"
atxA probe	FAM-CTT-TTA-TCT-CTT-GGA-AAT-TCT-ATT-ACC-ACA-Q			"
acpA-F	ATT-ATC-TTT-ACC-TCA-GAA-TCA-G	-	*acpA*	"
acpA-R	AAC-GTT-AAT-GAT-TTC-TTC-AG			"
acpA probe	FAM-CAA-TTT-CTG-AAG-CCA-TTT-CTA-ATC-TT-Q			"

Table 1: contd…

acpB-F	TTT-TTC-AAT-ACC-TTG-GAA-CT	-	*acpB*	"
acpB-F	AAT-GCC-TTT-TAG-AAA-CCA-C			"
acpB probe	CTT-GAA-GAA-TCA-TTA-GGA-ATC-TCA-TTA-CA			"
capB-F	TTT-GAT-TAC-ATG-GTC-TTC-C	-	*capB*	"
capB-R	CCA-AGA-GCC-TCT-GCT-AC			"
capB probe	ATA-ATG-CAT-CGC-TTG-CTT-TAG-C			"
16S rRNA-F	TTC-GGG-AGC-AGA-GTG	-	16S rRNA	"
16S rRNA-R	AAC-ATC-TCA-CGA-CAC-GAG			"
16S rRNA probe	CAG-GTG-GTG-CAT-GGT-TGT-C			"
CAP1	TAG-GAG-TTA-CAC-TGA-GCC	321	*cap2*	Brightwell *et al.,* (1998)
CAP2	TAA-TGG-TAA-CCC-TTG-TCT-TTG			"
CAP1242U	GAC-TTG-GAG-CTC-GCG-ATT-GGA-GAA-ACG-ACT-G	360	*capC*	"
CAPC1601L	GAC-TTG-GTC-GAC-GGG-CTG-ATT-AAA-AAC-GAG-TGC			"
CAPC1784U	CAC-ATA-GTC-GAC-CCA-TTT-GAG-ATT-TTT-GAA	258	*capC*	"
AAPC2041L	GAC-AAG-GCA-TGC-TAA-TAT-AAC-TGC-GAT-AAG-AGG			"
BA813 R1	TTA-ATT-CAC-TTG-CAA-CTG-ATG-GG	130	*BA813*	"
BA813 R2	AAC-GAT-AGC-TCC-TAC-ATT-TGG-AG			"
BA19U	GAC-TAC-GAG-CTC-CAA-AAC-TAA-CGA-ATC-TTT-CAT	122	*BA813*	"
BA140L	GAC-TTC-GTC-GAC-AAC-TCG-TTA-ATG-CTT-CAA-AT			"
BA171U	GAC-TTA-GTC-GAC-CTT-TTA-ATG-CCA-GGT-TCT-AT	81	*BA813*	"
BA251L	GAT-CTA-GCA-TGC-CTT-GCA-TAA-TAT-CCT-TGT-TTC			"
MO1	GCT-GAT-CTT-GAC-TAT-GTG-GGT-G	288	*capA*	Makino *et al.,* (1993)
MO2	GGC-TCA-GGA-TCT-GTC-CTT-CGG			"
Ba rpoB-F	CCA-CCA-ACA-GTA-GAA-AAT-GCC	89	rpoB	Sohni *et al.,* (2008)
Ba rpob-R	AAT-TTC-ACC-AGT-TTC-TGG-ATC-T			"
Ba rpob probe	FAM-ACT-TGT-GTC-TCG-TTT-CTT-CGA-TCC-AAA-GCG-MGB			"
AJ035 RAPD primer	AGC-ACC-TCG-TTC-ATG-CTC-ATA-ACG-G	-	AJ03	Levy *et al.,* (2005)
AJ036 RAPD primer	AGC-ACC-TCG-TCT-ACT-TCA-TTT-TGT-GC	-	AJ03	"

Table 1: contd…

AA032 RAPD primer	TTA-GCG-CCC-CCT-TGC-GTT-CC	-	AA03	"
AA033 RAPD primer	TTA-GCG-CCC-CTA-GAC-CAA-TTG-C	-	AA03	"
AT073 RAPD primer	CTC-CTC-AA-TTA-CTA-AAA-TGA-AAC-C	-	AT07	"
AT074 RAPD primer	TTG-GCA-TAG-ACG-TAT-ATT-GCG-GTC-C	-	AT07	"
Sir2-F	CAA-GAG-ATA-CTA-GAA-GAT-GCA-ATC-G	-	Sir2	Ariel *et al.*, (2002)
Sir2-R	TTA-CTG-ATT-TTG-CGT-TAA-AAA-TAA-A			"
MDH-F	CAA-AAT-ACA-GTA-ACA-ACA-GCG	-	MDH	"
MDH-R	TTA-ACG-GAA-ATA-GAT-TTG-TCC-CAT			"
SLP-F	ACT-GTT-AAG-GCT-GAT-AAC-ACA-GAC-TC	-	SLP	"
SLP-R	TTA-CTA-CTT-GCT-CTC-GAC-TTG-AAC-ATA-C			"
PH-F	GCA-ACA-ACA-GCT-GAT-ACA-AAA-CTA	-	PH	"
PH-R	TTA-TTA-AGG-TCC-TGT-TAC-AAT-AGA-ATA-TGT-CAG			"
IL-F	CTT-TTA-ACT-GCC-ACA-TGC	-	IL	"
IL-R	TTA-TCA-ACA-GGT-GTA-TTT-GGA-ATC-TG			"
MP-F	TTG-GAA-ATG-ATA-TAT-TTT-GCG	-	MP	"
MP-R	TTA-CGC-CAC-CCC-ATT-ATT-TG			"
CWRP-F	GAT-ACT-ATT-CCA-GTT-CTA-GTC-TTT-G	-	CWRP	"
CWRP-R	TTA-GTC-TAA-ATC-AGA-CAT-ATA-GAA-GTG-ACC			"
CAAX-LP-F	AGT-AGT-GGT-ATT-AGA-TAT-AAA-TCG-G	-	CAAX-LP	"
CAAX-LP-R	TCA-AGA-AAT-CTG-GAG-TAT-AAC-AAC-AAT			"
SLDI-F	GAA-AGT-GAT-TTA-AAA-AA-ATA-GTT-CCA-G	-	SLDI	"
SLDI-R	TTA-ACT-GTT-TGG-TAT-TCT-AAG-TTT-AGG			"
PB-F	TTT-AAT-TTT-GAT-GTA-TCA-TTC-C	-	PB	"
PB-R	CTA-TCT-ATG-AAC-ATG-AAA-AAT-ACG-G			"
AZP-F	CAA-GAG-CAA-ACA-TCT-GAA-AGT-CAT-AC	-	AZP	"

Table 1: contd...

AZP-R	TTA-ATG-CGC-CAT-CAT-CTC-G			"
F41	TGG-CGG-AAA-AGC-TAA-TAT-AGT-AAA-GTA	63	a/b hydrola se Kane *et al.,* (2009)	
R146	CCA-CAT-ATC-GAA-TCT-CCT-GTC-TAA-AA			"
Probe	FAM-ACT-TCT-AAA-AAG-CAG-ATA-GAA-AT-MGBNFQ			"
CAP6	TAC-TGA-CGA-GGA-GCA-ACC-GA	1035	*capBC*	Beyer *et al.,* (1995)
CAP103	GGC-TCA-GTG-TAA-CTC-CTA-AT			"
CAP9	ATG-TAT-GGC-AGT-TCA-ACC-CG	777	*capBC*	"
CAP102	ACC-CAC-TCC-ATA-TAC-AAT-CC			"
PAG8	GAG-GTA-GAA-GGA-TAT-ACG-GT	596	*pag*	"
PAG5	TCC-TAA-ACT-AAC-GAA-GTG-G			"
PAG7	ATC-ACC-AGA-GGC-AAG-ACA-CCC	210	*pag*	"
PAG6	ACC-AAT-ATC-AAA-GAA-CGA-CGC			"
BA57	ACT-CGT-TTT-TTA-TCA-GCC-CG	265	*capBC*	Sjostedt *et al.* (1997)
BA58	TGG-TAA-CCC-TTG-TCT-TTG-AAT			"
BA59	GAT-ATG-AAC-CCG-TAC-TTG-TAA-T	993	*lef*	"
BA60	TAA-ATC-CGC-ACC-TAG-GGT-TGC-G			"

[a]Underlined positions in the nucleotide sequence identify locked nucleic acids.
[b]S is a G/C nucleotide degeneracy.
[c]Bold letters indicate SNP location.
[d]From Qi *et al.,* (2001).
[e]From Stratilo *et al.,* (2006).
[f]From Helgason *et al.,* (2004).

Table 2: Transcriptional expression level of *capB* compared to wild type and extent of capsule formation in *atxA⁻*, *acpB⁻*, and *acpA⁻* mutant. Derived from data of Drysdale *et al.,* (2004).

Genotype	Level of *capB* Transcription	Extent of Capsule Formation
acpB-	No decrease	All cells encapsulated but capsules have reduced diameter.
acpA-	2-fold decrease	All cells fully encapsulated.
acpA- acpB-	>100-fold decrease	No capsules formed.
atxA-	32-fold decrease	Thin capsule on some cells & no capsule on majority of cells.
atxA- acpB-	100-fold decrease	Thin capsule on some cells and no capsule on majority of cells.
atxA- acpA-	>100-fold decrease	No capsules formed.

of the capsule, *dep,* and the genes that encode the anthrax protein toxins, *cya, lef, and pagA*. The three toxin genes are located at noncontiguous distant loci on pXO1, while the *cap* and *dep* genes are arranged in an apparent operon on pxO2. Additional genes include the spore germination operon *gerX,* the general transcription factor *sigB*, and the pXO1-encoded *trans*-activating gene *atxA* that positively controls expression of the toxin and capsule genes. Additional pXO1 genes include *pagR*, which is cotransduced with *pagA* and acts as a weak repressor of the *pagAR* operon (Hoffmaster and Koehler, 1999a), *gerXB, gerXA, and gerXC*, genes of an operon that affect germination with phagocytes (Guidi-Rontani *et al.,* (1999)), and *topX,* a type 1 topoisomerase gene thought to be involved in plasmid stability (Fouet *et al.*, 1994).

Cellular regulators, such as the transition state regulator AbrB and the sigma factor SigH, also influence virulence factor expression by regulating AtxA synthesis (Saile and Koehler, 2002; Hadjifrangiskou *et al.,* 2007). AbrB is itself regulated by the SpoOA master regulator of sporulation, which is controlled by a multitude of kinases and phophatases tuned to growth cues (Saile and Koehler, 2002). In a cascade of regulatory conrols, AtxA activates the transcription of the pxO2-encoded AcpA and AcpB, which in turn, activate capsule gene expression (Drysdale *et al.*, 2004). AtxA also regulates the expression of Pagr through the cotranscription of the *pagA-pagR* genes. Because PagR positively and negatively regulates the chromosomal *eag* and *sap genes*, respectively, for S-layer production, AtxA is again part of a regulatory network interconnecting virulence plasmids and chromosome gene expression (Mignot *et al.*, 2004).

Uchida *et al.,* (1993a) cloned into *B. subtilis* the *atxA* gene whose product acts in *trans* to stimulate the production of anthrax toxins. Their results demonstrated that the *atxA* gene product activates transcription of *pag, cya*, and *lef.* The activity of a *pag-lacZ* fusion was measured in several *B. anthracis* strains. The highest level of β-galactosidase synthesis was observed in a strain containing *atxA* on a multicopy plasmid and lacking a putative negative regulatory gene (*i.e.*, cured of pXO1). Bicarbonate had no effect on this strain, as might be expected if its normal action is to inhibit the activity of a negative regulatory gene product.

Production of PA has been shown to be transcriptionally regulated by bicarbonate in wild type cells of *B. anthracis* (Bartkus and Leppla, 1989). Koehler, Dai, and

Kaufman-Yarbray (1994) used an insertion mutant of *B. anthracis* in which the wild-type *pag* gene on pXO1 was replaced with a *pag*-lacZ transcriptional fusion to monitor promoter activity. Expression of the *pag*-lacZ fusion was induced five-to-eight-fold during growth in 5% CO_2 compared with growth in air. Growth in 20% CO_2 increased transcription up to 19-fold. The results of 5'-end mapping of *pag* transcripts indicated multiple sites of *pag* transcription initiation. Two major apparent start sites, designated P1 and P2 were identified. Analysis of total RNA from late-log-phase cells indicted comparable initiation from P1 and P2 in wild-type strains grown aerobically. However, initiation from P1 increases ~10-fold in cultures grown in 5% CO_2, with CO_2 having no effect on P2. The authors also identified a *trans*-acing regulatory locus on pXO1, more than 13- kb upstream from the *pag* gene which enhances *pag* transcription. The CO_2 effect on P1 was observed only in the presence of this activator locus.

Uchida *et al.,* (1993a) reported the cloning and sequencing of this locus functioning as a positive regulator of PA synthesis and designated it as the *atxA* gene. In the absence of *atxA* CO_2-enhanced transcription of *pagA* does not occur.

Hoffmaster and Koehler (1997) presented additional evidence that *atxA* is associated with CO_2 enhanced non-toxin gene expression in *B. anthracis*. When wild-type cells were grown in air and 5% CO_2 an increase in specific nontoxic proteins detected by two dimensional electrophoresis was found with 5% CO_2 compared to growth in air. An *stxA*-null derivative grown in 5% CO_2 produced lower levels of specific non toxic proteins quantitatively comparable to those resolved with growth in air.

Although the regulator gene *atxA* was found to be under the influence of CO_2 for expression of non-toxin proteins Dai and Koehler (1997) presented evidence that *atxA* regulation of the anthrax toxins is not influenced by CO_2 but is influenced by temperature. The relative amounts of *atxA* mRNA and AtxA protein were compared with cells grown to late log phase in air, 5% CO_2, and 20% CO_2 and were found not to differ with variations in atmospheric CO_2 levels. However, the steady-state level of *atxA* mRNA in cells grown under 5% CO_2 at 37 and 28 °C was five-to six-fold higher than the level in cells grown at 28 °C. In addition, steady state levels of the AtxA protein were found to be 10-fold higher with cells grown at 37 °C compared to cells grown at 28 °C. Interestingly, a strain of *B. anthracis* harboring an *atxA* deletion on pXO1 when complemented by the addition of *atxA* in *trans* on a multicopy plasmid overproduced AtxA but failed to restore PA synthesis to the wild-type level. The authors suggested that an additional factor is involved in regulation of *pag* and that the relative amounts of such a factor (s) and AtxA are important for optimum gene expression.

Hoffmaster and Koehler (1999a) found that a 300-bp gene located downstream of *pagA* is cotranscribed with *pagA* (formerly *pag*) and represses expression of the operon. This gene is designated *pagR* (for protective antigen repressor). A *pagR* null mutant resulted in an increase in the steady state level of *atxA* mRNA.

Activity of a *cap-lacZ* transcriptional fusion on pXO2 is induced approximately 25-fold during growth in a medium containing bicarbonate (Gigno *et al.*, 1997). The *capB, capC, capA,* and *dep* genes appear to be arranged in the same operon on pXO2. The pXO1-encoded gene *atxA* (anthrax toxin activator) is a global regulator of virulence gene expression (Uchida *et al.,* 1993a). In *B. anthracis*, *atx*-deleted strains produce no detectable EF, LF, or PA and exhibit reduced capsule synthesis (Guignot, Mock, and Fouet, 1997).

The structural toxin genes, *pag* (PA), *lef* (LF), and *cya* (EF), as well as those encoding their regulatory elements, AtxA and PagR, are carried on the pXO1 plasmid. AtxA has been described as a transcriptional activator of the toxin genes, and PagR has been proposed as a weak autorepressor (Hoffmaster and Koehler, 1999), while AcpA is considered a transcriptional activator of the cap BCAD capsule operon (Mignot *et al.*, 2004).

AtxA has major effects on the physiology of *B. anthracis*. It coordinates the transcription activation of the toxin genes with that of the capsule biosynthesis operon, located on the pXO2 virulence plasmid. In rich media, *B. anthracis* synthesizes alternatively two S-layer proteins (Sap and EA1) whose respective genes, *sap* and *eag* reside on the chromosome, where an exponential phase "Sap-layer" is subsequently replaced by a stationary phase "A1-layer". S-layer transcription is controlled by alternative σ factors and by Sap acting as a transcriptional repressor of *eag*. In addition, in the presence of CO_2 *in vitro* and *in vivo*, AtxA is part of the *sap* and *eag* regulatory network. Only *eag* is significantly expressed under these conditions which is due to AtxA activating *eag* and repressing *sap* transcription. Pagr, and not AtxA itself, is the direct effector of this regulation by binding to *sap* and *eag* promoter regions. PagR therefore mediates the effect of AtxA on *eag* and *sap* and is the most downstream element of a signaling cascade initiated by AtxA.

It has been recognized for several decades that pXO1-harboring strains of *B. anthracis* are more encapsulated than their pXO1 cured counterparts (Green *et al.*, 1985). To test a potential pXO1 influence on capsule gene transcription, Mignot *et al.*, (2004) constructed and assayed a *capB-lacZ* transcription fusion and found that the level of transcription of β-galactosidase (β-gal) with CO_2 grown cells was about 20-fold greater with a $pXO1^+$ strain compared to a $pXO1^-$ strain, clearly indicating that the pXO1 plasmid exerts quantitative transcriptional control of the *cap* operon (Table **3**). When the atxA gene was deleted from pXO1 ($pXO1^+$ ΔatxA) β-gal synthesis was reduced 10-fold, whereas a strain lacking the pXO1 plasmid and possessing the *atx* gene as a genomic insert ($pXO1^- + atxA$) exhibited the same elevated level of β-gal synthesis (Table **3**). It can therefore be concluded that AtxA is not only a toxin gene activator but also activates the capsule biosynthesis operon. In addition, Hoffmaster and Koehler (1977) found that pXO1-borne genes other than the toxin genes were regulated by AtxA.

Table 3: Influence of pXO1 harbored *atxA* on the activation of the *capBCAD* operon Transcription in *B. anthracis*. From Mignot *et al.*, (2004), with permission.[a]

	$pXO1^-$	$pX01^+$	$pXO1^+ + Δ\,atxA$[b]	$pXO1^- + atxA$[c]
- CO_2	2	75	3	64
+ CO_2	52	1485	145	1205

[a]Relative expression of *capB-lacz* transcriptional fusion in different genetic backgrounds with units of β-gal activity/mg protein.
[b]*atxA* gene deleted from pXO1
[c]atxA gene added to the chromosome of pXO1[-] strain cells grown with and without CO_2.

The surface of vegetative cells of *B. anthracis* grown in the presence of CO_2 is shrouded by the peptide capsule. The outer surface of the cell wall, immediately below the capsule harbors two S-layer (surface layer) proteins, Sap and EA1, chromosomally encoded by the *sap* and *eag* genes respectively (Mesnage *et al.*, 1997, 1998). Individual deletion of these genes has shown that Sap and Ea1 can each form an S-layer with distinct structural organization (Couture-Tasci *et al.,* 2002). Sap quantitatively increases during exponential growth and reaches a peak after entry into the stationary growth phase (Mignot *et al.,* 2002). *sap-lacZ* and *eag-lacZ* transcription fusions were constructed by Mignot *et al.,* (2002) and β-gal levels indicated that Sap is an *eag* transcription repressor. Similar studies with CO_2 grown cells indicated that the switch between *sap* and *eag* transcription was abolished (Mignot *et al.,* 2003). The absence of pXO1 was found to result in only *sap* being transcribed (Table **4**). The results indicated that pXO1 encodes a transcription factor repressing *sap* and activating *eag* transcription, namely AtxA. In addition, Mignot *et al.,* (2003) found that the absence of *pagR* in a pXO1^{+} containing host resulted in relief of AtxA repression of β-gal from the *sap-lac* fusion (Table **4**). PagR is therefore looked upon as an intermediate effector between AtxA and the *sap* and *eag* genes which is controlled and activated by AtxA.

Table 4: Influence of pXO1 harbored *atxA* on the activation of *capBCAD* operon transcription in *B. anthracis*. From Mignot *et al.,* (2003), with permission[a]

	pXO1^{-}	pX01^{+}	pXO1^{+} + Δ *atxA*[b]	pXO1^{+} + Δ*pagR*[c]
- CO_2	919	52	1209	1346
+ CO_2	33	1303	255	237

[a]Relative expression of *sap-lacz* and *eag-lacZ* transcriptional fusions in different genetic backgrounds with cells grown with CO_2. Numerical values are units of β-gal activity/mg protein.
[b]*atxA* gene deleted from pXO1.
[c]*pagR* gene deleted from pXO1.

Collectively, these results indicate that the *B. anthracis* transcriptional regulator AtxA controls the synthesis of the three toxin components and the surface elements (capsule and S-layer). Thus, AtxA residing on pXO1 is a master regulator that coordinates the response to host signals by orchestrating positive and negative controls over genes located on all three genetic elements (plasmids pXO1 and pXO2 and chromosomal DNA).

Expression of the structural genes for the anthrax toxin proteins is coordinately controlled by host-related signals such as elevated CO_2 and the *trans*-acting positive regulator AtxA. In addition to these factors, toxin gene expression is under growth phase regulation. During the late exponential phase of growth, when AbrB levels begin to decrease, toxin synthesis increases.

AbrB is considered to be a transition state regulator that plays a critical role in the suppression of post-exponential-phase gene expression during exponential phase growth (Phillips and Strauch, 2002). Toxin gene expression reaches a maximum during the late exponential growth phase, coinciding with a decrease in *abrB* transcription. A *B. anthracis abrB* deletion mutant produces higher levels of all three toxin proteins, and toxin gene expression was found to peak earlier during growth (Saile and Koehler, 2002). One AbrB target is *sigH* (Fujita and Sadaie, 1998); (Strauch, 1995), a gene encoding the alternative sigma factor σ^H, which has an important role in post-exponential growth phase gene expression. The activity of a reporter gene driven by the *atxA* promoter is reduced in *B. subtilis sigH* mutants (Strauch *et al.*, 2005). The σ^H RNA polymerase holoenzyme in *B. subtilis* recognizes and transcribes genes associated with the transition to the stationary phase of growth, including genes for cytochrome synthesis, generation of potential nutrient sources, transport, and cell wall metabolism (Britton *et al.*, 2002), as well as genes important for competence and sporulation initiation. Hajirangisko, Chen, and Koehler (2007) constructed an *sigH*-null mutant of *B. anthracis* and found it unable to sporulate, although the growth rate was not affected. In addition, the *sigH*-null mutant produced notably lower levels of PA, LF, and EF. Transcriptional analysis involving *pag-lacz, lef-lacZ,* and *cya-lacZ* revealed that β-gal production in the *sigH*-null mutant was 4% or less compared to the parent strains. Collectively, these observations indicate that σ^H controls toxin gene expression indirectly, *via* its indirect control of *atxA* gene transcription. Hajirangisko, Chen, and Koehler (2007) therefore proposed that control of toxin expression is fine tuned by the independent effects of σ^H and AbrB on the expression of atxA.

Wilson *et al.,* (2009) found that two haem-dependant, small c-type cytochromes, CccA and CccB, located on the extracellular surface of the cytoplasmic membrane, regulate *pagA* gene expression by affecting the expression of the master virulence regulator AtxA (Fig. **3**). Inactivation of *hemL, resB, resC*, or BAS3568 (haem defective mutants) resulted in increased transcription of *pagA* and *atxA* during exponential growth in response to the loss of the CccA-CccB signaling pathway. BAS3568 is a previously uncharacterized protein derived from a putative *hemL* gene that was found involved in cytochrome C activity and virulence regulation.

ResB and ResC are integral membrane proteins required for cytochrome C assembly and maturation. The *resB* and *resC* mutants did not produce active cytochrome C oxidase. HemL converts glutamic-1-semialdehyde to 5-aminolevulonic acid required for haem biosynthesis. The *hemL* mutant exhibited a reduced level of cytochrome C oxidase activity as did the BAS3568 mutant. A *pagA-lacZ* fusion was used to quantify the effect of various cyt mutants on *pagA* expression. The rate of induction of *pagA* was four-fold higher in the *resB, resC,* and BAS3568 mutants and 2.5-fold higher in the *hemL* mutant compared to the parent strain while the growth rate was unaffected.

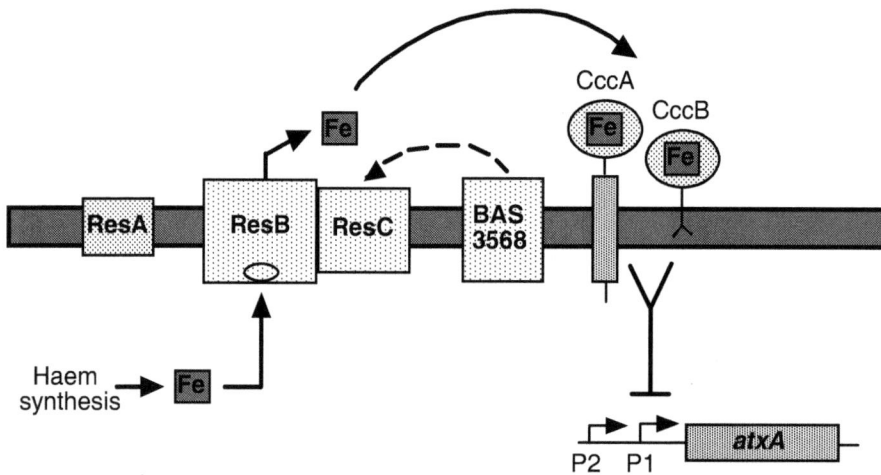

Figure 3: Schematic representation of the small c-type cytochrome pathway regulating virulence gene expression in *B. anthracis*. Intracellularly synthesized haem, through a pathway that requires the Hem proteins including HemL, is transported across the membrane by the resBC proteins and is covalently attached to CccA and CccB with the involvement of ResA. Either CccA or CccB can then act redundantly to indirectly repress *atxA* transcription at the P1 promoter. The BAS3568 protein, which is required for full cytochrome oxidase activity, is also involved in expression of *atxA* transcription in early exponential growth by an unknown mechanism (indicated by the broken arrow) likely to act on the cytochrome C maturation pathway. Adapted from Wilson *et al.,* (2009).

In *B. subtilis*, various environmental stresses induce the synthesis and activation of σ^B. σ^B then initiates the transcription of more than 100 stress genes that constitute the σ^B regulon. The *B. anthracis* operon encoding the general stress transcription factor σ^B and two proteins of its regulatory network Rsbv and RsbW were cloned by Fouet, Namy, and Lambert (2000). The associated *sigB* operon of *B. anthracis* was found to contain three genes, *rsbW, rsbV,* and σ^B rather than eight genes as in *B. subtilis.* The activation of σ^B itself involves a network of regulatory proteins encoded by *rsb* genes (for "regulator of sigma B"). The *B. anthracis sigB* operon is preceded by two σ^B-like promoter sequences similar to those of *B. subtilis,* the expression of which depends on an intact σ^B transcription factor. These two promoters in *B. anthracis* were induced during the stationary growth phase and induction required an intact *sigB* gene. The *sigB* operon was induced by heat shock. Mutants from which *sigB* was deleted were constructed in both a toxigenic and a plasmid free strain. These mutants differed from the parental strains morphologically by producing smaller colonies on BHI agar plates, flocculated during growth in liquid medium, and were more difficult to harvest by centrifugation, with cotton-like rather than sand-like pellets. Optical microscopy revealed that these mutants wee present as longer filaments than the parental strain.

When groups of 10 mice were injected with 10^5 CFU of the wild type strain of B anthracis 7702 and its $\Delta sigB$ deletion derivative SSB10, the wild type strain resulted in 50% deaths, whereas the deletion derivative strain resulted in no more than 20% deaths. These results indicate that the $\Delta sigB$ strain was less virulent than the parental strain. The *in vitro* production of PA was found to be identical in the mutant and parent strains. The authors concluded that *sigB* may be a minor virulence factor.

Barua *et al.,* (2009) identified the genes essential for intracellular germination. Eighteen divers genetic loci were identified for screening or mutants delayed in germination inside RAW264.7 NALB/c murine macrophages. Fourteen transposon mutants were identified in genes not previously associated with *B. anthracis* spore germination and included disruption of factors involved in membrane transport, transcriptional regulation, and intracellular signaling. Strain MIGD101, was the most highly represented mutant and revealed ungerminated cells within macrophages. In addition, this strain exhibited notably reduced lethality for infected macrophages.

Growth-Phase Dependant Toxin Expression

When grown in batch culture, *B. anthracis* toxin proteins are at the highest level during the transition from the exponential to the stationary growth phase. This transition is associated with the *abrB* gene, which is a transition state regulator (Hoffmaster and Koehler, 1999a). The *B. anthracis* chromosome encoded *abrB* gene negatively controls maximum gene expression in batch cultures by repressing all three toxin genes during early and mid-exponential growth.

Sail and Koehler (2002) identified two orthologues of the *B. subtilis* transition state negative regulator gene *abrB*. One was located on the chromosome and one on the pXO1 plasmid. Deletion of the pXO1 *abrB* did not affect toxin gene expression. However, a *B. anthraci*s mutant in which the chromosomal *abr* gene was deleted expressed *pagA* earlier during growth and at a higher level than the parent strain. Expression of a transcriptional *pagA-lacZ* fusion in the *abrB* chromosomal deletion mutant was increased up to 20-fold during exponential growth compared to the parent strain and peaked in the mid-exponential rather than the usual late exponential growth phase. In contrast to the strong effect of *abrB* on *pagA* expression, *lef-lacZ* and *cya-lacZ* expression during early log-phase growth was increased only two- to three-fold in the *abrB* deletion mutant. The mutant produced increased levels of PA, LF, and EF.

DNA ANALYSIS OF *B. anthracis* STRAINS

Variable Number Tandem Repeat (VNTR) Loci

The *vrrA* gene is highly conserved among *Bacillus* species and is thought to encode a putative protein of unknown function in *B. anthracis*. Jackson *et al.,* (1997) found that PCR analysis of 198 *B. anthracis* isolates of global and host diversity revealed a variable region of DNA sequence in the *vrrA* gene differing in length among isolates. Five polymorphisms differed by the presence of two to six copies of the 12-bp tandem repeat sequence 5'-CAA-TAT-CAA-CAA. PCR amplification of the five different VNTR sequences was accomplished with the primers EWA-1/EWA-2 (Table **1**).

Copy number differences in the VNTR region were used to define five different *B. anthracis* alleles. Both the Ames and Sterne strains were found to be located in VNTR category 4, reflecting four tandem repeats of the 12-bp VNTR sequence. There was a correlation between categories and geographic distribution of the isolates. The authors concluded that such molecular markers are ideally suited to monitor the epidemiology of anthrax outbreaks in domestic and herbivore populations.

Keim *et al.,* (1999, 2000) presented a novel molecular typing system based on rapidly evolving-variable number tandem repeat (VNTR) loci and involving nine novel VNTR loci which in combination with the *vrrA* locus which were used to genetically characterize an extensive collection of global *B. anthracis* isolates. The multiple-locus variable number tandem repeat (VNTR) analysis or multiple-locus variable-number tandem repeat analysis (MLVA) involves the use of multiple alleles of eight marker loci and eight pairs of primers (Table **1**), with one primer in each pair fluorescently labeled with FAM, HEX, or NED, and amplicons resolved in agarose gels. Each primer pair targets a different VNTR locus. Five of the eight loci were discovered by sequence characterization of molecular markers (*vrrC$_1$, vrrC$_2$, vrrBr$_1$, vrrB$_2$,* and CG3), two were discovered by searching complete plasmid sequences (pXO1-aat and pXO2-at), one was known previously (*vrrA*). Results (Keim *et al.,*1999), indicated that seven major diversity groups exist world-wide comprising two major groups A and B (Fig. **1**). Isolates from the "B" group are relatively rare and represent 10% of the total isolates characterized. The "A" group is divided into six distinct subgroups. A1 contains isolates similar to the well-known "Vollum" strain which are rare. A2 isolates are common but are found primarily in the Kruger National Park, South Africa. A3 isolates are rare and all were from Mozambique. A4 contains the vast majority of isolates from Western North America. A5 isolates are only rarely encountered and include the V770-NP1-R strain used in the U.S. for human vaccine production and an Argentinian isolate. A6 is a common strain type and includes the well-known "Ames" and "Sterne" strains. Isolates from this group appear to be the most widely distributed, being found in England, North America, South Africa, India, and Australia. Isolates from each of these seven groups are distinguishable and constitute numerous subtypes within a single group. The authors speculated that the common ancestor to the "A" and "B" groups may have been in the historically recent past and may have coincided with the human domestication of livestock which allowed the organism to become more successful and widely dispersed. The authors also speculated that it is also possible that the livestock domestication event is represented only in the "A" lineages and that the A1 – A6 lineages are dispersed clones corresponding to human activity. The "B" lineage would then represent a more ancient division in the evolution of *B. anthracis*, predating human activity.

MLVA characterization of 426 *B. anthracis* isolates by Keim *et al.,* (2000) resolved 89 distinct genotypes. Six genetically distinct groups were identified that appeared to be derived from clones (Keim *et al.,* (2002)). Some of these clones showed worldwide distribution, while others were restricted to specific geographic regions. The simple di-and trinucleotide tandem repeats (pX01-aat and pXO2-at) were the most diverse, while complex longer repeats such as with *vrrA* (12 nucleotide repeats) and *vrrC$_2$* (18 nucleotide repeats) yielded lower diversities. Slip strand repair mutations by DNA polymerase are thought to occur more frequently on short repeats (van Belkum *et al.*, 1998). Variation at a variable number tandem repeat locus such as *vrrA* is due to differing numbers of tandem repeats of the 12-bp sequence among different strains.

Ryu *et al.,* (2005) found that although AFLP yielded similarity levels of no less than 0.93 with Korean soil and clinical isolates of *B. anthracis,* MLVA using the 8 primer pairs and loci of Keim *et al.,* (2000) yielded nine MLVA types derived from the 34 Korean *B. anthracis* isolates, and 4 reference strains of *B. anthracis,* with similarity levels extending to 0.66. Three MLVA types were pathogenic *B. anthracis* isolates and were recognized as new genotypes (M1, M2, and M3) distinguished from 89 previously deduced genotypes by Keim *et al.,* (2000). An additional 4 types were nonpathogenic isolates. Among the 34 Korean isolates, 16 comprised the new M1 genotype. Because this new M1 genotype fell into the same cluster (A) as earlier isolated strains, this new genotype was suggested to represent evidence for historic adaptation to environmental conditions in Korea rather than recent introduction of a new genotype.

MLVA performed as described by Keim *et al.,* (2000) revealed the presence of five genotypes among a group of 10 *B. anthracis* isolates from epidemiologically unrelated cases of bovine anthrax in Eastern Poland (Gierczynski *et al.*, 2004). Eight of the isolates possessed the *pagA* and *capB* genes indicating the presence of both virulence plasmids pXO1 and pXO2, while two isolates revealed only *pagA* and lacked the pXO2 plasmid. MLVA and DNA sequence analysis indicated that seven of the isolates represented four novel genotypes. Five strains revealed a unique 144-bp *vrrB2* variant as well as a 220-bp variant of *vrrB1,* implying relatedness to the B2 lineage.

Keim *et al.,* (2000) observed only a single VNTR mutation in the *vvrA* locus of the Ames strain after 1000,000 passages. This suggests that the observed diversity among the Polish isolates is the result of genetic mutations, which could occur after a large number of generations resulting from anthrax outbreaks during the first half of the 20th century prior to vaccination in probable endemic regions rather than the acquisition of laboratory mutations.

Variable-number tandem repeats (VNTR) analysis and BOX-repeat-base PCR (BOX-PCR) genomic fingerprinting were performed by Kim *et al.,* (2001) as described by Anderson *et al.,* (1996) on 25 *Bacillus* strains to determine the generic relatedness of *B. anthracis* to the closely related species. Based on VNTR analysis, all *B. anthracis* strains could be assigned to VNTR, which is the most commonly found type in the world. Interestingly, a $VNTR_2$ type was also observed in *Bacillus cereus* KCTC 1661 and with an exact match to the tandem repeats in *B. anthracis*.

Single Nucleotide Repeat (SNR) Analysis of *B. anthracis* Strains

SNRs, also referred to as mononucleotide repeats, are a type of VNTR that display very high mutation rates (as high as 6.0×10^4 mutations per generation) (Keim *et al.,* 2004). Unlike some VNTR loci that can have complicated repeat structures, SNRs are stretches of one type of nucleotide that may vary in length between different bacterial isolates due to slip-strand mispairing (Levinson and Gutman, 1987). SNRs are more likely than other types of simple sequence repeats to undergo strand separation and base pair slippage, increasing the chance of slip–strand mispairing and causing a mutation at the SNR locus (Coenye and Vandamme, 2003; Gur-Arie *et al.,* 2000). In *Escherichia coli*, 93% of all mononucleotide repeats are A or T (Gur-Arie *et al.,* 2000). The lower melting temperature of A-T base pairs compared to G-C base pairs, is thought to increase the instability of the DNA helix, thereby increasing the incidence of slip-strand mispairing with A-T SNRs (Moxon and Rainey, 1995; van Belkum *et al.,* 1998).

Stratilo *et al.,* (2006) were the first to describe the SNR loci of *B. anthracis* which were used to discriminate between strains of *B. anthracis*. Primers for chromosomal clusters with at least six members and for plasmid clusters with four members were utilized for 22 SNR loci (Table 1). Among 19 *B. anthracis* strains, SNR analysis allowed strains with the same MLVA genotype to be distinguished from each other.

A multiplex four single nucleotide repeat (SNR) method for genotyping closely related isolates of *B. anthracis* has been described (Kenefic *et al.*, 2008). The forward primer of each of the four primer pairs was labeled with a different fluorescent dye at the 5'-end for identification of resolved fragments. Two of the primer pairs were derived from the pXO2 plasmid and two from the chromosomal DNA (Table 1). The assay system successfully identified the 89 genotypes described by Keim *et al.*, (2000).

Single Nucleotide Polymorphisms (SNPs)

In the Ames strain of *B. anthracis*, whole genome sequencing of the two plasmids and chromosome has revealed ~3,500 single-nucleotide polymorphisms (SNPs) among eight different strains (Pearson *et al.*, 2004; Read *et al.*, 2002). A "canonical SNP" (canSNP) is an SNP that can be used to effectively identify a point in the evolutionary history of a species. Mapping SNPs on a genetic population structure allows the identification of canSNPs that can be used diagnostically on a broad scale to define major genetic lineages in a species, or more narrowly, to define specific strains (Keim *et al.*, 2004).

Van Ert *et al.*, (2007) made use of two SNPs on the pXO1 plasmid (PS-1 and PS-32) and one SNP located on the pXO2 plasmid (PS-52) in addition to four chromosomal SNPs (Branch1-7, Branch1-26, Branch1-28, and Branch1-31) for establishing single-nucleotide polymorphism assays used in distinguishing the Ames strain from its various derivatives and related strains. For each of the primer pairs, two dual labeled probes were used that differed by one-base pair. One probe, was labeled with FAM at the 5'-end was designed to align itself only with the Ames strain (except for the PS-1 locus where it was labeled with VIC). The second probe was labeled at the 5'-end with VIC and designed to align with non-Ames strains. All probes were labeled at the 3'-end with a non fluorescent black hole quencher (BHQ). A total of 88 *B. anthracis* strains from diverse global sources were examined. The respective primer pairs and probes are listed in Table

1. Bold face letters in the probe sequences indicate the SNP location. The PS-52 SNP RTi-PCR assay separated the Ames MLVA type from all of its genetic relatives, whereas the PS-1 SNP Rti-PCR assay exhibited less specificity and grouped the Ames strain with four other isolates. One chromosomal SNP assay (Branch1-31) and two plasmid SNP assays (PS-1 and PS-52) were found suitable for discriminating Ames *versus* non-Ames strains.

Amplified Fragment Length Polymorphism Analysis (AFLP)

Both AP-PCR (arbitrary primed) also known as AFLP (amplified fragment length polymorphisms) are able to distinguish sequences of variable length in certain strains of *B. anthracis* and not in others. Both AP-PCR and AFLP variation among *B. anthracis* isolates are primarily due to insertion and deletion differences (Anderson *et al.,* 1996, Jackson *et al.*, 1997).

AFLP can provide a relatively rapid and thorough survey of a microbial genome. Purified genomic DNA is digested with two restriction endonucleases. AFLP variation among *B. anthracis* isolates are primarily due to insertion and deletion differences (Anderson *et al.,* 1996, Jackson *et al.*, 1997). Linkers containing a DNA sequence that can be used to prime DNA amplification are then ligated to the fragments which are then amplified. This results in a rapid screening of the entire genome for length polymorphism and evaluates about 0.4% of the genome for changes in DNA sequence (Jackson *et al.*, 1999). Analysis of the amplified DNA fragments is completed by electropohoresis through polyacrylamide DNA sequencing gels.

AFLP of microbial DNA generates fragments whose number is influenced by (1) the enzyme used to digest the DNA, (2) the A/T content of the genome, (3) secondary modification of the DNA (*i.e.,* methylation of restriction sites), and (4) genome size (Jackson *et al.,* (1999)). AFLP analysis of *B. anthracis* using 16 different "+1" ALP primers was found by Jackson *et al.,* (1999) to generate 388 different fragments. Keim *et al.,* (1997) found that AFLP profiles generated from a large battery of *B. anthracis* strains showed very little difference between isolates. In contrast, a significant difference was found between AFLP profiles generated for any *B. anthracis* strain and even the most closely related *Bacillus* species. This observation was later confirmed by Ryu *et al.,* (2005) who applied AFLP to 34 *B. anthracis* isolates from soil and clinical samples from Korea which yielded a high level of similarity (0.93%) allowing AFLP to be used to differentiate *B. anthraci*s from other *Bacillus* species.

Although *B. anthracis* and *B. cereus* have been found to share 98% DNA sequence homology, comparison of AFLP profiles revealed marked differences. A comparison of all AFLP fragments produced using 16 different primer combinations revealed that only 45% of the *B. anthracis* fragments were shared by *B. cereus* (Keim *et al.,* (1997)). Subtraction of fragments derived from the pXO1 and pXO2 plasmids allows a direct comparison of genomic DNA among different species.

Velappen *et al.,* (2001) subjected a number of Gram-positive and Gram-negative species including *B. anthracis* to single-enzyme amplified fragment length polymorphism (SE-AFLP) analysis. Following extraction, DNA (100 ng) was digested with *Hind*III. The *Hind*III adapter (5'-GTA-GAC-TGC-GTA-CCA-GCT) was ligated to the digestion fragments. Non-selective amplification was achieved using primer *Hind*III+0 primer (5'-GTA-GAC-TGC-GTA-CCA-GCT) corresponding to the adapter and restriction enzyme recognition sequences. Selective amplification of AFLP fragments was achieved by use of the primer *Hind*III 5'-GTA-GAC-TGA-GTA-CCA-GCT-XX), where XX represents one or two additional specific nucleotides used to amplify a subset of the restriction fragments. The five isolates of *B. anthracis* examined in the study yielded much greater DNA band pattern similarity than isolates of *B. cereus, B. thuringiensis*, and *B. mycoides*, reflecting the monomorphic nature of *B. anthracis* isolates. The *Hind*III primers with AC and G extensions respectively were able to differentiate *B. cereus* and *B. anthracis* isolates. Fifteen µL of diluted Hind+0 amplification products were used as target DNA with primer *Hind*III + AC which was able to clearly distinguish between the Ames strain and the four other *B. anthracis* strains, whereas bands that differentiated the other four strains of *B. anthracis* were more subtle. Visual examination of the subtle band distinctions between these four strains suggests that may have been enhanced by using 150 or 200 ng of digested DNA in place of 100 ng.

A total of 25 strains of *B. anthracis*, 62 strains of *B. cereus*, and 244 strains of *B. thuringiensis* were subjected to AFLP analysis. Extracted DNA (100 ng) was digested with *Eco*RI and *Mse*I restriction nucleases and the resulting fragments ligated to double-stranded adapters. The digested and ligated DNA was then amplified by PCR using *Eco*RI and *Mse*I primers. The resulting PCR products were then analyzed by agarose gel electrophoresis. Each isolate yielded about 40 fragments between 100 and 500-bp in length with simultaneous digestion by *Eco*RI + *Mse*I. Results of AFLP analysis indicated that these type 1 bacilli are heterogeneous. AFLP using only one primer combination did not provide sufficient resolution to discriminate among the 25 *B. anthracis* isolates.

The authors concluded that the concept of horizontal gene transfer of plasmid and/or extrachromosomal markers is an important factor in defining the phenotypes of type 1 bacilli and that *B. anthracis* isolates, unlike *B. cereus* and *B. thuringiensis*, form a distinct clade within the diverse group of *B. cereus* and *B. thuringiensis* isolates, reflecting the extreme monomorphic nature of this subgroup. The authors assumed that the ancestral *B. anthracis* was derived from a single *B. cereus-like* isolate that acquired the pXO1 and pXO2 plasmids by a genetic exchange event or by gene transfer. Comparison of the sequence genomes of *B. cereus* ATCC 14579 and *B. anthracis* Ames indicates that a large core set of genes are conserved between these two species (Ivanova *et al*., 2003). Approximately 4,500 out of 5,366 open reading frames in *B. cereus* ATCC 14579 have 80 to 100% identity to corresponding homologs in *B. anthracis* (Ivanova *et al*., 2003). The authors further concluded that *B. anthracis* is most probably a clonal derivative of an ancestral *B. cereus* that acquired and maintained the specific pathogenic properties as a result of advantageous selection pressure. In addition, the authors further emphasized that it cannot be assumed that simply inserting pXO1 and pXO2 into a randomly chosen *B. cereus* isolate will produce a successfully pathogenic *B. anthracis* strain and that clearly there are chromosomally encoded factors that are critical to the success of *B. anthracis* as a lethal pathogen.

Random Amplified Polymorphic DNA (RAPD) Analysis

RAPD analysis, also known as amplified polymorphic PCR (AP-PCR) analysis is able to distinguish sequences of variable length in certain strains of *B. anthracis* and not in others. RAPD variation among *B. anthracis* isolates is primarily due to insertion and deletion differences (Anderson *et al.*, 1996, Jackson *et al.*, 1997). The use of a single random M13 bacteriophage derived primer (Table **1**) was found by Anderson *et al.,* (1996) to yield a 1,480-bp genomic DNA fragment from the Sterne strain that contained four consecutive repeats of CAA-TAT-CAA-CAA. Sequencing indicated that the same fragment from the Vollum strain was identical except that two of these repeats were deleted. The Ames strain of *B. anthracis* differed from the Sterne strain by a single-nucleotide deletion. More than 150 nucleotide differences separated *B. cereus* and *B. mycoides* from *B. anthracis* in pairwise comparisons. The nucleotide sequence of the variable fragment form each species contained one complete open reading frame (ORF) designated vvrA, for variable region with repetitive sequences. The primers EWA2/EWA1 (Table **1**) generated amplicons of 167 and 168 for the Ames and Sterne strains respectively and were therefore indistinguishable by agarose gel electrophoresis (AGE), whereas the Vollum strain yielded a 143-bp amplicon which readily distinguished this strain from the Ames and Sterne strains. The primers EWA2/EWA1 yielded an amplicon of ~100-bp with a *B. cereus* strain while a ~150-bp amplicon resulted with a *B. mycoides* strain.

No difference was found between the Ames and Vollum strains of *B. anthracis* using PFGE and sequences of the 16S-23S and *gyr*B-*gyr*A intergeneric spacer regions (Harrell, Andersen, and Wilson, 1995). Henderson, Duggleby, and Turnbull (1994) reported no differences in restriction fragment length patterns among 37 isolates of *B. anthracis* for 18 restriction enzymes. They did however note a slight pattern variation in PCR fragments generated from an M13 bacteriophage-based primer (Table **1**) in arbitrary primed-PCR (RAPD).

Levy *et al.,* (2005) subjected 8 strains of *B. anthracis* varying in pX01/pXO2 plasmid content (-/-, +/-,+/+, a*nd -/+)* to RAPD analysis using six RAPDs primers Strains of *B. thuringiensis, B. cereus* and *B. licheniformis* were also examined. The AJ035, AJ036, AA032, and AA033 primers (Table **1**) were used in a multiplex PCR assay. The AT073 and AT074 primers were used in a separate multiplex PCR assay because the bands were similar in size to the bands derived from the former multiplex assay with four primers. The eight strains of *B. anthracis* could be divided into five genotypes on the basis of the multiplex

RAPD analyses which yielded a total of three DNA bands with each strain (two bands with the quadruplex assay and one band with the duplex assay).

Multilocus Sequence Typing (MLST)

Helgason *et al.,* (2004) developed a MLST typing scheme for members of the *B. cereus* group. Primers were designed for conserved regions of seven housekeeping genes (*adk, ccPA, fts, glp, pyrE, reF,* and *sucC*) and the resulting amplicons sequenced for all 77 strains examined. The number of alleles at each loci ranged from 25 to 40, and a total of 53 allelic profiles or sequence types were distinguished. All five *B anthracis* isolates had identical alleles at all seven loci Some *B. cereus* isolates shared one or more alleles with *B. anthracis.*

Olsen *et al.,* (2007) used the MLST assay system of Helgason *et al.,* (2004) utilizing the same seven housekeeping genes and primer pairs (Table **1**) to assess the genetic distribution of 296 *B. cereus* group members including four strains of *B. anthracis.* A total of four clusters were recognized, with all four *B. anthracis* isolates falling into a single cluster (A) with identical allelic profiles in agreement with previous MLST studies (Daffonchio *et al.,* 2006, Helgason *et al.,* 2004, Ko *et al.,* 2004). Cluster A was found not only to contain all four strains of *B. anthracis* but also 27 *B. cereus* and *B. thuringiensis* strains. In addition, the primer pair BA5510-1/BA5510-2 (Table **1**) was used to target the putative *B. anthracis* "specific" chromosomal gene *BA5510* that encodes the techoic acid ABC transporter (ATP-binding protein). This primer pair yielded a 162-bp amplicon from all four *B. anthracis* isolates. In addition, the primers BAlef-f/BAlef-r (Table **1**) targeting the *lef* and *cap* genes located on the *B. anthracis* plasmids pXO1 and pXO2, respectively were also used and yielded amplicons with only the *B. anthracis* isolates. The BA5510 primers yielded amplicons with the four *B. anthracis* isolates in addition to *two B. cereus* isolates. In addition, the frequently used *B. anthracis* marker BA813 was amplified from all four *B. anthracis* strains in addition to 31 of 289 (11%) non-*B. anthracis* strains (see section below entitled *The Ba813 chromosomal sequence*). Sequencing of the seven MLST amplicons was able to distinguish all four strains of *B. anthracis* from one another.

Long-Range Repetitive Element-PCR (LR REP-PCR)

LR REP-PCR makes use of the fact that the presence of multiple copies of highly conserved repetitive extragenic palindromic (REP) sequences may be found seemingly randomly distributed throughout the genomic DNA of prokaryotes. Brumlik *et al.,* (2000) made use of LR REP-PCR for allocating each of 105 strains of *B. anthracis* into one of five groups. The method made use of a single primer LL-Rep1LR (Table **1**) and to that extent resembles random amplified polymorphic DNA (RAPD) analysis which also uses a single primer. A high level of discrimination was achieved by using a "hot start" PCR after initial thermal denaturation. In addition, the annealing temperature was incrementally increased over the first 17 thermal cycles from 54.0 to 62.5 °C. This gradual increase in stringency eliminated non-specific amplification. A far greater diversity in banding patterns was obtained among a collection of 10 different *B. cereus* and *B. thuringiensis* strains, than with the *B. anthracis* isolates which readily allowed all 105 *B. anthracis* isolates to be distinguished from the other *Bacillus* spp.

Conventional PCR

B. licheniformis, B. megatherium, and *B. Subtilis* all express capsule-like substances that are immunologically cross-reactive with the capsule of *B. anthracis* but exhibit no homology with the *B. anthracis cap* region at the nucleotide sequence level (Makino *et al.,* 1988; Makino *et al.,* 1989). Makino *et al.,* (1993) developed a PCR assay for detection of *B. anthracis* in animal tissue and blood. The primers MO1/MO2 (Table **1**) amplified a 288-bp sequence of the *capA* gene required for capsule synthesis. The primers were found to be highly specific for *B. anthracis.* Amplicons were not obtained from *B. cereus, B. licheniformis, B. megatherium, B. subtilis,* or *B. thuringiensis* nor from 33 additional non-*Bacillus* cultures. Boiling of infected murine blood samples followed by centrifugation and direct incorporation of the supernatant in the PCR required DNA from 10^5-10^6 CFU per PCR reaction. In contrast, phenol purification of DNA released after enzymatic treatment of infected spleen tissue or blood and detergent lysis of *B. anthracis* cells allowed detection of DNA from 10^3 CFU per PCR reaction.

Reif *et al.,* (1994) described the development of a dual probe PCR assay for capsule forming *B. anthracis* strains using spores. Spore suspensions were heated at 60 °C for 1 hr. to destroy vegetative cells. Spore suspensions were then allowed to partially germinate for 4.5 hrs. and were then boiled for 30 min. to release DNA. Alternatively spores were disrupted with a mini-bead beater for 10 min. After disruption, supernatants were used directly in PCR reactions. The primers BACA1FI/BACA6RI (Table 1) amplified a 623-bp sequence of the *capB* gene of *B. anthracis*. The 5'-biotin-labeled probe PACABL1 (Table 1) and the 5'-fluorescein-labeled probe PACAFL1 (Table 1) which hybridize to different portions of the amplicon were added to a portion of the denatured product and incubated at 55 °C for 15 min. to achieve probe hybridization. Streptavidin was then added and the preparation passed through a biotinylated membrane. Polyclonal anti-fluorescein antibody conjugated to urease was then incubated for 30 min. with the streptavidin-captured-probe-PCR product complex. Excess-anti-fluorescein antibody was then washed off and the membrane exposed to urea and the change in pH recorded using a pH-sensitive silicon sensor. The method was able to detect a minimum of 1 spore per PCR reaction mixture.

The detection of *B. anthracis* in soil samples is significantly more difficult than with clinical specimens. This is primarily due to the lack of effective selective enrichment procedures, the presence of high numbers of other related aerobic spore-forming bacilli of the *B. cereus* group which may readily overgrow *B. anthracis*, and the content of chemical compounds in soils inhibiting spore germination and the PCR. Beyer *et al.,* (1995) developed a nested PCR assay to detect *B. anthracis* spores in natural soil and soil samples heavily contaminated with tannery waste agents. One hundred grams of soil from the location of a former tannery were seeded with varying numbers of *B. anthracis* spores and incubated with 200 ml of d.H_2O and 60 g of glass beads at room temperature with rotary agitation at 150 rpm for 15 hrs. The samples were then passed through 500μ and 250μ stainless steel sieves. The filtrates were centrifuged for 25 min at 37 °C overnight with rotary agitation at 150 rpm. One hundred μL of the enrichment broth was then inoculated into 10 of fresh TSB and incubated at 37 °C for 6 hrs. to avoid spore formation. H_2O_2 was then added to a final concentration of 3.0% and the sample further incubated at 37 °C to destroy vegetative cells. Cells were then lysed after treatment with mutanolysin, lysozyme, and proteinase K and DNA purified with phenol.

The primers CAP6/CAP103 (Table **1**) amplified a 1035-bp sequence of the *capBC* genes on pXO2 and were used in the first stage of the nested PCR. The primers CAP9/CAP102 (Table **1**) amplified a 777-bp sequence of the *capBC* genes and were used in the second stage of the nested PCR. The primers PA8/PA5 (Table **1**) amplified a 596-bp sequence of the *pag* gene on pXO1. The primers PA7/PA6 (Table **1**) amplified a 210-bp sequence of the *pag* gene in the second stage of the nested PCR. The limit of detection was less than 10 *B. anthracis* spores per 100 g of soil. The major problem encountered in this study that had to be overcome was the high content of PCR inhibitors consisting of humic substances in addition to a high content of heavy metals in the tannery soil samples. The addition of T4 gene 32 protein at a concentration of 1 μg/100 μL of PCR reaction mix notably reduced PCR inhibition.

A PCR assay for detection of *B. anthracis* spores in various types of soils was developed by Sjöstedt *et al.,* (1997). The primers BA57/BA58 (Table **1**) amplified a 265-bp sequence of the capBCA operon. The primers BA59/BA60 (Table **1**) amplified an 893-bp sequence of the *lef* gene. The detection limit for the primer pair BA57/BA58 was 10 CFU and for BA59/BA60 was 100 CFU for pure culture suspensions of spores. Ten microliters of spore suspensions were added to 100 mg of various soil samples which were allowed to incubate overnight at room temperature. Samples were then subjected to three freeze-thaw cycles and DNA was released from the spores enzymatically and then purified using phenol-chloroform-isoamyl alcohol. The detection limit varied from 10^3 to 10^4 spores per sample. A major problem encountered was the high level of PCR inhibitors present in the soil and the possible binding of DNA to soil particles. The use of a detergent to separate the spores from soil particles followed by differential centrifugation, and the addition of carrier DNA from an extraneous organism might have reduced the level of detectable spores.

A PCR-ELISA assay to detect *B. anthracis* in soil samples was developed by Beyer, Pocivalsek, and Böhm (1999). The application of streptavidin-coated microtiter plates in conjunction with covalently linked oligonucleotides as capture probes resulted in a sensitivity of detection of less than 10 spores seeded into 100 g of soil. Some uninoculated soil samples yielded positive results with previously published primers or probes targeting the B or C gene of pXO2 and with primers targeting the B813 chromosomal sequence. Sequencing of the resulting amplicons led to the assumption that there is at least one unknown organism in soil samples having high sequence similarity to *B. anthracis* with highly conserved capsule-encoding genes.

The protocol developed by Beyer, Pocivalsek, and Böhm (1999) for detection of *B. anthracis* in soil samples consisted of inoculating 200 ml of Tryptic Soy Broth (TSB) with 100 of soil held at 70 °C for 30 min to destroy vegetative cells and then the volume was made up to a final volume of 1L with TSB. One hundred μL of the enrichment broth were then inoculated into 10 ml of fresh TSB and incubated at 37 °C for 6 hrs to avoid spore formation. H_2O_2 was added to a final concentration of 30% with incubation for 1 hr. to destroy vegetative cells. DNA was purified with a commercial kit.

Perez, Hohn, and Higginsk (2005) developed a concentration protocol for detection of *B. anthracis* spores from 100 ml and 1L volumes of tap water and source waters using filtration through 0.45μ nitrocellulose filter membranes, followed by heating at 65 °C for 15 min. and overnight culture on blood agar plates at 37 °C. A nested PCR targeting the variable repeat region A (*urrA*) gene and primers for the *lef* gene were utilized in the PCR. When 100 viable spores were spiked into 100 ml of finished water, a mean of 48 were recovered on blood agar plates. With source water and spiking doses >250 spores, recoveries from 100 ml were successful; however, reduction of the spiking dose to 100 spores or less gave comparatively poor results due to the overnight growth of the plates with other species of *Bacillus* native to the water. When 35 or 10 spores were spiked into 1L volumes of tap water, mean recoveries were 15.3 and 7.2 spores, respectively. A nested PCR involving DNA extraction from spores using a spiking dose of 534 spores in 100 ml of tap water yielded positive PCR results. Detection of spores spiked into 1L of tap water could be achieved with the nested PCR. When the spiking dose was reduced to 100 spores in 100 ml of tap water assays were negative.

Ramisse *et al.,* (1996) developed a multiplex PCR assay involving the use of nine primer pairs (Table **1**) for amplifying a 152-bp sequence of the Ba813 chromosomal region *B. anthracis* in addition to sequences derived from the *pag,* *cya,* and *lef* genes on the pXO1 plasmid and for cap genes on the pXO2 plasmid. The system unambiguously identified fully virulent (pxO1$^+$ pxO2$^-$) isolates and attenuated isolates (pxO1$^+$ pxO2$^-$), (pxO1$^-$ pxO2$^+$), and (pxO1$^-$ pxO2$^-$) and distinguished "anthrax-like" strains from other *B. cereus* group isolates.

Brightwell, Pearce, and Leslie (1998) reported on the development and evaluation of a multiplex PCR assay for detection of *B. anthracis* strains harboring the pXO2 plasmid, the multiplex PCR assay also incorporated an internal control (IC) to avoid false negative reactions. Internal controls consisted of plasmids containing modified PCR target sequences, corresponding to the *capC* and BA813 genes of *B. anthracis* which were then co-amplified with the original target sequences using the same set of primers as the IC. The BA813 chromosomal gene was chosen as a PCR target because it is considered unique to all *B. anthracis* strains, although the primers used have been found to cross-react with other *Bacillus* species (Table **1**). Its function is unknown. The *capC* gene carried on the pXO2 plasmid encodes one of several genes involved in capsule synthesis and was chosen as a PCR target on the basis that naturally occurring pXO1$^-$ pXO2$^+$ strains do not occur. Sensitivity of detection was the DNA from 10^2 - 10^3 organisms. The IC PCRS product size was found not to effect the native PCRS sensitivity. The concentration of the IC was critical, excessive IC outcompeted the genomic DNA template since the same pair of primers were involved with the IC in a competitive PCR format. The initial IC constructs were comprised of an internally deleted form of the genomic target sequences (*capC* or *BA813*) cloned into plasmid pUC19 respectively. A series of nested DNA fragments corresponding to the 23S rRNA sequences of *B. cereus* were then subcloned into the point of deletion, producing a number of IC constructs with identical flanking sequences homologous to the same forward and reverse primers but increasing in product size on PCR amplification. The primers CAP1/CAP2 (Table **1**) amplified a 321-bp sequence of the *cap2* gene. Primers BA813 R1/BA813 R2 (Table **1**) amplified a 130-bp sequence of the *BA813* putative gene. Additional primer pairs are listed in Table **1**. The reader is referred to Rupf, Merte, and Eschrich (1999) for an alternative method of constructing competitive internal PCR standards that utilize the same forward and reverse primers used for amplifying the genomic target DNA sequence.

Radnedge *et al.,* (2003) made use of suppression subtractive hybridization to identify 28 *B. anthracis* chromosomal sequence loci not present in strains of *B. cereus* or *B. thuringiensis*. In addition, these 28 loci were found to be absent from a variety of enteric pathogens. From these 28 loci, four were selected with respective to primer pairs so as to yield amplicons of different sizes (Table **1**) for multiplex PCR allowing agarose gel resolution of the four amplicons (133-, 163-, 196-, and 241-bp). In addition, a pair of primers 23S-F/23S-R (Table **1**) was used to amplify a 288-bp sequence of the 23S rDNA as an internal positive control for the multiplex PCR assay. The study utilized ten strains fo *B. anthracis*, six of *B. cereus*, and four of *B. thuringiensis*. A far more extensive list of test strains would be required to assess the extent of any exceptions to the observation that the four amplicons are generated from all isolates of *B. anthracis* and that all four loci are absent from in all non *B. anthracis* isolates of the *B. cereus* group.

Klichko, *et al.,* (2003) made use of reverse transcription of RNA followed by PCR for confirming the expression of hemolysin in *B. anthracis* that are expressed only under strictly anaerobic conditions on human but not sheep red blood cells (RBCs). These hemolysins were designated anthraysins (Anls) The primer pairs along with several positive control primers pairs *are* listed in Table **1**. See chapter 2 for a more detailed discussion of these hemolysins.

Dauphin *et al.,* (2008) exposed spores from 8 genetic groups of *B. anthracis* at density of 10^7 CFU/ml to 2.5 x 10^6 rads of gamma irradiation. PCR analysis of irradiated and nonirradiated spores indicated that radiation treatment significantly enhanced the PCR detection of *B. anthracis* chromosomal targets but had no significant effect on the detection of targets on the pXO1 and pXO2 plasmids of *B. anthracis*. When analyzed by an ELISA, irradiation affected the detection of *B. anthracis* spores in a direct ELISA assay but had no effect on the limit of detection of a sandwich ELISA. These results indicate that gamma irradiation-inactivated spores can be tested by real-time PCR or sandwich ELISA without decreasing the sensitivity of either type of assay. The exposure of spores to gamma irradiation is therefore ideally suited for the inactivation of spore samples prior to PCR identification for purposes of laboratory safety.

Kane *et al.,* (2009) developed a rapid-viability-polymerase chain reaction (RV-PCR) method to determine the presence of viable spores of *B. anthracis*. The method combined high-throughput sample processing with 96-well real-time PCR (Rti-PCR) analysis and detected 1 to 10 live spores in a background of 10^6 dead spores. Destruction of spores was achieved with gaseous ClO_2. Filtered surface swab samples were MPN diluted and incubated for 14 hrs at 30 $^\circ$C in 2 ml wells containing 1.8 ml of Tryptic Soy Broth for detection of resulting vegetative cells. Samples from turbid wells were then heated at 95 $^\circ$C to lyse the vegetative cells (releasing DNA) but not spores and 5 µl used for Rti-PCR analysis The PCR primers used were F41/Ri46 (Table **1**). Targeted a 63-bp sequence of the abhydrolase (alpha/beta-hydrolase) encoded on the chromosome and previously documented by Bode, Hurtle, and Norwood (2004) to be specific for *B. anthracis*. A dual labeled probe with FAM at the 5'-end and a nonfluorescent quencher at the 3'-end was used as the reporter molecule (Table **1**). In studies with spores exposed to ClO_2 producing partially killed spore populations, the results from 10 hr. incubation did not completely match that from 24 r. incubation. This observation suggested that some viable spores were delayed in germination and/or growth relative to non-exposed spores. A conservative endpoint of 14 hrs. incubation was therefore chosen to accommodate variability introduced by ClO_2 exposure.

Real-Time PCR (Rti-PCR)

A Rti-PCR assay for *B. anthracis* was developed by Lee *et al.,* (1999) using the fluorescent dye SYBR gold which was found to exhibit greater stability than SYBR green. Internal controls were developed for detection of false negative assays, each with a distinct Tm value. An internal control with 20% GC content was found to be optimum. Non specific artifacts were reduced to a minimum using a "hot start" technique. A 347-bp sequence of the *cap-C* gene located on the pXO2 plasmid was amplified with the primers CAP-1/CAP2 (Table **1**).

Qi *et al.,* (2001) developed a Rti-PCR assay for *B. anthracis* utilizing fluorescent resonance energy transfer (FRET) probes. The primers rpoBF1/rpoBR1 (Table **1**) amplified a 175-bp sequence of the *rpoB* gene, which encodes the β-subunit of RNA polymerase, and has been found useful for discrimination of bacterial species. The BaP1 probe was labeled at the 3' end with fluorescein and the BaP2 probe was labeled at the 5' end with Cy5 and was blocked at the 3' end with a phosphate group (Table **1**). When the FRET assay was applied to 144 *B. anthracis* strains, all tested positive and also yielded the expected 175-bp amplicon. An additional group of 175 closely related *Bacillus* strains including representatives of the *B. cereus* group and closely related Ba813⁺strains were FRET negative except for one of 72 Ba813⁺ strains.

Ellerbrok *et al.,* (2002) developed a Rti-PCR assay for distinguishing between pathogenic and apathogenic isolates of *B. anthracis*. The primers rpoBF1/rpoBR1 (Table **1**) from Qi *et al.,* (2001) amplified a 175-bp sequence of the *rpoB* gene. The primers PA-S/PA-R (Table **1**) amplified a 204-bp sequence of the *pag* gene on plasmid pXO1. The primers Cap-S/Cap-R (Table **1**) amplified a 291-bp sequence of the *cap-C* gene involved in capsule synthesis. Detection of the amplicons was with dual labeled fluorescent probes designated rpoB-TM, PA-TM, and CAP-TM respectfully (Table **1**). Although the chromosomal *rpoB* gene that encodes for the β-subunit of RNA polymerase was previously found to be specific for *B. anthracis* by Qi *et al.,* (2001), Ellerbrok *et al.,* (2002) found that four of five *B. cereus* strains yielded the *rpoB* amplicon, as did one of three *B. megatherium* isolates. The Ct values occurred with these non-B. anthracis isolates over 10 cycles later than those seen for *B. anthracis* using the same number of CFU. When 800 spores were added directly to Rti-PCR mixes, 20% (160) could be detected using the rpoB primers. Eight spores added directly to Rti-PCR mixes were undetected.

A Rti-PCR assay hardware unit can be utilized for detection of a specific species based on the DNA melting point (Tm) of the resulting amplicons of related species utilizing SYBR Green as the fluorophore in a multiplex format. Differences in Tm values can be expected to be derived from minor sequence differences in amplicons of the same size but from different species. Kim *et al.,* (2005) amplified *B. anthracis* sequences from the chromosomal gene *sspE* that encodes a spore structural component, in addition to the sequences derived from the plasmid genes *lef* and *capC* using primer pairs (Table **2**) that yielded amplicons of 188, 475, and 318-bp with *B. anthracis* having Tm values of 83.85, 80.57, and 77.65 °C respectively. In addition, the *sspE* primers yielded a 71-bp amplicon from both *B. anthracis* and other *Bacillus* species isolates. The 71-bp amplicon from non *B. anthracis Bacillus* isolates had a much broader Tm value than that from *B. anthracis* strains in addition to the absence of the *lef* and *capC* amplicons.

Antwerpen *et al.,* (2008) developed a Rti-PCR system for *B. anthracis* based on a 645-bp region of the chromosomal DNA (locus dhp61.1.183) that is thought to encode a hypothetical protein. The primers dhp61_183_113F/dhp61_183-2008R (Table **1**) amplified a 96-bp sequence of this hypothetical gene. The dual labeled probe dhp_183_1435 (Table **1**) was labeled at the 5'-end with FAM and the 3'-end with TAMRA. All 92 isolates of *B. anthracis* were from diverse sources were detected whereas 236 strains belonging to 19 *Bacillus* species other than *B. anthracis* were PCR negative. The limit of detection was 12.7gene copies per Rti-PCR reaction. The authors suggested that a PCR strategy consisting of Rti-PCR assays targeting the dhp61.183 96-bp sequence and primers targeting sequences of the pXO1 and pXO2 plasmids would be of use in detecting or confirming intentional plasmid transfer to non-*B. anthracis Bacillus* species.

An internal noncompetitive amplification control (IAC) was developed by Sohni, Kanjilal, and Kapur (2008) for detection of the *B. anthracis* chromosomal gene *rpOB* in Rti-PCR assays to decrease the incidence of false-positive and false-negative results. Synthetic IAC oligonucleotides were subcloned into a *B. subtilis* strain. Separate pairs of primers for the *rpoB gene* (Table **1**) and IAC with differentially labeled target (Table **1**) and IAC probes were used in the Rti-PCR assays the limit of detection for both target and IAC was 5 fg, corresponding to a singe gene copy.

The approved confirmatory tests of the Center for Disease Control (CDC) for *B. anthracis* require isolation of individual colonies from an overnight culture. Suspected colonies are then confirmed either by a direct fluorescence assay (DFA) for a *B. anthracis*-specific capsular protein and plate lysis by gamma phage or by a capsule DFA and *B. anthracis*-specific cell wall component DFA (Popovic *et al.,* 2005). The gamma phage is considered highly specific for all strains of *B. anthracis* with the exception of a rare susceptible non-*B. anthracis* strain (Abshire *et al.,* 2005) and is widely used for identifying *B. anthracis* isolates. Reiman *et al.,* (2007) developed a Rti-PCR assay combined with primers designed for highly specific gamma phage DNA amplification to indirectly detect *B. anthracis* in less than 5 hrs. Gamma phage (9 x 10^7 PFU/ml) were added to varying numbers of *B. anthracis* vegetative cells in 5 ml of Brain Heart Infusion (BHI). Culture tubes were then incubated at 37 °C for 4 hrs with rotary agitation at 120 rpm. After 4 hrs. incubation, 50 μl were used to extract DNA. Purified DNA was incorporated into Rti-PCR reaction mixtures with SYBR Green as fluorophore. The primers gamma-F/gamma-R (Table **1**) amplified a 117-bp sequence of the gamma phage DNA. Ct values were then correlated with the number of CFU of B. anthracis seeded into the BHI broth cultures. The minimum level of detection of *B. anthracis* was 207 CFU/ml equivalent to 10 CFU per Rti-PCR reaction vial.

Wattiau *et al.,* (2008) studied the distribution of *B. anthracis* at a commercial wool and goat hair cleaning plant in Belgium where the import of nondisinfected wool is allowed. *B. anthracis* was found in air filter dust, wastewater and goat hairs, where it accounted for ~10% of the total counts of viable bacteria. The presence of the *capC* and *pagA* viruence genes was detected using Rti-PCR. Primers CAPC-F/CAPC/R (Table **1**) amplified a 96-bp sequence of the *capC* gene in conjunction with the dual labeled probe capC labeled at the 5'-end with HEX and at the 3'-end with the black hole quencher BHO1 (Table **1**). Primers PAGA-F/PAGA-R amplified a 97-bp sequence of the *pagA* gene in conjunction with the dual labeled probe PAGA-1 labeled at the 5'-end with FAM and at the 3'-end with BHQ1 (Table **1**). VNTR typing was performed with eight primer pairs as described by Keim *et al.,* (2000). SNR analysis typing was performed with the use of four sets of primers for loci CL7, CL10, CL12, and CL33 (Table **1**). The forward or reverse primers for SNR analyses were labeled with Cy5 at the 5'-end and amplicons resolved by capillary electrophoresis. All of the 41 *B. anthracis* isolates harbored both the pXO1 and pXO2 plasmids as assessed by RTi-PCR. In addition, all 41 isolates belonged to VNTR cluster A4 of Keim *et al.,* (2000). Five closely related genotypes differing at only two VNTR loci were identified. Among the SNR primer pairs, three (for loci CL10, CL12, and CL13) yielded substantial variability among goat hair isolates resulting in eight different SNR subtypes.

The Ba813 Chromosomal Sequence

Patra *et al.,* (1996) identified a 277-bp chromosomal sequence designated Ba813 present in the chromosome of a *B. anthracis* pXO1⁻ pXO2⁻ strain which was missing in other *Bacillus* species. The primers R1/R2 (Table **1**) amplified a 152-bp sequence of the Ba813 region. Detection of amplicons involved a sandwich hybridization microwell assay utilizing wells coated with a capture probe designated C1, phosphorylated at the 5' end (Table **1**). The detection probe designated D3 (Table **1**) was labeled with biotin and final processing involved alkaline phosphatase conjugated to extravidine and para-nitrophenyl phosphate as colorimetric substrate. The primers appeared to be highly specific for strains of *B. anthracis* with no amplicons obtained from isolates of 10 other *Bacillus* spp.

Ramisse *et al.,* (1999) evaluated the distribution of the Ba813 chromosomal sequence in closely related *Bacillus* species. Ba813 was found present in the chromosomal DNA of 47 strains of *B. anthracis*. Among 60 strains of closely related *Bacillus* spp. examined, 3 of 3 *B. cereus* and 1 of 11 *B. thuringiensis* strains were found to harbor Ba813. These results and others indicate that the Ba813 sequence is present in all *B. anthracis* strains examined in addition to being present in strains of other species of the *B. cereus* group.

DNA Microarrays

Wang *et al.,* (2004) developed a qualitative DNA chip assay for identification and characterization of *B. anthracis* using multiplex PCR combined with a biotin-avidin phosphatase indicator system. Conventional PCR was used for a first round of multiplex DNA amplifications for regeneration of amplicons. Primers BA1/BA2 (Table **1**) were used to amplify a 168-bp sequence of the Ba813 chromosomal marker, present in all *B. anthracis* isolates. Primers PAG1/PAG2 (Table **1**) were used to amplify a 140-bp sequence of the *pag* gene on the pXO1 plasmid, and primers CP1/CAP2 (Table **1**) were used to amplify a 428-bp sequence of the *cap* gene on the pXO2 plasmid. Forward primers (BA11, PAG11, and CAP11) were extended at the 5'-end with a 10 unit poly T linker (Table **1**) for covalent binding to silanized glass slides. Denatured amplicons were then applied to the immobilized forward primers on the glass slides for hybridization to the immobilized forward primers followed by PCR reagents including biotin-11-dUTP and a cover slip applied for the second round of amplification. After amplification, the reaction areas of the chips were blocked with blocking buffer and alkaline phosphatase conjugated to avidin added. A purple color was developed for positive samples with the addition of 5-bromo-4-chloro-3-indolyl phosphate plus nitro blue tetrazolium. Fig. **4** presents the DNA chip process involved. About 1pg of specific DNA fragments on the chip wells could be detected after the PCR. The avirulent pXO1$^+$pXO2$^-$, pXO1$^-$pXO2$^+$, and pXO1$^-$pXO2$^-$ isolates were unambiguously identified. The reader is reminded that although the Ba813 chromosomal marker sequence is present in all *B. anthracis* isolates examined the Ba813 marker sequence has also been detected in isolates of other *Bacillus* species (Olsen *et al.*, 2007; Ramisse *et al.*, 1999; Qi *et al.*, 2001) and that the presence of the Ba813 chromosomal marker along with the pXO1$^-$pXO2$^-$ genotype cannot be used to definitively identify an avirulent strain of *B. anthracis*.

Nübel *et al.*, (2004) developed a DNA microarray for identification of *B. anthracis* and other members of the *B. cereus* group. Nucleotide sequences of the 16S-23S rDNA internal intergeneric transcribed spaces (ITS) containing genes for tRNA^{Ile} from 52 B. anthracis strains were found to be identical to sequences from seven strains published previously and different for all other bacteria outside the *B. cereus* group. The primers 1099-F/ILE-R (Table **1**) were used to amplify a 560-bp sequence of the ITS region containing tRNA^{Ile} genes from all members of the *B. cereus* group. The 1099-F primer was labeled at the 5'-end with Cy5. Glass slides were silanized and coated with neutravidin. A microarray consisting of 25 DNA probes, labeled at th 5'-end with biotin was immobilized onto the glass slides. Amplicons from individual strains were hybridized to the immobilized probes and fluorescence detected. Three probes Ba-112, Ba113, and Ba113-B (Table **1**) were found to be specific for *B. anthracis*. Oligonucleotide RC-BWW-111 (Table **1**) labeled at the 5'-end with Cy5 was used as a positive control. Probe Bact-1242 (Table **1**) is considered specific for the entire *B. cereus* group.

Volokhov *et al.*, (2004) developed a rapid microarray assay involving analysis of PCVR amplicons for reliable identification of *B. anthracis* isolates for discrimination from other species of the *B. cereus* group. PCR reactions targeted six *B. anthracis*-specific plasmid-associated virulence genes *cyaA, pagA, lef, capA, capB,* and *capC* in addition to one chromosomal gene BA-5449 specific for *B. anthracis* (Table **1**). Amplicons derived from these genes were denatured and subjected to a second round of 40 thermal cycles in the presence of a mixture of reverse primers and Cy5-dCP resulting in the synthesis of Cy5-labeled single stranded amplicons which were hybridized to probes immobilized on silanized glass slides. Identification was based on hybridization with 47 individual microarray oligonucleotide probes specific for an individual targeted *B. anthracis* gene. Results indicated that all six *B. Anthracis* strains and none of the 56 non-*B. anthracis Bacillus* isolates yielded the expected amplicons from pXO1 and pxO2. The BA-5449 amplicon was highly specific for the *B. anthracis* isolates with none of the 56 non-*B. anthracis* isolates yielding the 1-1017-bp amplicon. The chromosomal DNA sequence BA813 was previously proposed for identification of B. anthracis isolates. However, as described above, the utility of the BA813 sequence as a highly specific *B. anthracis* chromosomal marker has been found to be limited, because of the presence of sequences homologous to BA813 in other *Bacillus* species (Olsen *et al.*, 2007; Ramisse *et al.*, 1999; Qi *et al.*, 2001). The BA-5449 sequence is longer (1017-bp) and contains the BA813 sequence (168-bp) and was found to be more highly specific for *B. anthracis* isolates.

Pulsed-Field Gel Electrophoresis (PFGE)

A (PFGE) method was developed by Zhong *et al.*, (2007) for discriminating *B. anthracis* from *B. cereus* and *B. thuringiensis*. A collection of 25 *B. anthracis* isolates from diverse global origins exhibited high-profile homology. All of the *B. anthracis* isolates were unambiguously distinguished from *B. cereus* and *B. thuringiensis* isolates using the single restriction nuclease NitI. Other commonly used restriction nucleases, SmaI, XbaI, and XmaI failed to perform as well as NotI by either producing too many bands or too few. *B. cereus* and *B. thuringiensis* strains were not segregated by the method.

Although 16s rRNA sequence analysis has been successfully used for the identification and discrimination of many microorganisms, the high level of similarity of 16S and 23S rRNA sequences of the *B. cereus* group (99-100%) and attempts to develop DNA sequences to discriminate between species has repeatedly failed to allow discrimination between species of this group. Bavykin *et al.*, (2008) however were successful in developing a 3D gel element microarray of oligonucleoide probes targeting 16S and 23S rRNA markers for distinguishing seven subgroups derived from isolates of *B. anthracis*, *B. cereus*, *B. thuringiensis*, and *B. mycoides*. Eleven pairs of oligonucleotide probes targeted the 16S RNA sequence and four pairs of oligonucleotide probes targeting the 23S rRNA sequences. Each probe was 20 nucleotides in length. The microarray bearing perfect match/mismatch probe pairs was specific enough to discriminate single nucleotide polymorphism and was able to identify targeted organisms in 5 min.

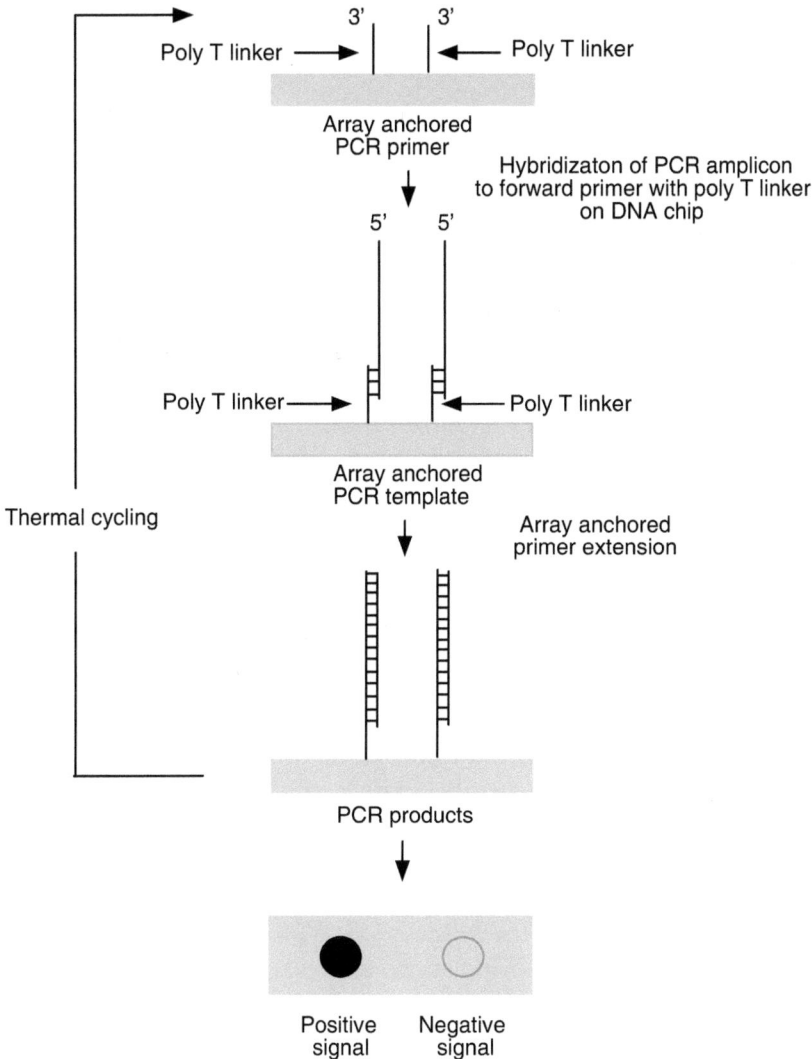

Figure 4: Diagrammatic representation of DNA chip PCR amplification resulting in biotin labeled amplicons. Adapted from Wang *et al.,* (2004).

QUORUM SENSING BY *B. anthracis*

The genome sequence of *B. anthracis* contains an open reading frame (BA50487) predicted to encode an ortholog of *lux*S, the gene required for synthesis of the quorum-sensing signaling molecule autoinducer-2 (AI-2). To determine whether *B. anthracis* produces AI-2, the *Vibrio harveyi* luminescence bioassay was used. Cell-free conditioned media from the Sterne 34F2 vaccine strain was found to induce luminescence in the *V. harveyi* reporter strain BB170, indicating its production of AI-2 (Bassler, Wright and Silverman, 1994). A *luxS* deletion mutant of the Sterne strain was found to exhibit noticeable growth defects compared to the wild-type. The *luxS* mutant exhibited a brief but significant delay (30 to 60 min) in the transition between the lag and early exponential growth phase compared to the wild type strain. Subsequent exponential growth of the wild type and mutant were similar but the mutant was found to enter the stationary growth phase at a much lower cell density. AI-2 function therefore appears necessary to achieve maximum cell density *in vitro*.

MOLECULAR ENGINEERING OF *B. anthracis*

Overcoming Methylation Restriction Nucleases

B. anthracis is refractory to transformation by DNA extracted from *dam$^+$ dcm$^+$ E. coli,* requiring that DNA for transformation be prepared from Dam methylation-deficient strains of *E. coli* (Marrero and Welkos, 1995). Therefore, it is probable that *B. anthracis* encodes methylation-dependant restriction nucleases (MDRs). Pomerntsev *et al.,* (2006) reported on the sequential inactivation of the *mrr* and *mcrB* genes in *B. anthracis*, two of the three genes proposed as encoding MDRs. An antibiotic resistance marker was temporarily inserted into a selected region of the genome, and subsequently removed, leaving the target region permanently mutated. For this purpose, a spectinomycin resistance cassette flanked by bacteriophage P1 *loxP* sites oriented as direct repeats was inserted within a selected gene. A thermo-sensitive plasmid expressing Cre recombinase was introduced at the permissive temperature. Cre recombinase acting at the *loxP* sites excised the spectinomycin marker, leaving a single *loxP* site within the targeted gene or genomic segment. The Cre-expressing plasmid was then removed by growth at the restrictive temperature. The procedure could then be repeated to sequentially mutate additional genes. In this manner, two pairs of genes were subsequently mutated: *pepM* (encodes a secreted, zinc-dependant metaloprotease) and *spoOA* (encodes a global regulator essential for spore formation), and *mcrB* and *mrr*. In addition, *loxP* sites introduced at distant genes could be recombined by Cre recombinase to cause deletion of large intervening regions. In this way, the *capBCAD* region of the pXO2 plasmid and the entire 30 kb chromosomal sequence encompassing the *mcrB* and *mrr* genes was deleted, with the result that the 32 intervening ORFs were found not to be essential for growth in a rich medium. The methodology employed readily allowed (1) the deletion of *spoOA* gene and the construction of nonsporeforming strains, (2) the deletion of the *pepM* gene to enhance the stability of secreted proteins, (3) deletion of the *capBCAD* genes resulting in nonencapsulated strains, and (4) deletion of the *mrr* and *mcrB* genes may enhance successful transformation of DNA from *dam⁻ dcm⁻* strains of *E. coli*.

Engineering Allelic Exchange

Janes and Stibitz (2006) developed improved genetic methodology suitable for routine markerless allelic exchange in *B. anthracis*. The utility of the methodology was demonstrated by the introduction of insertions, deletions, and missense mutations on the chromosome and pXO1 plasmid of the Sterne strain of *B. anthracis*. The methodology involves the use of the intron-encoded homing restriction enzyme I-SceI. The ability of this enzyme, which recognizes an 18-bp sequence to cleave an introduced site unique to a genome was utilized for the establishment of homologous recombination in *B. anthracis*. In such a scheme, the integration of a suicide plasmid by a cloned region of homology containing the desired genetic change results in one of the two crossovers required to effect allelic exchange. To promote the second, the synthesis of the I-SceI enzyme results in cleavage at the unique I-SceI site within the vector. This double-stranded cut is a potent substrate for host recombination systems that can repair the break by homologous recombination of he region of sequence homology that flank the ends of the cut as a result of the initial plasmid cross-in. The method used two plasmids, pBK236 and pBKJ223. Gene replacement constructs were first cloned into plasmid pBKJ236 for integration into the *B. anthracis* chromosome by homologous recombination. This plasmid contained an erythromycin resistance gene and temperature-sensitive replication origin for conditional maintenance in Gram-positive organisms. In addition, the OReT from RP4 was added to facilitate conjugative transfer from *E. coli* to *B. anthracis* and the 18-bp recognition site for I-SceI. Plasmid integrants were then isolated by a shift to the replication-nonpermissive temperature after conjugative transfer and growth at the permissive temperature. The second plasmid, pBKJ223 was then introduced by electroporation and selection for tetracycline resistance. This plasmid encoded the I-SceI enzyme under the control of a hybrid amylase promoter and a Gram-positive ribosome-binding site. Transformants were streaked onto agar plates containing tetracycline, and then single colonies were scored for loss of erythromycin resistance. Following screening by PCR for incorporation of the desired mutation, the pBKJ223 plasmid was lost spontaneously by streaking onto agar plates lacking tetracycline and scoring colonies for tetracycline sensitivity.

A Lethal Gene in *B. anthracis*

Prokaryotic toxin-antitoxin (TA) loci are usually organized as an operon, where the first cistron encodes a small (65 – 85 as) labile antitoxin and the second one encodes a large stable toxin (95 – 135 aa) (Brown and Shaw, 2005). Most of these TA modules are associated with plasmids. When the cell loses the plasmid during a segregational event, the degradation of antitoxin by cellular protease renders the toxin free to exert its lethal effect on the bacterial cell. These modules have therefore been implicated in maintaining the stability of extra-chromosomal elements in the bacterial host so as o ensure propagation of only plasmid-inherited cells. Such a genetic unit has been termed an "Addiction Module" because the cells become addicted to the short-lived antitoxin and its *de novo* synthesis is essential for cell survival. Agarwal *et al.,* (2007) found a 351-bp open reading frame (ORF) in the genome of *B. anthracis* that encodes a putative protein of 116 amino acids designated PemK (K, killer protein) and a second ORF (PemI) (I, inhibitor protein). When the PemK protein of *B. anthracis* was overexpressed in *E. coli* using a recombinant plasmid designated pETSAT, growth was notably inhibited for the first 5 hrs. then the *pemI* gene was cloned to yield the pGEXSAAT plasmid and inserted into *E. coli* containing the PemK expression plasmid, pETSAT growth was normal. When *pemK* and *pemI* were cloned into the plasmid vector pHCMCO5 independently and simultaneously to generate pSpacpemk pSpacpemI, and pSpacpemIK under the control of an IPTG inducible promoter and β-D-thiogalactopyranoside and then inserted into *B. anthracis* by electroporation, the expression of pemK on plasmid pSpacpemK was severely toxic to cell growth and no transformants were obtained. In contrast, transformants containing pSpacpemI or pspacpemIK were obtained which grew normally. These results were confirmed with the use of polyclonal antiserum to PemK and PemI and western blots of sonically disrupted cells indicating independent production of PemK and PemI derived from the respective transformed plasmids and the simultaneous production of both PemK and PemI in *B. anthracis* derived from the pSpacpemIK plasmid. The mechanism of the TA system relies on the differential lifespan of toxin and antitoxin. In addition, the direct physical interaction of both PemK and PemI was demonstrated by their copurification and ELISA assays confirming that this TA module is a Typical addiction system which refers to neutralization of the PemK toxin by the binding of PemI forming an inactive complex. Structural instability of the antitoxin, making it susceptible to proteolysis, combined with the well-ordered stable structure of the toxin is required for the appropriate functioning of the system on loss of the plasmid. The *pemK-pemI* operon is assumed to be silent in *B. anthracis* when harbored strictly on the chromosome and is presumably activated when transferred by recombination to an acquired plasmid.

MICROBIAL FORENSICS APPLIED TO *B. anthracis* SPORES OF POST 9/11 MAILINGS

With micro-organisms having notably higher mutation rates than *B. anthracis* such as *Escherichia coli, Salmonella enterica, Campyobacter, Shigella,* and *Listeria monocytogenes*, the CDC uses restriction DNA analysis that cuts genomic DNA at specific but infrequent sites (restriction fragment length polymorphism, RFLP). The resulting DNA fragments are then resolved by pulsed field gel electrophoresis. The resulting DNA band-profiles provide rapid typing for tracking outbreaks of food borne diseases. The author has found that the DNA banding patterns derived from random amplified polymorphism (RAPD) analysis using three different mono primers is very effective for distinguishing between different strains of *L. monocytogenes* (Gu *et al.*, 2006). However, these methods are not suitable for distinguishing strains of *B. anthracis*. Keim *et al.,* (2008) have emphasized that *B. anthracis* is a highly clonal species with no evidence of recombination or horizontal genetic transfer since its evolution from a *B. cereus* ancestor. This clonal characteristic is common for recently emerged pathogens that rapidly acquire a new environment that furnished the organism with great and unrestricted reproductive capacity. The success of *B. anthracis* as a warm blooded animal pathogen has served to ecologically isolate the vegetative state of the organism from other members of the *B. cereus* group, that are native to soil. *B. anthracis* can now be considered ecologically distinct from its soil relatives and therefore has little opportunity for genetic exchange (Keim *et al.,* 2008). The clonality of this species coupled with the extreme stability of its SMPs, means that a single nucleotide polymorphism (SNP) can be used to define an entire lineage of *B. anthracis* (Keim *et al.,* 2008). Single nucleotide polymorphisms (SNPs) selected to represent phylogenetic branches for genotyping assays are termed canonical SNPs (canSNPs). The Ames branch can now be defined as consisting of 32 SNPs (Keim *et al.,* 2008). These 32 SNPs are found only in isolates that are clear relatives of the Ames strain. Van Ert *et al.,* (2007) found that among 11 non-Ames isolates of *B. anthracis* that had been previously placed in the Ames canSNP cluster five of the 32 SNPs are unique to the ancestral Ames strain.

Most virulent cultures of *B. anthracis* used in new animal vaccine studies in the U.S.A. are derived from the original Ames strain which is highly virulent. On September 18, 2001, exactly one week after the terrorist attack on the Word Trade Center five letters containing weaponized spores of *B. anthracis* were placed into the U.S. mail. Three weeks later, two more anthrax spore laden letters entered the U.S. mail service. A total of 22 individuals contracted anthrax; 11 individuals were stricken with cutaneous anthrax and 11 with inhalation anthrax. Amongst the 11 inhalation victims, five succumbed, primarily as a result of delayed diagnosis. In addition, 31 individuals were found positive for exposure to *B. anthracis* spores. Th FBI was faced with the task of tracking a genetically homogeneous organism to its point of origin. When spores derived from the powder of the tainted letters were plated on sheep blood agar, some of the colonies had unusual morphological features that differed from the majority in size, shape, color, and texture suggesting that the Ames spores did not consist of a homogeneous population, but included rare variant morphotypes (Koblentz and Tucker, 2010). This observation indicated that the strain used had a unique colony morphology profile that could be compared to the Ames strains used in various laboratories. The DNA from each colony morphotype was sequenced (at enormous cost in 2002) for detection of the unique and subtle differences between these clones and the sequence of a reference *B. anthracis* Ames strain. The FBI then asked three outside laboratories to develop genetic assays to test samples of *B. anthracis* Ames regarding the four most stable morphotype mutations. A total of 1,070 isolates of the Ames strain were obtained by the FBI from 16 laboratories in the U.S.A, Canada, Sweden, and the U.K. Only eight contained all four mutations present in the post 9/11 mailed spores. The genetic profile of all eight isolates matched that of a liquid suspension of *B. anthracis* spores stored in a large flask for testing the efficacy of experimental animal vaccines. This spore suspension designated RMR-1029 had been created in 1997 by a staff scientist at Fort Detrick, MD (Enserink, 2008). Interestingly, the spores in the flask had been pooled from more than 30 production runs at two different U.S. military facilities, each of which had created the opportunity for mutations to arise. The fact that *B. anthracis* is one of the most thoroughly characterized pathogens greatly facilitated the microbial forensic investigation.

The anthrax strain derived from spinal fluid of the first victim of the anthrax contaminated letters were found to be molecularly identical to the standard "wild type" Ames strain and did not yield the unusual morphotypes of the dry-spores in the mailed envelopes (Ravel *et al.*, 2009). This discrepancy was due to the fact that the morphotypes associated with the RMR-1029 strain made up only about 1% of the spores, and the resulting infection was therefore due primarily to the majority strain. Over 100 people were found to have access to the RMR-1029 preparation at a total of three U.S. military facilities. At this point the FBI began to establish personality profiles of these individuals. The suicide shortly before indictment, of the staff scientist at Fort Detrick who had prepared the RMR-1029 spore suspension, hampered further investigation, making absolute identity of the perpetrator impossible to determine.

The cost of sequencing the entire genome of an organism like *B. anthracis* has dropped sharply from about $500,000 in 2002 to no more than $500 in 2013. In addition to a comparison of genomic sequences the manner in which the weaponized spores are dispersed to prevent clumping is of significance. The use of bentonite or an alternate agent for reduction of particle size to no more than 3 microns is of forensic value.

REFERENCES

Abshire, T., Brown, J., Ezzell, J. (2005). Production and validation of the use of gamma phage for identification of *Bacillus anthracis*. J. Clin. Microbiol. 43:47809-4788.

Agarwal, S., Agarwal, S., Bhatnagar, R. (2007). Identification and characterization of a novel toxin-antitoxin module from *Bacillus anthracis*. FEBS Lett. 581:1727-1734.

Anderson, G., Simchock, J., Wilson, K. (1996). Identification of a region of genetic variability among *Bacillus anthracis* strains and related species. J. Bacteriol. 178:377- 384.

Antwerpen, M., Zimmerman, P., Bewly, K., Frangoulidis, D., Meyer, H. (2008). Real- time PCR system targeting a chromosomal marker specific for *Bacillus anthracis*. Molec. Cell. Probes. 22:313-315.

Bartkus, J., Leppla, S. (1989). Transcriptional regulation of the protective antigen gene of *Bacillus anthracis*. Infect. Immun. 57:2295-2300.

Barua, S., McKevitt, M., Deguisti, K., Hamm, E., Labarbee, J., Shakir, S., Bryant, K., Koehler, T., Blanke, S., Dyer, D., Gillaspy, A., Ballard, J. (2009). The mechanism of *Bacillus anthracis* intracellular germination requires multiple and highly diverse genetic loci. Infect. Immun. 77:23-31.

Bassler, B., Wright, M., Silverman, M. (1994). Multiple signaling systems controlling expression of luminescence in *Vibrio harveyi*: sequence and function of genes encoding a second sensory pathway. Mol. Microbiol. 13:273-286.

Bavykin, A., Mikhailovich, V., Zakharyev, V., Lysov, Y., Kelly, J., Alferov, O., Gavin, I., Kukhtin, A., Jackman, J., Stahl, D., Chandler, D., Mirzabekov, A. (2008). Discrimination of *Bacillus anthracis* and closely related microorganisms by anal- ysis of 16S and 23S rNA with oligonucleotide microarray. Chemico-Biol. Inter- actions. 171:212-235.

Beyer, W., Glöckner, P. Otto, J., Böhm, R. (1995). A nested PCR method for the detection of *Bacillus anthracis* in environmental samples collected from former tannery sites. Microbiol. Res. 150:179-186.

Beyer, W., Pocivalsek, S., Böhm, R. (1999). Polymerase chain reaction-ELISA to detect *Bacillus anthracis* from soil samples— limitations of present published primers. J. Appl. Microbiol. 87:229-236.

Bode, E., Hurtle, W., Norwood, D. (2004). Real-time PCR assay for a unique chromosomal sequence of *B. anthracis*. J. Clin. Microbiol. 42:5825-5831.

Bourgogne, A., Drysdale, M., Hilsenbeck, G., Petrson, N., Koehler, T. (2003). Global effects of virulence gene regulators in a *Bacillus anthracis* strain with both virulence plasmids. Infect. Immun. 71:2736-2743.

Bowen, J., Quinn, C. (1999). The native virulence plasmid combination affects the segregational stability of a theta-replicating shuttle vector in *Bacillus anthracis* var. New hampshire. J. Appl. Microbiol. 87:270-278.

Brightwell, G., Pearce, M., Leslie, D. (1998). Development of internal controls for PCR detection of *Bacillus anthracis*. *Molec. Cell. Probes*. 12:367-377.

Britton, R., Eichenberger, P., Gonzalez-Pastor, J., Fawcett, P., Monson, R., Losick, R., Grossman, A. (2002). Genome-wide analysis of the stationary-phase sigma factor (sigma-H) regulon of *Bacillus subtilis*. J. Bacteriol. 184:4881-4890.

Brown, J., Shaw, K. (2005). A novel family of *Escherichia coli* toxin-antitoxin gene pairs. Nucleic Acids. Res. 33:966-976.

Brumlik, M., Szymajda, U., Zakowski, D., Liang, X., Redkar, R., Patra, G., Vecchio, V. (2001). Use of long-range repetitive element polymorphism-PCR to differentiate *Bacillus anthracis* strains. Appl. Environ. Microbiol. 67:3021-3028.

Coenye, T., Vandamme, P. (2003). Simple sequence repeats and compositional bias in Bipartite *Ralstonia solanaceanum* GMI1000 genome. BMC Microbiol. 4:10. [Online,] http://biomedcenral.com/1471-2164/4/10. GMI1000 genome. BMC Microbiol. 4:10.

Couture-Tosi, Delacroix, H., Mignot, T., Mesnage, S., Chami, M., Fouet, A., Mosser, G., (2002). Structural analysis and evidence for the dynamical emergence of *Bacillus anthracis* S-layer networks. J. Bacteriol. 184:6448-6456.

Daffonchio, D., Raddadi, N., Merabishvili, M., Cherif, A., Carmagdnola, L., Brusetti, L., Rizzi, A., Chanishvili, N., Visca, P., Borin, (2006). *Appl. Environ. Microbiol.* 72:1295- 1301.

Dai, Z., Sirard, J., Mock, M., Koeler T. (1995). The *atxA* gene product activates transcription of anthrax toxin genes and is essential for virulence. Molec. Microbiol. 16:1171-1181.

Dai, Z., Koehler, T. (1997). Regulation of anthrax toxin activator gene (*atxA*) expression In *Bacillus anthracis*: temperature, not CO2/bicarbonate, affects AtxA synthesis. Infect. Immun. 65:2576-2582.

Dauphin, L., Newton, B., Rasmussen, M., Meyer, R., Bowen, M. (2008). Gamma irradiation can be used to inactivate *Bacillus anthracis* spores without compromising the sensitivity of diagnostic assays. Appl. Environ. Microbiol. 74:4427-4433.

Drysdale, M., Borgogne, A., Hilsenbeck, S., Koehler, T. (2004). atxA controls *bacillus anthracis* capsule synthesis *via acpA* and a newly discovered regulator, *acpB*. J. Bacteriol. 186:307-315.

Ellerbrok, H., Natternann, H., Ozel, M., Betin, L., Appel, B., Pauli, G. (2002). Rapid and sensitive identification of pathogenic and apathogenic *Bacillus anthracis* by real-time PCR. FEMS Microbiol. Lett. 214:51-59.

Enserink, M., (2008). Sequencing paved the way from spores to a suspect. Science. 321:898-899.

Fouet, A., Namy, O., Lamnbert, G. (2000). Characterization of the operon encoding the alternative σ^B factor form *Bacillus anthracis* and its role in virulence. J. Bacteriol. 182:5036-5045.

Fjita, M, Sadie, Y. (1998). Feedback loops involving SpoA and AbrB in *in vitro* transcription of the genes involved in the initiation of sporulation in *Bacillus subtilis*. J. Biochem. (Tokyo). 124:98-104.

Gierxzynski, R., Kaluzewski, S. Rakin, A., Jagielski, M. Zasada, A., Jakubczak, A., Borkowska-Opacka, B. Rastawicki,W. (2004). Intriguing diversity of *Bacillus anthracis* in easten Poland—the molecular echoes of the past outbreaks. FMS Microbiol. Lett. 239:235-240.

Green, B., Battista, L., Loehler, T., Torne, C., Ivins, B. (1985). Demonstration of a capsule plasmid in *Bacillus anthracis*. Infect. Immun. 49:291-297.

Gu, W., Gonzalez-Rey, C., Krovacek, K., Levin, R. (2006). Genetic variability among isolates of *Plesiomonas shigelloides* from fish, human clinical sources and fresh water, determined by RAPD typing. Food Biotechnol. 20:1-12.

Guidi-Rontani, C., Pereira, Y., Ruffie, S., Sirard, J., Weber-Levy, M., Mock, M. (1999). Identification and characterization of a germination operon on the virulence plasmid pXO1 of *Bacillus anthracis*. Mol. Microbiol. 33:407-414.

Gur-Arie, R., Cohen, C., Eitan, Y., Shelef, L., Hallerman, M., Kashi, Y. (2000). Simple sequence repeats in *Escherichia coli*: abundance, distribution, composition and poly- morphism. Genome Res. 10:62-71.

Guignot, J., Mock, M., Fouet, A. (1997). AtxA activates the transcription of genes harbored by both *Bacillus anthracis* virulence plasmids. FEMS Microbiol. Let. 147:203- 207.

Hadjifrangiskou, M., Chen, Y., Koehler, T. (2007). The alternative sigma factor s^H is required for toxin gene expression by *Bacillus anthracis*. J. Bacteriol. 189:1874-1883.

Harrell, L., Andersen, G., and Wilson, K. (1995). Genetic variability of *Bacillus anthracis* and related species. J. Clin. Microbiol. 33:1847-1850.

Helgason, E., Okstad, O., Caugant, D., Johansen, H., Fouet, A., Mock, M., Hegna, I., Kolsto, A. (2000). *Bacillus anthracis, Bacillus cereus*, and *Bacillus thuringiensis* — One species on the basis of genetic evidence. Appl. Environ. Microbiol. 66:2627-2630.

Helgason, E., Tourasse, N., Meisal, R., Caugant, D., Kolsto, A. (2004). Multilocus sequence typing scheme for bacteria of the *Bacillus cereus* group. Appl. Environ. Microbiol. 70:191-201.

Henderson, I., Duggleby, C., Turnbull, P. (1994). Differentiation of *Bacillus anthracis* from other *Bacillus cereus* group bacteria with the PCR. Int. J. Syst. Bacteriol. 44:99-105.

Hill, K., Ticknor, L., Okinaka, R., Assay, M., Blair, H., Bliss, K., Laker, M., Pardington, P., Richardson, A., Tonks, M., Beecher, D., Kemp, J., Kolsto, A., Wong, A., Keim, P., Jackson, P. (2004). Appl. Environ. Microbiol. 70:1068-1080.

Hoffmaster, A., Koehler, T. (1997). The anthrax toxin activator gene *atxA* associated with CO_2-enhanced non-toxin gene expression in *Bacillus anthracis*. Infect. Immun. 65:3091-3099.

Hoffmaster, A., Koehler, T. (1999a). Autogenous regulation of the *Bacillus anthracis pag* operon. J. Bacteriol. 181:4485-4492.

Hoffmaster, A., Koehler, T. (1999b). Control of virulence gene expression in *Bacillus anthracis*. J. Appl. Microbiol. 87:279-281.

Ivanova, N., Sorokin, A., Anderson, I., Galleron, N., Candelon, R., Kapatral, V., Bhattaharya, A., Reznik, G., Mikhailova, N., Lapidus, A., d'Souza, M., Walunus, T., Grechkin, Y., Pusch, G., Haselkorn, R., Fonstein, M., Ehrich, S., Overbeck, R., Kyrpides, N. (2003). Genome sequence of *Bacillus cereus* and comparative analysis with *Bacillus anthracis*. Nature. 423:87-91.

Jones, M., Blaser, M. (2008). Detection of *luxS*-signaling molecule in *Bacillus anthracis*. Infect. Immun. 71:3914-3919.

Jackson, P., Hill, K., Laker, M., Ticknor, L., Keim, P. (1999). Genetic comparison of *Bacillus anthracis* and its close relatives using amplified fragment length polymorph- ism and polymerase chain reaction analysis. J. Appl. Microbiol. 87:263-269.

Jackson, P. Walthers, E., Kalif, A., Richmond, K., Adair, D., Hill, K., Kuske, C, Andersen, G., Wilson, K., Hugh-Jones, M., Keim, P. (1997). Characterization of the variable- number tandem repeats in vrrA from different *Bacillus anthracis* isolates. Appl. Environ. Microbiol. 63:1400-1405.

Janes, B., Stibitz, S. (2006). Routine markerless gene replacement in *Bacillus anthracis*. Infect. Immun. 74:1949-1953.

Kane, A., Leant, S., Murphy, G., Alfaro, T., Krauter, P., Mahnke, R., Legler, T., Raber, E. (2009). Rapid, high-throughput, culture-based methods to analyze samples for viable spores of *Bacillus anthracis* and its surrogates. J. Microbiol. Meth. 76:278-284.

Klichko, V., Miller, J., Wu, A., Popov, S., Alibek, K. (2003). Anaerobic induction of *Bacillus anthracis* hemolytic activity. Biochem Biophys. Res. Comm. 303:855-862.

Koehler, T., Dai, Z., Kaufman-Yarbray, M. (1994). Regulation of the *Bacillus anthracis* protective antigen gene: CO_2 and a *trans*-acting element activate transcription from one of two promoters. J. Bacteriol. 176:586-595.

Keim, P., Kalif, A., Scupp, J., Hill, K., Travis, S., Richmond, K., Adair, D., Hugh-Jones, M., Kuske, C., Jackson, P. (1997). Molecular evolution and diversity in *Bacillus anthracis* as detected by amplified fragment polymorphism markers. J. Bacteriol. 179:818-824.

Keim, P., Smith, K. (1992). *Bacillus anthracis* evolution and epidemiology. Current topics in Microbiology and Immunology. 271:20-32.

Keim, P., Klevytska, A., Price, L., Schupp, J., Zinser, G., Smith, K., Hugh-Jones, M., Okinaka, R., Hill, K., Jackson, P. (1999). Molecular diversity in *Bacillus anthracis*. J. Appl. Microbiol. 87:215-217.

Keim, P., Pearson, T., Okinaka, R. (2008). Microbial Forensics: DNA fingerprinting of *Bacillus anthracis* (Anthrax). Anal. Chem. 80:4791-P., Keim, P., Price, I., Klevytska, A, Smith, K, Schupp, J, Okinaka, R, Jackson, P, Hugh-Jones, M. (2000). Multiple-locus variable-number tandem repeat analysis reveals genetic relationships within *Bacillus anthracis*. J. Bacteriol. 182:2928-2936.

Keim, P., van Ert, N., Pearson, T., Vogler, A., Huynh, L., Wagner, D. (2004). Anthrax molecular epidemiology and forensics: using the appropriate marker for different evolutionary scales. Infect. Genet. Evol. 4:205-213.

Kenefic, L., Beaudry, J., Trim, C., Huynh, L., Zanecki, S., Mahews, M., Schupp, J., Van Ert, M., Keim, P. (2008). A high resolution four-locus multiplex single nucleotide r epeat (SNR) genotyping system in *Bacillus anthracis*. J. Microbiol. Meth. 73:269-272.

Kim, W., Hong, Y., Yoo, J. Lee, W., Choi, C., Chung, S. (2001). Genetic relationships of *Bacillus anthracis* and closely related species based on variable-number tandem repeat analysis and BOX-PCR genomic fingerprinting. FEMS Microbiol. Lett. 207:21-27.

Kim, K., Seo, J., Wheeler, K., Park, C., Kim, D., Park, S., Kim, W., Chung, S., Leighton, T. (2005). Rapid genotypic detection of *Bacillus anthracis* and the *Bacillus cereus* group by multiplex real-time PCR melting curve analysis. FEMS Immunol. Med. Microbiol. 43:301-310.

Ko, K., Kim, J., Kim, J, Kim, W., Chung, S., Kim, I. Kock, Y. (2004). Population structure of the *Bacillus cereus* group as determined by sequence analysis of six housekeeping genes and the *plcR* gene. Infect. lmmun. 72:5253-5261.

Koblentz, G., Tucker, J. (2010). Tracing an Attack: The Promise and Pitfalls of Microbial Forensics. Survival. 52:159-186.

Koehler, T., Dai, Z., Kauman-Yarbray, M. (1994). Regulation of the *Bacillus anthracis* protective antigen gene: CO_2 and a *trans*-acting element activate transcrip- tion from one of two promoters. J. Bacteriol. 176:586-595.

Koehler, T. (1999). *Bacillus anthracis* genetics and virulence gene regulation. Current topics in Microbiology and Immunology. 271:142-164.

La Duk, M., Satomi, M., Agata, N., Venkateswaran, K. (2004). *gyrB* as a phylogenetic discriminator for members of the *Bacillus anthracis—cereus —thuringiensis* group. J. Microbiol. Meth. 56:383-394.

Lamonica, J., Wagner, M., Eschenbrenner, M., Williams, L., Miller, T., Patra, G., Delvecchio, V. (2005). Comparative secretome analyses of three *Bacillus anthracis* strains with variant plasmid contents. Infect. Immun. 73:3646-3658.

Lee, M., Brightwell, G., Leslie, D., Bird, H, Hamilton, A. (1999). Fluorescent detection techniques for real-time multiplex strand specific detection of *Bacillus anthracis* using rapid PCR. J. Appl. Microbiol. 87:218-223.

Levi, H., Fisher, M., Ariel, N., Altboum, Z., Kobiler, D. (2005). FEMS Microbiol. Lett. 244:199-205.

Levinson, G., Gutman, G. (1987). Slipped-strand mispairing: a major mechanism for DNA sequence evolution. *Mol. Biol. Evol.* 4:203-221.

Liang, X., Yu, D. (1999). Identification of *Bacillus anthracis* strains in China. J. Appl. Microbiol. 87:200-203.

Luna, V., King, D. Peak, K., Reeves, F., Heberlein-Larson, L., Veguilla, W., Heller, L., Duncan, K., Cannons, A., Amuso, P., Cattani, J. (2006). *Bacillus anthracis* virulent plasmid pXO2 genes found in large plasmids o f two other *Bacillus* s pecies. J. Clin. Microbiol. 44:2367-2377.

Makino, S, Sasakawa, C., Uchida, I., Terakado, N., Yoshikawa, M. (1988). Cloning and CO_2 expression of the genetic region for encapsulation from *Bacillus anthracis*. Mol. Microbiol. 2:371-376.

Makino, S, Uchida, I., Terakado, N., Sasakawa, C., Yoshikawa, M. (1989). Molecular characterization and protein analysis of the *cap* region, which is essential for encapsulation in *Bacillus anthracis*. J. Bacteriol. 171:722-730.

Makino, S., Iinuma-Okada, Y., Muruyama, T., Ezaki, T., Sasakawa, C., Yoshikawa, M. (1993). Direct detection of *Bacillus anthracis* DNA in animals by polymerase chain reaction. J. Clin. Microbiol. 31:547-551.

Marrero, R., Welkos, S. (1995). The transformation frequency of plasmids into *Bacillus anthracis* is affected by adenine methylation. Gene. 152:75-78.

Mesnage, S., Tosi-Couture, E., Gounon, P., Mock, M., Gounon, P., Fouet, A. (1997). Molecular characterization of *Bacillus anthracis* main S-layer component, evidence that it is the major cell-associated antigen. Mol. Microbiol. 23:1147-1155.

Mesnage, S., Tosi-Couture, E., Gounon, P., Mock, M., Fouet, A. (1998). The capsule and S-layer, two independent and yet compatible macromolecular structures in *Bacillus anthracis*. J. Bacteriol. 180:52-58.

Mignot, T., Couture-Tosi, E., Mesnage, S., Mock, M., Fouet, A. (2004). *In vivo Bacillus anthracis* gene expression requires PagR as an intermediate effector of the AtxA signalling cascade. Int. J. Med. Microbiol. 293:619-624.

Mignot, T., Mock, M., Fouet, A. (2003). A plasmid-encoded regulator couples the synthesis of toxins and surface structures in *Bacillus anthracis*. Mol. Microbiol. 47:917- 927.

Moxon, E., Rainey, P. (1995). Pathogenic bacteria: the wisdom of their genes, p. 255-268. In B. Vander Zeijst, L. van Alphen, W. Hokstra, and J. van Embden (ed.), Ecology of Pathogenic Bacteria. Royal Dutch Academy of Sciences, 2nd series, no. 96. Royal Dutch Academy of sciences. Amsterdam, The Netherlands. Nübel, U., Schimdt, P., Reib, E., Bier, F., Beyer, W., Naumann. (2004). Oligonucleotide microarray for Identification of *Bacillus anthracis* based on intergeneric transcribed spacers in ribosomal DNA. FEMS Microbiol. Lett. 240:215-223.

Okinaka, R., Cloud, K., Hampton, O., Hoffmaster, A., Hill, K., Keim, P., Koehler, T., Lamke, G., Kumano, S., Mahillon, K., Manter, D., Martinez, Y., Rick, D., Svensson, R., Jackson, P. (1999a). Sequence and organization of pXO1, the large *Bacillus anthracis* plasmid harboring the anthrax toxin genes. J. Bacteriol. 181:6509-6515.

Okinaka, R., Cloud, K., Hampton, O., Hoffmaster, A., Hill, K., Keim, P., Koehler, T., Lamke, G., Kumano, S., Manter, D., Martinez, Y., Rick, D, Svensson, R., Jackson, P. (1999b). Sequence, assembly and analysis of pXO1 and pXO2. J. Appl. Microbiol. 87:261-262.

Olsen, J., Skogam, G., Fykse, E., Rawlinson, E., Tomaso, H., Granum,P., Blatny, J. (2007). Genetic distribution of 295 *Bacillus cereus* group members based on *adk*-screening in combination with MLST (Multilocus Sequence Typing) used for validating a primer targeting a chromosomal locus in *B. anthracis*. J. Microbiol. Meth. 71:265-274.

Patra, G., Sylvesdtre, P., Ramisse, V., Therasse, J., Guesdon, J. (1996). Isolation of a specific chromosomic DNA sequence of *Bacillus anthracis* and its possible use in diagnosis. FEMS Immunol. Med. Microbiol. 15:223-231.

Pearson, T., Busch, J., Ravel, J., Read, T., Rhoton, S., U'ren, J., Simonson, T., Kachur, S., Leadem, R., Cardon, M., Van Ert, M., Huynh, L., Fraser, C., Keim, P. (2004). Phylogenetic discovery bias in *Bacillus anthracis* using single-nucleotide polmorphisms from whole-genome sequencing. Proc. Natl. Acad. Sci. USA. 101:13536-13541.

Perez, A., Hohn, C., Higgins, J. (2005). Filtration methods for recovery of *Bacillus anthracis* spores spiked into source and finished water. Water Res. 39:5199-5211.

Phillips, Z., Strauch, M. (2002). *Bacillus subtilis* sporulation and stationary phase gene expression. Cell. Mol. Life Sci. 59:392-402.

Pomerantsev, A., Sitaraman, R., Galloway, C., Kivovich, V., Leppla, S. (2006). Genome engineering in *Bacillus anthracis* using Cre recombinase. Infect. Immun. 74:682-693.

Pomerantsev, A., Shishkova, N., Soronin, I., Sukovatova, L., Marinin, L. (1993). Interaction of *Bacillus anthracis* with benzylpenicillin *in vivo* and *in vitro*. Antibiot. Khimioter. 38:30-33.

Popovic, T., Hoffmaster, A., Ezzell, J., Abshire, T., Brown, J. (2005). Validation of methods for confirmatory identification of presumptive isolates of *Bacillus anthracis*. J.A.O.A.C. Intl. 88:175-177.

Qi, Y., Patra, G., Liang, A., Williams, L., Rose, S., Redkar, R., Del Vecchio. (2001). Utilization of the *rpoB* gene as a specific chromosomal marker for real-time PCR detection of *Bacillus anthracis*. Appl. Environ. Microbiol. 67:3720-3727.

Radnedge, L., Agron, P., Hill, K., Jackson, P., Ticknor, L., Keim, P., Anderson, G. (2003). Genome differences that distinguish *Bacillus anthracis* from *Bacillus cereus* and *Bacillus thuringiensis*. Appl. Environ. Microbiol. 69:2755-2764.

Ramisse, V., Patra, G., Garrigue, H. Guesdon, J., Mock, M. (1996). Identification and characterization of *Bacillus anthracis* by multiplex PCR analysis of sequences on plasmids pXO1 and pXO2 and chromosomal DNA. FEMS Microbiol. Lett. 145:9-16.

Ramisse, V., Patra, G., Vaissaire, J., Mock, M. (1999). The Ba813 chromosomal DNA sequence effectively traces the whole *Bacillus anthracis* community. J. Appl. Microbiol. 87:224-228.

Rasko, D., Worsham, P., Abshire, T., Stanley, S., Bannand, J., Wilson, M., Langham, R., Decker, R., Jianga, L., Reade, T., Phillippy, A., Salzberg, S., Pop, M., Van Ertg, M., Kenefic, L., Keim, P., Fraser-Liggetti, C., Ravela, J. (2011). B*acillus anthracis* comparative genome analysis in support of the Amerithrax investigation. Proc. Natl. Acad. Sci. 108:5027-5032.

Ravel, J., Jiang, L., Stanley, S., Wilson, M., Decker, R., Read, T., Worsham, P., Keim, P., Salzberg, S., Fraser-Liggett, C., Rasko, D. (2009). The complete genome sequence of *Bacillus anthracis* "Ames ancestor". J. Bacteriol. 191:445-446.

Read, T., Salzberg, S., Pop, M., Shumway, M., Umayam, L., Jiang, L., Holtzapple, E., Busch, J., Smith, K., Schupp, J., Solomon, D., Keim, P., Fraser, C. (2002). Comparative genome sequencing for discovery of novel poymorphisms in *Bacillus anthracis*. Science. 296:2028-2033.

Read, T. Read, Peterson, S., Tourasse, N. Baillie, L., Paulsen, I. Nelson, K., Tettelin, H., Fouts, D., Eisen, J., Gill, S., Holtzapple, E., Økstad, O., Helgason, E., Rilstone, J., Wu, M., Kolonay, J., Beanan, M., Dodson, R., Brinkac, L., Gwinn, M., DeBoy, R., Madpu, R., Daugherty, S., Durkin, A., Haft, D., Nelson, W., Peterson, J., Pop, M., Khouri, H., Radune, D., Benton, J., Mahamoud, Y., Jiang, L., Hance, I., Weidman, J., Berry, K., Plaut, R., Wolf, A., Watkins, K., Nierman, W., Hazen, A., Cline, R., Redmond, C., Thwaite, J., White, O., Salzberg, S., Thomason, B., Friedlander, A., Koehler, T., Hanna, P., Kolstø, A., Fraser. C. (2003). The genome sequence of *Bacillus anthracis* Ames and comparison to closely related bacteria. Nature. 423:81-86.

Reif, T., Johns, M., Pillai, S., Carl, M. (1994). Identification of capsule-forming *Bacillus anthracis* spores with the PCR and a novel dual-probe hybridization format. Appl. Environ. Microbiol. 60:1622-1625.

Reiman, R., Atcley, D., Voorhees, K. (2007). Indirect detection of *Bacillus anthracis* using real-time PCR to detect amplified gamma phage DNA. J. Microbiol. Meth. 68:651-653.

Rupf, S., Mert, K., Eschrich, K. (1999). Quantification of bacteria in oral samples by competitive polymerase chain reaction. J. Dent. Res. 78:850-856.

Ryu, C., Le, K., Hawng, H., Yoo, C., Seong, W., Oh, H. (2005). Molecular characterization of Korean *Bacillus anthracis* isolates by amplified fragment length polymorphism analysis and multilocus variable-number tandem repeat analysis. Appl. Environ. Microbiol. 71:4664-4671.

Saile, E., Koehler, T. (2002). Control of anthrax toxin gene expression by the transition state regulator *abrB*. J. Bacteriol. 184:370-380.

Sjöstedt, A., Erikssn, U., Ramisse, V., Garrigue, H. (1997). Detection of *Bacillus anthracis* spores in soil by PCR. (1997). FEMS Microbiol. Ecol. 23:159-168.

Sohni, Y., Kanjilal, S., Kapur, V. (2008). Cloning and development of synthetic internal amplification control for *Bacillus anthracis* real-time polymerase chain reaction assays. Diag. Microbiol. Infect. Dis. 61:471-475.

Stratilo, C., Lewis, C., Bryden, L., Mulvey, M., Bader, D. (2006). Single-nucleotide repeat analysis for subtyping *Bacillus anthracis* isolates. J. Clin. Microbiol. 44:777-782.

Strauch, M. (1995). Delineation of AbrB-binding sites on the *Bacillus subtilis spoOH, kinB, ftsAZ*, and *pbpE* promoters and use of a derived homology to identify a previously unsuspected binding site in the *bsuB1* methylase promoer. J. Bacteriol. 177:6999-7002.

Strauch, M., Ballar, P., Rowsan, A., Zoller, K. (2005). The DNA-binding specificity of he *Bacillus anthracis* ABrB protein. Microbiol. 151:1751-1759.

Thorne, C. (1993). B*acillus anthracis*. P. 113-124. In A. Sonensheim (ed.), *Bacillus subtilis* and other gram-positive bacteria. American Society for Microbiology, Washington, D.C. Tinsley, E., Khan, S. (2007). A *Bacillus anthracis*-based *in vitro* system supports replication of plasmid pXO2 as well as rolling-circle-replicating plasmids. Appl. Environ. Microbiol. 73:5005-5010.

Uchida, I., Hornung, J., Thorne, C, Klimpel, K., Lepla, S. (1993a). Cloning and characterization of a gene whose product is a *trans*-activator of anthrax toxin synthesis. J. Bacteriol. 175:5329-5338.

Uchida, I., Makino, S., Sasakawa, M., Yoshikawa, M., Sugimoto, C., Terakado, N. (1993b.) Identification of a novel gene *dep*, associated with depolymerization of the capsular polymer of *Bacillus anthracis*. Mol. Microbiol. 9:487-496.

Uchida, I., Makino, S., Sekizaki, T., Terakado, N. (1977). Cross-talk to the genes for *Bacillus anthracis* capsule synthesis by *atxA*, the gene encoding the *trans*-activator of anthrax toxin synthesis. Molec. Microbiol. 23:1229-1240.

van Belkum, A., Scherer, S., Alphen, L., Verbrugh, H. (1998). Short-sequence DNA repeats in prokaryoteic genomes. Microbiol. Mol. Biol. Rev. 62:275-293.

Van Ert, M., Easterday, W., Simonson, S., U'Ren, J., Pearson, T., Kenefic, L., Busch, J., Huynh, L., Dukerich, M., Trim, C., Beaudry, J., Welty-Bernard, A., Read, T., Fraser, C., Ravel, J., Keim, P. (2007). Strain specific single-nucleotide polymorphism assays for the *Bacillus anthracis* Ames strain. J. Clin. Microbiol. 45:47-53.

Velappan, N., Snodgrass, J., Hakovirta, J., Marrone, B., Burde, S. (2001). Rapid identification of pathogenic bacteria by single-enzyme amplified fragment length polymorphism analysis. Diag. Microbiol. Infect. Dis. 39:77-83.

Vietri, N., Marrero, R., Hoover, T., Welkos, S. (1995). Identification and characterization of a *trans*-activator involved in the regulation of encapsulation by *Bacillus anthracis*. Gene. 152:1-9.

Volokhov, D., Pomerantsev, A., Kivovicvh, V., Rasooly, A., Chizhikov, V. (2004). Identification of *Bacillus anthracis* by multiprobe microarray hybridization. Diag. Microbiol. Infect. Dis. 49:163-171.

Wang, A., Wen, J., Zhou, Y., Zhang, Z., Yang, R., Zhang, J., Chen, J., Zhanag, X. (2004). Identification and characterization of *Bacillus anthracis* by multiplex PCR on DNA chip. Biosens. Bioelectron. 20:807-813.

Wattiau, P., Klee, S., Fretin, D., Van Hessche, M., Ménart, M., Franz, T., Chasseur, C., Butaye, P., Imberechts, H. (2008). Occurrence and genetic diversity of *Bacillus anthracis* strains isolated in an active wool-cleaning factory. Appl. Environ. Microbiol. 74:4005-4011.

Wilson, A., Hoch, J., Perego, M. (2009). Two small c-type cytochromes affect virulence gene expression in *Bacillus anthracis*. Molec. Microbiol. 72:109-123.

Zhong, W., Shou, Y., Yoshida, T., Marrone, B. (2007). Differentiation of *Bacillus anthracis, B. cereus,* and *B. thuringiensis* by using pulse-field gel electrophoresis. Appl. Environ. Microbiol. 73:3446-3449.

Index

A

Acapsular sterne strain   178

AcpA- strain   213, 217

Active anthrax infection   81

Adaptive immune responses   84, 89-90, 139

Adverse events (AEs)   127, 129-30, 138

Aerosol   21, 89, 120, 122, 140, 143, 154, 161, 164-5, 168, 179, 188

Aerosolized spores   136, 149, 154, 171, 176

Aerosolized Stern strain spores   172

AFLP   208, 240, 243-5

AFLP analysis   243, 245

AFLP profiles   243-4

Africa   3-4, 12

African green monkeys (AGMs)   98, 177

Alanine   55, 134

Albania   3, 5, 7, 13

Allergic reactions   128-30

Aluminum hydroxide   120-1, 126, 134, 142, 161, 164

Alveolar macrophages   50, 57, 59, 91

Alveoli   50, 58-9, 189

America (U.S.A.)   3

Ames and Sterne strains   238, 246

Ames pXO2 plasmid   103

Ames spores   101, 143, 161, 266

Ames strain   19, 37, 89, 113, 120, 136, 140, 149, 158, 164, 167, 187, 190-2, 241-4, 266-7

Amino acids   55, 71, 79, 145-6, 157-8, 160, 264

Amino terminus   78, 80, 84

Amplicons   239, 246-9, 251, 253, 255-9

Animals surviving anthrax infection   114

Anl genes   37-9

B

www.ingramcontent.com/pod-product-compliance
Lightning Source LLC
Chambersburg PA
CBHW050814220326
41598CB00006B/210